Visit us at

www.syngress.com

Syngress is committed to publishing high-quality books for IT Professionals and delivering those books in media and formats that fit the demands of our customers. We are also committed to extending the utility of the book you purchase via additional materials available from our Web site.

SOLUTIONS WEB SITE

To register your book, visit www.syngress.com/solutions. Once registered, you can access our solutions@syngress.com Web pages. There you may find an assortment of value-added features such as free e-books related to the topic of this book, URLs of related Web sites, FAQs from the book, corrections, and any updates from the author(s).

ULTIMATE CDs

Our Ultimate CD product line offers our readers budget-conscious compilations of some of our best-selling backlist titles in Adobe PDF form. These CDs are the perfect way to extend your reference library on key topics pertaining to your area of expertise, including Cisco Engineering, Microsoft Windows System Administration, CyberCrime Investigation, Open Source Security, and Firewall Configuration, to name a few.

DOWNLOADABLE E-BOOKS

For readers who can't wait for hard copy, we offer most of our titles in downloadable Adobe PDF form. These e-books are often available weeks before hard copies, and are priced affordably.

SYNGRESS OUTLET

Our outlet store at syngress.com features overstocked, out-of-print, or slightly hurt books at significant savings.

SITE LICENSING

Syngress has a well-established program for site licensing our e-books onto servers in corporations, educational institutions, and large organizations. Contact us at sales@syngress.com for more information.

CUSTOM PUBLISHING

Many organizations welcome the ability to combine parts of multiple Syngress books, as well as their own content, into a single volume for their own internal use. Contact us at sales@syngress.com for more information.

SYNGRESS®

SYNGRESS®

BUILDING ROBOTS WITH LEGO® MINDSTORMS® NXT

Dave Astolfo Lead Author & Technical Editor
Mario Ferrari
Giulio Ferrari

CLINTON-MACOMB PUBLIC LIBRARY

Special Foreword by Former Space Shuttle Astronaut Dan Barry

Elsevier, Inc., the author(s), and any person or firm involved in the writing, editing, or production (collectively "Makers") of this book ("the Work") do not guarantee or warrant the results to be obtained from the Work.

There is no guarantee of any kind, expressed or implied, regarding the Work or its contents. The Work is sold AS IS and WITHOUT WARRANTY. You may have other legal rights, which vary from state to state.

In no event will Makers be liable to you for damages, including any loss of profits, lost savings, or other incidental or consequential damages arising out from the Work or its contents. Because some states do not allow the exclusion or limitation of liability for consequential or incidental damages, the above limitation may not apply to you.

You should always use reasonable care, including backup and other appropriate precautions, when working with computers, networks, data, and files.

Syngress Media®, Syngress®, "Career Advancement Through Skill Enhancement®," "Ask the Author UPDATE®," and "Hack Proofing®," are registered trademarks of Elsevier, Inc. "Syngress: The Definition of a Serious Security Library"™, "Mission Critical™," and "The Only Way to Stop a Hacker is to Think Like One™" are trademarks of Elsevier, Inc. Brands and product names mentioned in this book are trademarks or service marks of their respective companies.

KEY	SERIAL NUMBER
001	HJIRTCV764
002	PO9873D5FG
003	829KM8NJH2
004	BPOQ48722D
005	CVPLQ6WQ23
006	VBP965T5T5
007	HJJJ863WD3E
008	2987GVTWMK
009	629MP5SDJT
010	IMWQ295T6T

PUBLISHED BY
Syngress Publishing, Inc.
Elsevier, Inc.
30 Corporate Drive
Burlington, MA 01803

Building Robots with LEGO Mindstorms NXT

Copyright © 2007 by Elsevier, Inc. All rights reserved. Printed in the United States of America. Except as permitted under the Copyright Act of 1976, no part of this publication may be reproduced or distributed in any form or by any means, or stored in a database or retrieval system, without the prior written permission of the publisher, with the exception that the program listings may be entered, stored, and executed in a computer system, but they may not be reproduced for publication.

Printed in the United States of America
1 2 3 4 5 6 7 8 9 0
ISBN-13: 9781597491525

Publisher: Amorette Pedersen
Acquisitions Editor: Andrew Williams
Technical Editor: Dave Astolfo
Cover Designer: Michael Kavish
Project Manager: Gary Byrne
Page Layout and Art: Patricia Lupien
Copy Editor: Audrey Doyle
Indexer: J. Edmund Rush

For information on rights, translations, and bulk sales, contact Matt Pedersen, Commercial Sales Director and Rights, at Syngress Publishing; email m.pedersen@elsevier.com.

CLINTON-MACOMB PUBLIC LIBRARY

Lead Author and Technical Editor

Dave Astolfo (GIS A.S.) is a Project Manager/Business Analyst with the technical arm of a North American environmental consulting company. He currently provides project management, technical consulting, database design, and software architecture with a focus on geographic information systems ranging from desktop mapping software to Web mapping portals and mobile software applications. His specialties include database modeling and software design/architecture. Dave is a key contributor to the business development and implementation of products and services. As such, he develops enterprisewide technology solutions and methodologies focused on client organizations.

Dave holds a bachelor's degree from Trent University and is a certified Geographic Information Systems Applications Specialist holding a certificate from Sir Sandford Fleming College. In early 2006, Dave was invited by LEGO to participate in the LEGO MINDSTORMS Developer Program (MDP) to help LEGO beta test the prerelease of the LEGO MINDSTORMS NXT robotics system. After the release of the NXT, Dave was invited back to participate in the LEGO MINDSTORMS Community Partner Program (MCP) to work with LEGO in developing the product further while providing an ear to the community. Dave has been involved with LEGO all of his life, and he has been working with the MINDSTORMS product since the initial release of the LEGO MINDSTORMS Robotics Invention System (RIS) in 1998. Dave is well known for building a MINDSTORMS domino-placing robot that was published in a previous Syngress book (*10 Cool Lego Ultimate Builders Lego Mindstorms Robots*, ISBN: 1-931-836-60-4). Since then, he has created newer versions with a more recent NXT version being used by LEGO marketing staff in Europe. Visit Dave's Web site, www.plastibots.com, for more information on his work.

Lead Authors

Called the "DaVincis of LEGOS," Mario and Giulio Ferrari are world-renowned experts in the field of LEGO MINDSTORMS robotics.

Mario Ferrari received his first LEGO box around 1964, when he was four. LEGO was his favorite toy for many years, until he thought he was too old to play with it. In 1998, the LEGO MINDSTORMS RIS set gave him reason to again have LEGO become his main addiction. Mario believes LEGO is the closest thing to the perfect toy. He is Managing Director at EDIS, a leader in finishing and packaging solutions and promotional packaging. The advent of the MINDSTORMS product line represented for him the perfect opportunity to combine his interest in IT and robotics with his passion for LEGO bricks, which started during his early childhood. Mario has been a very active member of the online MINDSTORMS community from the beginning and has pushed LEGO robotics to its limits. Mario holds a bachelor's degree in Business Administration from the University of Turin and has always nourished a strong interest for physics, mathematics, and computer science. He is fluent in many programming languages, and his background includes positions as an IT manager and as a project supervisor. Mario estimates he owns more than 60,000 LEGO pieces. Mario works in Modena, Italy, where he lives with his wife, Anna, and his children, Sebastiano and Camilla.

Giulio Ferrari is Sales Manager at EDIS, a company that specializes in premiums and promotional packaging. He studied economics and engineering at the University of Modena and Reggio Emilia. He is fond of computer programming and mathematical sciences, as well as puzzles and games in general. He has a collection of 1,500 dice of all kinds and shapes. Giulio coauthored the bestselling *Building Robots with LEGO Mindstorms* (Syngress Publishing, ISBN: 1-928994-67-9) with his brother Mario and Ralph Hempel. The book has quickly become a fundamental reference and source of ideas for many LEGO robotics fans. He is also a contributor to

Programming LEGO MINDSTORMS with Java (Syngress Publishing, ISBN: 1-928994-55-5) and LEGO MINDSTORMS Masterpieces (Syngress Publishing, ISBN: 1-931836-75-2). Giulio has been playing with LEGO bricks since he was very young, and his passion for robotics started in 1998, with the arrival of the MINDSTORMS series. From that moment on, he held an important place in the creation of the Italian LEGO community, ItLUG, now one of the most important LEGO user groups worldwide. He works in Modena, where he lives with his girlfriend, Claire, and his son, Zeno.

Contributors

Bryan Bonahoom is a LEGO MINDSTORMS enthusiast. He is a member of the Lafayette LEGO Robotics Club and one of the original team that developed the Great Ball Contraption. Bryan is also cofounder of Brickworld™. Bryan was selected by LEGO in 2006 as a member of the MDP and later as a member of the MINDSTORMS Community Partners (MCP). Bryan was also awarded the Best Robot Design Trophy at the 2005 AFOL Tournament at LEGO headquarters as a member of Team Hassenplug. Bryan is possibly most well known for the creation of an NXT-based robot that plays tic-tac-toe with a human opponent.

John Brost had a passion for LEGO and all things mechanical at an early age. However, the interest waned, and like most adult LEGO fans, John went through his "dark ages" in high school and while attending Purdue University. The release of the Star Wars LEGO sets brought a renewed interest in LEGO to John. But it was a chance encounter with an announcement of a LEGO Robotic Sumo competition being held locally that brought John back to LEGO 100 percent. It took only this event to get John hooked. Less than two weeks later, John had his first MINDSTORMS RIS kit, and he has been busily building robots and all sorts of mechanical LEGO contraptions ever since then.

John has participated in all types of MINDSTORMS competitions, winning a few here and there. He has also been a coach for LafLRC's (Lafayette LEGO Robotics Club) FIRST LEGO League team for the past five years. In 2006, John was lucky enough to become a member of the LEGO MINDSTORMS Developer Program to test the MINDSTORMS NXT. Currently, John is a moderator on LEGO's NXTlog Web site and a coordinator for the 2007 Brickworld convention being held in Chicago.

Rebeca Dunn-Krahn is a member of a working group at the University of Victoria dedicated to increasing awareness and understanding of computer science among children and young adults. She is currently producing a short documentary film about these outreach efforts that include free robotics festivals using LEGO MINDSTORMS. Rebeca has also worked in quality assurance and as a Java developer.

Rebeca holds a bachelor's degree in computer science and biochemistry from the University of Victoria and lives in Victoria, Canada, with her family. Rebeca would like to thank her husband, Tobias, and her children, Sophia and Sebastian, for their support.

Richard Li is one of two nonadult contributors to *Building Robots with LEGO MINDSTORMS NXT*. He is currently a seventh-grader at Beck Academy and heads two award-winning FIRST LEGO League (FFL) teams. They have been honored with the Champion's Award on several occasions and are ranked as two of the top teams yearly. When not working with his FLL teams, he experiments with his own robots at home. He would like to thank his parents, Lin and Liang-Hong, for buying him his LEGOs and dealing with him as he stayed up late several nights to meet his deadline for this book. Richard currently resides in Simpsonville, SC.

Christopher Dale Minamyer (Bachelor of Science in Mechanical Engineering, University of Arizona, 2007) began studying mechanical engineering during the fall of 2002 and graduated from the University of Arizona in May 2007. A lifelong LEGO fan, Chris has been building with LEGO for 20 years.

For the past four years Chris was an instructor of LEGO Robotics at Ventana Vista Elementary School in Tucson, AZ. During this time he instructed more than 300 students in grades one through five in LEGO MINDSTORMS RCX and NXT. In addition Chris is a founding member of the Tucson LEGO Club Masters Group and the head coach. Chris has been a FIRST LEGO League coach for the past four years. In that time the teams he coached have won the Robot Performance award (2004) and the Research Quality award (2005) for the state of Arizona. In addition Chris received the Adult Coach/Mentor award (2005) and the Appreciation award (2006).

Chris would like to thank his mother, Martha, and father, Rodger, for always supporting his LEGO building, and giving him his first set at the age of three. In addition he would like to thank Misha Chernobelskiy of the Tucson LEGO Club for providing an ideal environment for the instruction of LEGO enthusiasts. Finally, Chris would like to thank Caryl Jones of Ventana Vista Elementary School for her support and her continuing dedication to teaching LEGO robotics.

Deepak Patil developed interest in LEGO robotics when the original LEGO MINDSTORMS kit was introduced in 1998. Since then Deepak has coached FLL teams and has conducted several robotics workshops with LEGO MINDSTORMS.

Deepak has a master's degree in Industrial Design from IIT Bombay and has designed user interfaces for diverse products, including programmable logic controllers, telephony software, and multimedia systems. Deepak has worked for Cisco and other leading technology companies, and he has led technology projects with globally interspersed teams of engineers.

Deepak lives in Richmond, VA, with his wife, Priti, daughter, Vibha, and son, Uday, an avid LEGO Robotics fan.

Mac Ruiz is a retired construction superintendent. His work entailed problem solving and coordinating of off-site engineers and subcontractors with the projects' realities. He also has experience in fabrication of farm equipment from his family's dealership. This included steel fabrication and mechanicals.

Christian Siagian is working toward a Ph.D. degree in Computer Science at the University of Southern California (USC). He is involved in the Beobot Project that develops a biologically inspired vision-based mobile-robot localization and navigation system. His research interests include robotics and computer and biological vision.

As a teaching assistant for CS445 Introduction to Robotics at USC, Christian develops laboratory curriculum to prepare undergraduates for research in robotics. Christian also volunteers for after-school programs at St. Agnes Parish School and EPICC in Los Angeles. These programs use robotics to promote interest in science and mathematics in elementary and middle school students.

Christian holds a bachelor's degree from Cornell University in Computer Science and is a member of the IEEE.

Dick Swan is an embedded software consultant. He partnered with the Robotics Academy at Carnegie Mellon University in developing the RobotC programming environment for the NXT. He also codeveloped with Tufts University the Robolab programming environment for the NXT. Dick has 30 years' experience in software and hardware projects, including embedded systems, telephone systems, and compilers. Dick has both a bachelor's and master's degree in Computer Science from the University of Waterloo and is a member of the IEEE.

Sivan Toledo is Associate Professor of Computer Science in Tel-Aviv University in Israel. He holds a BSc degree in Math and Computer Science and an MSc degree in Computer Science, both from Tel-Aviv University, and a PhD in Computer Science from the Massachusetts Institute of Technology. He authored more than 50 scientific papers and one textbook. He serves on the editorial boards of the *SIAM Journal of Scientific Computing* and of *Parallel Computing*.

Joshua Whitman is a home-schooled eighth-grader from Wichita, KS. He has been building with LEGO bricks for as long as he can remember. He received his first MINDSTORMS RCX kit at age eight. He has participated in numerous robotics groups, clubs, and classes. He competed on a team in WSU's MINDSTORMS Robotics Challenge for two years. His team won the first-place trophy both times. As an experienced member of the team, he had to learn how to help teach the newer kids about MINDSTORMS. He was a part of the first winning team to use an NXT to compete in the challenge.

His favorite (and most impressive) creation is a robot that can actually lock and unlock his room through a rotation sensor combination lock. The system is surprisingly secure, and 99.9 percent foolproof. He loves programming more than anything else in robotics. His current project involves using the NXT display as a screen for simple videogames like Pong.

Larry Whitman (Ph.D., P.E.) is an Associate Professor of Industrial and Manufacturing Engineering at Wichita State University. Larry promotes engineering in every context possible. He is especially interested in promoting the technical literacy of all citizens, not just those who intend to be engineers. To this end, he and several colleagues at Wichita State have developed a course using LEGO MINDSTORMS in a hands-on environment to demonstrate basic engineering skills to nonengineering undergraduates. He also coordinates a LEGO MINDSTORMS challenge competition for middle school students. Finally, he promotes engineering by training his two sons, Joshua and David, to love building LEGO robots.

Larry holds bachelor's and master's degrees from Oklahoma State University. After spending 10 years in the aerospace industry as a practicing engineer, he completed his Ph.D. from the University of Texas at Arlington.

Larry, Joshua, and David are forever indebted to Larry's wife, Heidi, for putting up with pieces of robots around the house and dinner conversations about robots and engineering.

Guy Ziv is now finishing his graduate studies in biological physics at the Weizmann Institute of Science in Israel. He holds a bachelor's degree in math and physics from the Hebrew University of Jerusalem and a master's degree in physics from the Weizmann Institute. Guy has been working in the field of measurement and automation for several years. He is an experienced LabView™ programmer and was a beta tester of NI LabView™ toolkit for NXT and MINDSTORMS NXT v. 1.1. Guy is the author and editor of NXTasy.org, the second largest NXT community site, and he moderates NXTasy.org's repository and forums.

Foreword Contributor

Daniel T. Barry (M.D., Ph.D.) is a former NASA astronaut who was a crew member aboard the Space Shuttles Discovery and Endeavor. He logged more than 734 hours in space, including four spacewalks totaling 25 hours and 53 minutes. He holds a BS degree in electrical engineering from Cornell University; a master's of engineering degree and a master of arts degree in electrical engineering/computer science from Princeton University; a doctorate in electrical engineering/computer science from Princeton University; and a doctorate in medicine from the University of Miami in 1982.

Organizations to which he belongs include the Institute of Electrical and Electronic Engineers (IEEE), the American Association of Electrodiagnostic Medicine (AAEM), the American Academy of Physical Medicine and Rehabilitation (AAPMR), the Association of Academic Physiatrists (AAP), and the Association of Space Explorers. He holds five patents, has written 50 articles in scientific journals, and has served on two scientific journal editorial boards.

Dr. Barry retired from NASA in April 2005 to start his own company, Denbar Robotics, where he currently builds robots. Dr. Barry currently lives in South Hadley, MA.

Contents

Foreword

I was always building stuff as a kid. I didn't have LEGO, but I was pretty happy with an erector set and whatever accessories were lying about the house. I still remember the day that I got a motor to run a crane built from flimsy metal spars. The simple process of lifting wooden blocks kept me entertained for hours. To me, that crane was not just a rickety pile of metal and string; it was a massive construction machine, and I was building skyscrapers, airports, entire cities. It was not so much about the things I built as it was about the things I imagined I was building. That was true when my kids starting building stuff, too, although they had the advantage of having LEGO, so we could build many more things. Together, we built hundreds of airplanes (the more engines, the better!) and used them to run around the house dogfighting. We built cities and destroyed them from missile bases. We built spaceships and ocean liners, trucks and buses, bulldozers and backhoes, motorcycles and race cars, coliseums and fortresses. What a blast! However, even with all that variety, it was basically the same activity that I had done when I was a child. Our stuff had no smarts.

And then along came MINDSTORMS. Suddenly, our creations were no longer completely predictable. Before MINDSTORMS, a car went where you pointed it, at whatever speed your hand (or its motor) pushed it, and fairly quickly crashed. At best, you could buy a remote control car, but even that just went where it was told. A MINDSTORMS car navigates by itself, avoiding obstacles (or picking them up), and keeps going.

However, building things is only half the battle; *programming* them is key. Furthermore, these creations can actually do useful stuff! Right away my kids and I built a robot that cleared the table. It ran around, detecting objects, scooping them up, and dumping them off the table, while being careful not to fall off the table itself. Lights! Sensors! Action! A single bot can do a million different things by just having you change the software.

At first I didn't like the graphical programming environment that came with MINDSTORMS. I was used to standard programming languages, and having to move blocks around felt slow and clumsy. However, I found that, with no prior programming experience, my kids could program intuitively with the blocks. And then I found I could understand their programs with just a glance, not even having to read the code to follow the logic.

Cool! Ultimately, the kids built and programmed a working model of the Space Shuttle's robotic arm, and I took it along with me into orbit on board shuttle Discovery.

The FIRST LEGO League encourages kids to build robots as teams, and I am happy to participate, both as a judge and in the design of the contests. One day I was backstage while the competitions were running out front, and I saw a fifth-grader all alone, intently programming her MINDSTORMS brick. I asked her if she had a moment to talk, and she replied, "Well, OK, but just a minute because I have to get this working right now." We talked for a bit about her project, and I realized that her grasp of concepts like torque, friction, and acceleration was at the level of a high school physics student. Finally I said, "You seem really into this project; what got you interested in robotics?" She said, "I'll tell you, but you have to keep it secret, OK?" "You bet," I replied. "Well, in our school there's a deal where, if you do the LEGO league, you don't have to do *any* science for the whole year, and I hate science, so I signed up for this—and it is so great! Anyway, I have to get back to programming since I'm the only one who knows how this part works." So I left, amused to have met someone who "hated science" and loved being a scientist!

With the advent of MINDSTORMS NXT, the motors, sensors, and programming environment improved so much that I am incorporating NXT robots into the robots of my company, Denbar Robotics. I have not duplicated with NXT the types of robots we are making at Denbar. Instead, I decided to make robots that the Denbar robots command. For example, I built an NXT robot that has two drive motors and uses the third motor to move its light and ultrasound sensors. This robot can find light or dark directions, avoid obstacles, and navigate around the house. Our big robot, Neel, can turn the lights in the house on and off. So he can call the light-seeking robot by turning room lights on and off in a pattern that results in the light seeker getting to him. My idea is to have groups of robots that include leaders and followers. The LEGO robots can help Neel get access to spots that are too small for him and can do tasks for him while he is busy doing something else.

As robotic tasks become more interesting, the robots themselves must become mechanically more robust, able to withstand crashes, recover from upsets, steer accurately, and balance loads well. The authors of this book, led by Technical Editor Dave Astolfo, a member of the LEGO MINDSTORMS Developer Program/Community (MDP, MCP), have updated Mario and Giulio Ferrari's content so that it is current with the NXT. They take us through the basics of gears, motors, and sensors and then move on to pneumatics, grabbers, and navigation, and eventually to tasks such as solving mazes and racing against time. The accurate descriptions and precise images have already helped me to make my robots stronger and more versatile. So gather your gear and let the inventing begin!

—Dan Barry, M.D., Ph.D.
NASA Astronaut (retired)
Denbar Robotics

Preface

LEGO has been a part of my life since I was about four years old. My first sets were basic LEGO SYSTEM sets. However, I soon jumped to the early TECHNIC sets that were beginning to appear on the market. Because I was one of those kids who had to take everything apart to figure out how it worked, TECHNIC seemed like a good fit for me. The best part was that I was no longer breaking toys—a relief for my parents.

I remember sets such as the now-classic 856, 853, 855, and 8865. I don't seem to have the original parts for any of these sets now. Like many other adult fans, I have gone through dark years during which some of my LEGO parts were sold, others were thrown out, and the rest were stored.

My interest in LEGO was rekindled in the late '90s, however, when I read about the MINDTORMS Robotics Invention System (RIS) 1.0. The moment it was available for sale, I ordered mine, and I now find myself where I am now. I had wished for this sort of thing many years ago. Now with the advent of the NXT system, a whole new era of fun with robotics has begun.

In early 2006, I was honored to be one of the 100 testers chosen by LEGO as part of its MINDSTORMS Developer Program (MDP) for the beta testing of the new NXT system. Once the product went to market in fall of 2006, I was also invited by LEGO to be part of its MINDSTORMS Community Partner (MCP) Program, which has allowed a core group of adult fans to keep involved (with a great deal of excitement) with LEGO on upcoming features/releases for the NXT.

Since my initial RIS purchase, I have built up an inventory of more than 50,000 pieces, including three NXT sets, five RIS sets, three DDKs, one RDS, countless motors and sensors, and a whole slew of other TECHNIC pieces. Oddly enough, however, I still struggle to find parts when building robots.

A few years back, Syngress, now an imprint of Elsevier Inc., asked me to author building instructions in its now popular book, *10 Cool Lego Mindstorms Ultimate Builders Projects* (ISBN: 1-931836-60-4). My chapter of the book provided details and instructions on how to build my RCX-based DominoBot. So when Syngress asked me in November 2006 to become the technical editor of this book, my answer, of course, was yes.

As many of you may know, this book is a revision to the bestseller written by Mario and Giulio Ferarri. My goal was to revise and update the content and make it specific to the NXT system. In addition to the revised content, you will notice a significant shift from the traditional brick-and-plate building approach to studless building techniques—the chapters are rife with ideas and approaches to help guide you and ensure that your robot-building experience is enjoyable! On behalf of myself and the rest of the authors of this book, we hope you enjoy the diverse and plentiful information within it.

—*David Astolfo*
Technical Editor

Chapter 1

Understanding LEGO® Geometry

Solutions in this chapter:

- Expressing Sizes and Units
- Squaring the LEGO World: Vertical Bracing
- Tilting the LEGO World: Diagonal Bracing
- TECHNIC Liftarms: Angles Built In

Introduction

Before you enter the world of LEGO robotics, we want to be sure you know and understand some basic geometric properties of the LEGO bricks and beams. Don't worry; we're not going to test you with complex equations or trigonometry. We'll just discuss some very simple concepts and explain some terminology that will make assembling actual systems easier from the very beginning.

You will discover which units LEGO builders use to express sizes, the proportions of the bricks and beams, and how this affects the way you can combine them with different orientations into a solid structure.

In the past few years, there has been a shift from building with TECHNIC bricks and beams to building with studless beams, pins, and connectors. After we introduce some basic concepts, you will be exposed to these new ideas and see examples of how you can use studless building.

We encourage you to try to reproduce all the examples we show in this chapter with your own LEGO parts. If for any reason, you feel that what we present is too complex or boring, don't force yourself to read it. Skip the chapter and go to another one. You can always come back and use this chapter as a sort of glossary whenever you need it.

Expressing Sizes and Units

LEGO builders usually express the size of LEGO parts with three numbers, representing *width*, *length*, and *height*, in that order. The standard way to use LEGO bricks is "studs up." When expressing sizes, we always refer to this orientation, even when we are using the bricks upside down or rotating them in 3D space.

Height is the simplest property to identify. It's the vertical distance between the top and bottom of the basic brick. Width, by convention, is the shorter of the two dimensions which lie on the horizontal plane (length is the other one). Both width and length are expressed in terms of *studs*, also called *LEGO units*. Knowing this, we can describe the measurements of the most traditional brick, the one whose first appearance dates back to 1949, which is 2 x 4 x 1 (see Figure 1.1).

Figure 1.1 The Traditional LEGO Brick

LEGO bricks, although their measurements are not expressed as such, are based on the metric system: A stud's width corresponds to 8 mm and the height of a brick (minus the stud) to 9.6 mm. These figures are not important to remember. What's important is that they do not have equal values, meaning you need two different units to refer to length and height. Their *ratio* is even more important: Dividing 9.6 by 8 you get 1.2 (the vertical unit corresponds to 1.2 times the horizontal one). This ratio is easier to remember if stated as a proportion between whole numbers: It is equivalent to 6:5. Figure 1.2 shows the smallest LEGO brick, described in LEGO units as a 1 x 1 x 1 brick. For the reasons explained previously, this LEGO "cube" is not a cube at all.

Figure 1.2 Proportions in a 1 x 1 x 1 LEGO Brick

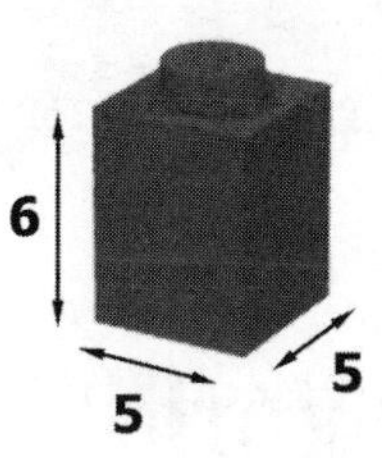

The LEGO system includes a class of components whose height is one-third of a brick. The most important element of this class is the *plate*, which comes in a huge variety of rectangular sizes, and in some special shapes too. If you stack three plates, you get the height of a standard brick (see Figure 1.3).

Figure 1.3 Three Plates Make One Brick in Height

The advent of studless building has thrown a wrench into our understanding and use of the classic brick-and-plate-construction approaches. The new studless components provide a different way to construct LEGO models. The jury is still out regarding which method is better or more preferred—there are proponents on both sides of the fence with this. Figure 1.4 shows the traditional plates and bricks next to a newer TECHNIC beam and a set of three liftarms that are stacked, all to provide some reference on size. You will notice that their stacked heights are not compatible across the studded (left) and studless (right) parts.

The one thing that is undeniable is that LEGO has made a significant shift with its TECHNIC and MINDSTORMS lines toward studless components. If you are a die-hard studded builder, you are encouraged to take the plunge and try building studless. Many found the change in approach a challenge at first, but use it almost exclusively now. Studless building offers countless options for connectivity and even allows for building at odd angles that was difficult to accomplish with traditional beams.

Figure 1.4 Comparing Bricks to Studless Beams and Liftarms

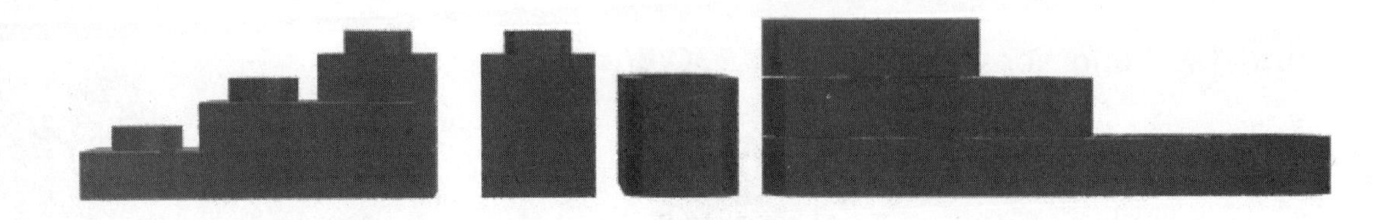

Squaring the LEGO World: Vertical Bracing

Why do we care about all these relationships? To answer this, we must travel back to the late 1970s when the LEGO TECHNIC line was created. Up to that time, LEGO was designed and used to build things made of horizontal layers: Bricks and plates integrate pretty well when stacked together. Every child soon learns that three plates count for a brick, and this is all they need to know. But in 1977, LEGO decided to introduce a new line of products targeting an older audience: LEGO TECHNIC. It turned the common 1 x *N* brick holes into what we call a TECHNIC brick, or a *beam* (Figure 1.5, left). These holes allow *axles* to pass through them, and permit the beams to be connected to each other via *pegs*, thus creating a whole new world of possibilities.

In the late 1990s, the advent of studless ~~beams~~ (Figure 1.5, right) opened the door to alternative building options. One of the best sets in TECHNIC history is undoubtedly the 8448 Super Street Sensation, which is built almost entirely from studless parts. LEGO was clever with its approach here. Instead of using beams to construct the chassis and plates to provide the "shell" or "form" for the model, the chassis was built using studless beams and its style was handled by fairing panels, allowing the curves of the car to "flow" with the design. LEGO did this to reduce costs: Less material required equals a cost savings in production. You can see a great example of this if you compare the 8448 Super Street Sensation to the classic TECHNIC 8880 Supercar. Compare their approaches to construction, and their weight. You will notice significant differences. The 8880 gets all its design cues from classic TECHNIC beams, whereas the 8448 uses fairings and flex axles for its design.

Figure 1.5 The LEGO TECHNIC Beams

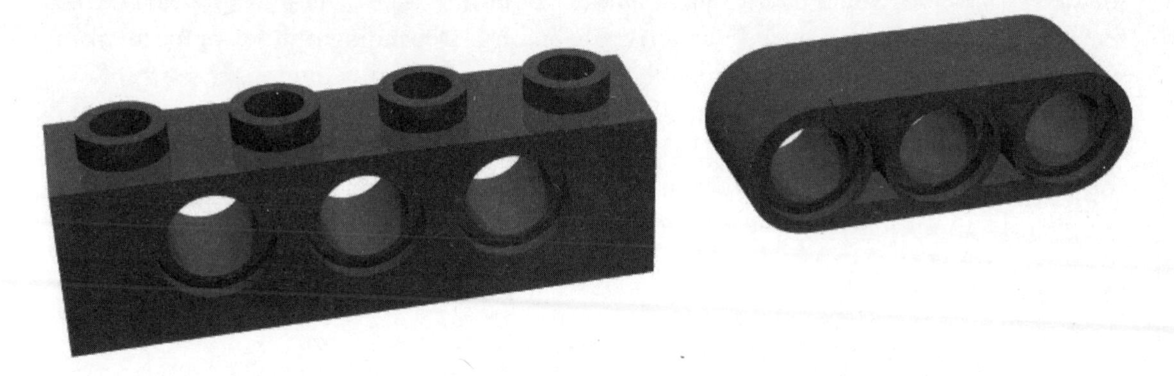

Suppose you want to mount a beam in a vertical position to brace two or more layers of horizontal beams. Here's where you must remember the 6:5 ratio. The holes inside a beam are spaced at exactly the same distance as the studs, but are shifted over by half a stud. So, when we stand the beams up, the holes follow the horizontal units and not the vertical ones. Consequently, they don't match the corresponding holes of the layered beams. In other words, the holes in the vertical beam cannot line up with the holes in the stack because of the 6:5 ratio. At least not with all the holes. But let's take a closer look at what happens. Count the vertical units by multiples of 6 (6, 12, 18, 24, 30…) and the horizontal ones by multiples of 5 (5, 10, 15, 20, 25, 30…). Don't count the starting brick and the starting hole—that is your reference point; you are measuring the *distances* from that point. You see? After counting five vertical units you reach 30, which is the same number you reach after counting six horizontal units (see Figure 1.6).

Figure 1.6 Matching Horizontal and Vertical Beams

Now suppose you want to construct a robot that needs to be strong but light. With the studded beams, you would have to use a number of beams, plates, and pins to create the frame, and potentially you would need to cross-brace it. Depending on the approach taken, you may even use the aforementioned stacking technique, which would make your robot strong but heavy. With some of the newer parts that are now available, creating a *strong and light* chassis is quite simple and straightforward. Figure 1.7 shows a sample chassis that you could use as a base for your robot. It employs very few pieces, and in fact, it uses only four unique parts (in quantity) to make for a solid structure.

Figure 1.7 Sample Chassis

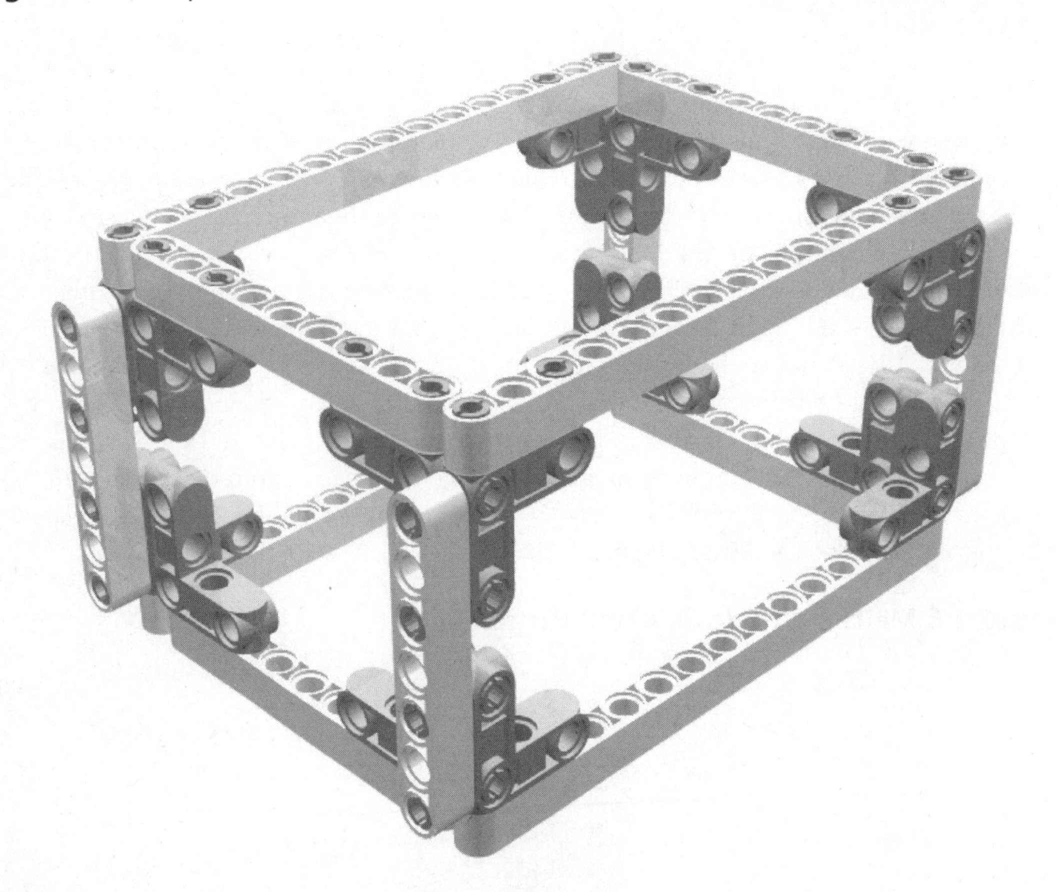

Tilting the LEGO World: Diagonal Bracing

Who said that the LEGO beams *must* connect at a right angle to each other? The very nature of LEGO is to produce squared things, but with the advent of studless parts, diagonal connections are mainstream now, making our world a bit more varied and interesting, and giving us another tool for problem solving.

You now know that you can cross-connect a stack of plates and beams with another beam. And you know how it works in numerical terms. So how would you brace a stack of beams with a diagonal beam?

You must look at that diagonal beam as though it were the hypotenuse of a right-angled triangle. Continuing from the previous sample, Figure 1.8 adds a cross-brace to support the structure and provides a sample for this next bit. Now proceed to measure its sides, remembering not to count the first holes, because we measure lengths in terms of distances from them. The base of the triangle is eight holes. Its height is six holes: Remember that in a standardized grid, every horizontal beam is at a distance of two holes from those immediately below and above it. In regard to the hypotenuse, it counts 10 holes in length.

For those of you who have never been introduced to Pythagoras, the ancient Greek philosopher and mathematician, the time has come to meet him. In what is probably the most famous theorem of all time, Pythagoras demonstrated that there's a mathematical relationship between the length of the sides of right-angled triangles. The sides composing the right angle are the catheti—let's call them A and B. The diagonal is the hypotenuse—let's call that C. The relationship is:

$$A^2 + B^2 = C^2$$

Now we can test it with our numbers:

$$8^2 + 6^2 = 10^2$$

This expands to:

$$(8 \times 8) + (6 \times 6) = (10 \times 10)$$

$$64 + 36 = 100$$

$$100 = 100$$

Yes! This is exactly why the example works so well. It's not by chance; it's the good old Pythagorean theorem. Reversing the concept, you might calculate whether any arbitrary pair of base and height values brings you to a working diagonal. This is true only when the sum of the two lengths, each squared, gives a number that's the perfect square of a whole number. Let's try some examples (see Table 1.1).

Figure 1.8 Pythagoras' Theorem

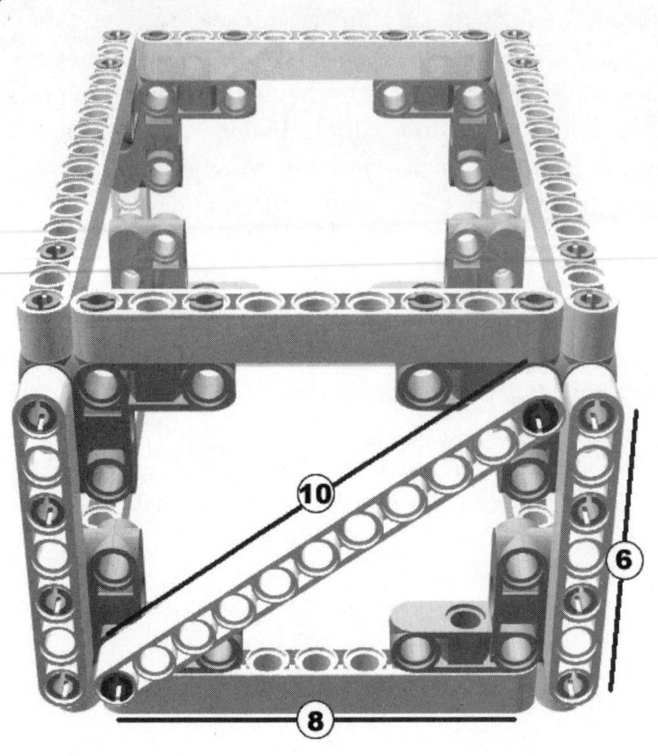

Table 1.1 Verifying Working Diagonal Lengths

A (Base)	B (Height)	A^2	B^2	$A^2 + B^2$	Comments
5	6	25	36	61	This doesn't work.
3	8	9	64	73	This doesn't work.
3	4	9	16	25	This works! 25 is 5 x 5.
15	8	225	64	289	This works too. Although 289 is 17 x 17, this would come out a very large triangle.
9	8	81	64	145	Note that 145 is not the square of a whole number, but it is so close to 144 (12 x 12) that if you try to make it your diagonal beam, it will fit with no effort at all. After all, the difference in length is less than 1 percent.

At this point, you're probably wondering whether you have to keep your pocket calculator on your desk when playing with LEGO, and maybe dig up your old high school math textbook to reread. Don't worry; you won't need either, for many reasons:

- You won't need to use diagonal beams very often.
- Most of the useful combinations derive from the basic triad 3-4-5 (see the third line in Table 1.1). If you multiply each side of the triangle by a whole number, you still get a valid triad—by 2: 6-8-10, by 3: 9-12-15, and so on. These are by far the most useful combinations, and they are very easy to remember.
- We provide a table in Appendix B with many valid side lengths, including some that are not perfect but so close to the right number that they will work very well without causing any damage to your bricks.

We suggest you take some time to play with triangles, experimenting with connections using various angles and evaluating their rigidity. This knowledge will prove precious when you start building complex structures.

TECHNIC Liftarms: Angles Built In

As noted earlier, over the past several years LEGO has introduced a number of new TECHNIC parts that divert from the concept of straight beams and 90-degree connectivity. We could review numerous parts here, but there simply is not enough room in the book for this. Some of the more popular ones fit in the common group of studless beams, called *liftarms*. They come in many shapes and sizes, you can use them to connect parts at differing angles, and you often see them in robot grabbers, fingers, ball casters, and so on. Figure 1.9 shows a sample of liftarms from the TECHNIC line.

Figure 1.9 A Variety of Liftarms

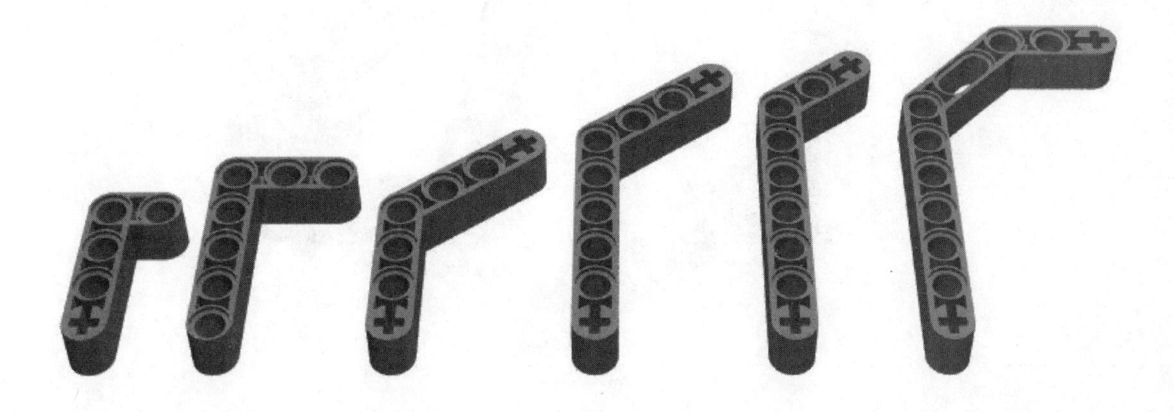

Liftarms are quite versatile parts that often come to the rescue when you're trying to connect components in odd ways. As you browse through the chapters of this book, keep an eye out for examples of this. You will see several samples that use liftarms in differing ways. Some use them to connect motors at odd angles, and others use them for bracing. Figure 1.10 shows some examples of how you can use liftarms to brace a structure.

Using your own parts, try to re-create this sample. You will notice that you can connect the liftarms at only certain holes and angles; not every combination works. However, by using different types of liftarms, you can see how each one connects a little differently, thus providing a number of ways to brace your robot.

It is also important to think outside the box here. With many of the newer TECHNIC parts, your models do not have to follow the traditional square or rectangular building approach. For example, using Figure 1.10, you could extend one of the liftarms upward to mount a sensor, or use it to connect a servo motor to provide a drive mechanism mounted at an angle. Try to experiment with connecting liftarms and beams together and see how you can brace your structure or extend components of your robot at odd angles.

The important thing to remember here is that you don't always have to follow the traditional approach of connecting beams and bricks at 90-degree angles.

Figure 1.10 Liftarm Bracing

Summary

Did you survive the geometry? You can see it doesn't have to be that hard once you get familiar with the basics. First, it helps to know how to identify the beams by their proportions, counting the length and width by studs, and recognizing that the vertical unit to horizontal unit ratio is 6 to 5. Thus, according to the simple ratio, when you're trying to find a locking scheme to insert axles or pins into perpendicular beam holes, you know that every five bricks in height, the holes of a crossed beam match up. Also, because three plates match the height of a brick, the most compact locking scheme is to use increments of two plates and a brick, because it gives you that magic multiple of 5. If you stay with this scheme, the standard grid, everything will come easy: one brick, two plates, one brick, two plates...

To fit a diagonal beam, use the Pythagorean theorem. Combinations based on the triad of 3-4-5 constitute a class of easy-to-remember distances for the beam to make a right triangle, but there are many others. You also were exposed to the TECHNIC liftarm, which offers countless connectivity options for your robots. Remember to think outside the box and not assume that you have to build via the traditional square or rectangular approach. Explore with the parts you have in your kit, and discover new ways to connect parts using studless building techniques. You will soon find that this way of building offers great flexibility in design.

Chapter 2

Playing with Gears

Solutions in this chapter:

- **Counting TeethGearing Up and Down**
- **Riding That Train: The Geartrain**
- **Worming Your Way: The Worm Gear**
- **Limiting Strength with the Clutch Gear**
- **Placing and Fitting Gears**
- **Using Pulleys, Belts, and Chains**
- **Making a Difference: The Differential**

Introduction

You might find yourself asking "Do I really *need* gears?" Well, the answer is yes, you do. Gears are so important for machines that they are almost their symbol: Just the sight of a gear makes you think *machinery*. In this chapter, you will enter the amazing world of gears and discover the powerful qualities they offer, transforming one force into another almost magically. We'll guide you through some new concepts—velocity, force, torque, friction—as well as some simple math to lay the foundations that will give you the most from the machinery. The concepts are not as complex as you might think. For instance, the chapter will help you see the parallels between gears and simple levers.

We invite you once again to experiment with the real things. Prepare some gears, beams, and axles to replicate the simple setups of this chapter. No description or explanation can replace what you learn through hands-on experience.

Counting Teeth

A single gear wheel alone is not very useful—in fact, it is not useful at all, unless you have in mind a different usage from that for which it was conceived! So, for a meaningful discussion, we need at least two gears. In Figure 2.1, you can see two very common LEGO gears: The right one is an 8t, and the left is a 24t. The most important property of a gear, as we'll explain shortly, is its *teeth*. Gears are classified by the number of teeth they have; the description of which is then shortened to form their name. For instance, a gear with 24 teeth becomes "a 24t gear."

Figure 2.1 8t and 24t Gears

Let's go back to our example. We have two gears, an 8t and a 24t, each mounted on an axle. The two axles fit inside holes in a beam at a distance of two holes (one empty hole in between). Now, hold the beam in one hand, and with the other hand gently turn one of the axles. The first thing you should notice is that when you turn one axle, the other turns too. The gears are *transferring motion from one axle to the other.* This is their fundamental property, their very nature. The second important thing you should notice is that you are not required to apply much strength to make them turn. Their teeth match well and there is only a small amount of friction. This is one of the great characteristics of the LEGO TECHNIC system: Parts are designed to match properly at standard distances. A third item of note is that the two axles turn in opposite directions: one clockwise and the other counterclockwise.

A fourth, and subtler, property you should have picked up on is that the two axles revolve at different speeds. When you turn the 8t, the 24t turns more slowly, whereas turning the 24t makes the 8t turn faster. Let's explore this in more detail.

Gearing Up and Down

Let's start turning the larger gear in our example. It has 24 teeth, each one meshing perfectly between two teeth of the 8t gear. While turning the 24t, every time a new tooth takes the place of the previous one in the contact area of the gears, the 8t gear turns exactly one tooth too. The key point here is that you need to advance only eight teeth of the 24 to make the small gear do a complete turn (360 degrees). After eight teeth more of your 24, the small gear has made a second revolution. With the last eight teeth of your 24, the 8t gear makes its third turn. This is why there is a difference in speed: For every turn of the 24t, the 8t makes three turns! We express this relationship with a ratio that contains the number of teeth in both gears: 24 to 8. We can simplify it, dividing the two terms by the smaller of the two (8), so we get 3 to 1. This makes it very clear in numerical terms that one turn of the first corresponds to three turns of the second.

You have just found a way to get more speed! (To be technically precise, we should call it *angular velocity*, not *speed*, but you get the idea.) Before you start imagining mammoth gear ratios for race car robots, sorry to disappoint you—there is no free lunch in mechanics; you have to pay for this gained speed. You pay for it with a decrease in *torque*, or, to keep in simple terms, a decrease in strength.

Bricks & Chips...

What Is Torque?

When you turn a nut on a bolt using a wrench, you are producing torque. When the nut offers some resistance, you've probably discovered that the more the distance from the nut you hold the wrench, the less force you have to apply. Torque is in fact the product of two components: force and distance. You can increase torque by either increasing the applied force, or increasing the distance from the center of rotation. The units of measurement for torque are thus a unit for the force, and a unit for the distance. The International System of Units (SI) defines the newton-meter (Nm) and the newton-centimeter (Ncm).

If you have some familiarity with the properties of levers, you will recognize the similarities. In a lever, the resulting force depends on the distance between the application point and the fulcrum: The longer the distance, the higher the force. You can think of gears as levers whose fulcrum is their axle and whose application points are their teeth. Thus, applying the same force to a larger gear (that is, to a longer lever) results in an increase in torque.

So, our gearing is able to convert torque to velocity—the more velocity we want the more torque we must sacrifice. The ratio is exactly the same: If you get three times your original angular velocity, you reduce the resulting torque to one-third.

One of the nice properties of gears is that this conversion is symmetrical: You can convert torque into velocity or vice versa. And the math you need to manage and understand the process is as simple as doing one division. Along common conventions, we say that we *gear up* when our system increases velocity and reduces torque, and that we *gear down* when it reduces velocity and increases torque. We usually write the ratio 3:1 for the former and 1:3 for the latter.

When should you gear up or down? Experience will tell you. It largely depends on the motor you start with and the robot you want to end up with. The MINDSTORMS NXT servo motors are already geared down significantly within their plastic case, so they turn at a relatively slow velocity and produce quite a bit of torque (see Chapter 3). Without gearing up or down, they will provide a good match of speed and torque for many robots. If your vehicle will climb steep slopes or your robotic arm will lift some load, you may still want to gear down. If your vehicle will be lightweight and you want more speed, you can gear up.

Older LEGO motors have less internal gearing than MINDSTORMS NXT servo motors. They rotate at a higher velocity but produce less torque. When using them, you will generally want to gear down to reduce speed and increase torque.

One last thing before you move on to the next topic. We said that there is no free lunch when it comes to mechanics. This is true for this conversion service as well: We have to pay something to get the conversion done. The price is paid in *friction*—something you should try to keep as low as possible—but it's unavoidable. Friction will always eat up some of your torque in the conversion process.

Riding That Train: The Geartrain

The largest LEGO gear is the 40t, and the smallest is the 8t (used in the previous discussion). Thus, the highest ratio we can obtain is 8:40, or 1:5 (Figure 2.2).

Figure 2.2 A 1:5 Gear Ratio

What if you need an even higher ratio? In such cases, you should use a *multistage reduction* (or multiplication) system, usually called a *geartrain*. Look at Figure 2.3. In this system, the result of a first 1:3 reduction stage is transferred to a second 1:3 reduction stage. So, the resulting velocity is one-third of one-third, which is one-ninth, and the resulting torque is three times three, or nine. Therefore, the ratio is 1:9.

Figure 2.3 Geartrain with a Resulting Ratio of 1:9

Geartrains give you incredible power, because you can trade as much velocity as you want for the same amount of torque. Two 1:5 stages result in a ratio of 1:25, whereas three of them result in a 1:125 system! All this strength must be used with care, however, because your LEGO parts may get damaged if for any reason your robot is unable to convert it into some kind of work. In other words, if something gets jammed, the strength of a LEGO motor multiplied by 125 is enough to deform your beams, wring your axles, or break the teeth of your gears. We'll return to this topic later.

Designing & Planning...

Choosing the Proper Gearing Ratio

We suggest you perform some experiments to help you make the right decision in choosing a gearing ratio. Don't wait to finish your robot to discover that some geared mechanics don't work as expected! Start building a very rough prototype of your robot or just of a particular subsystem, and experiment with different gear ratios until you're satisfied with the result. This prototype doesn't need to be very solid or refined, and it doesn't even need to resemble the finished system you have in mind. It is important, however, that it accurately simulates the kind of work you're expecting from your robot, and the actual loads it will have to manage. For example, if your goal is to build a robot capable of climbing a slope with a 50 percent grade, put on your prototype all the weight you imagine your final model is going to carry: additional motors for other tasks, the NXT itself, extra parts, and so on. Be generous

Continued

with extra parts. It is always better to add extra weight, more than you think the final robot will have. Then see if your prototype can climb a test slope. Don't test it without load, as you might discover it doesn't work as a finished robot.

NOTE

Remember that in adding multiple reduction stages, each additional stage introduces further *friction*, the bad guy that makes your world less than ideal. For this reason, if you're aiming for maximum efficiency, you should try to reach your final ratio with as few stages as possible.

Worming Your Way: The Worm Gear

In your NXT box, you've probably found another strange gear, a black one that resembles a sort of cylinder with a spiral wound around it. Is this thing really a gear? Yes, it is, but it is so peculiar we have to give it special mention.

In Figure 2.4, you can see a worm gear engaged with more familiar gears. The assembly on the right uses a special LEGO part. It is called a *worm gear block*. With this single piece you can connect the worm gear to a 24t gear. In just building these simple assemblies, you will discover many properties. Try to turn the axles by hand. Notice that although you can easily turn the axle connected to the worm gear, you can't turn the one attached to the other gears. We have discovered the first important property: The worm gear leads to an *asymmetrical system*; that is, you can use it to turn other gears, but it can't be turned *by* other gears. The reason for this asymmetry is, once again, friction. Is this a bad thing? Not necessarily. It can be used for other purposes.

Figure 2.4 Worm Gears Engaged with Other Gears

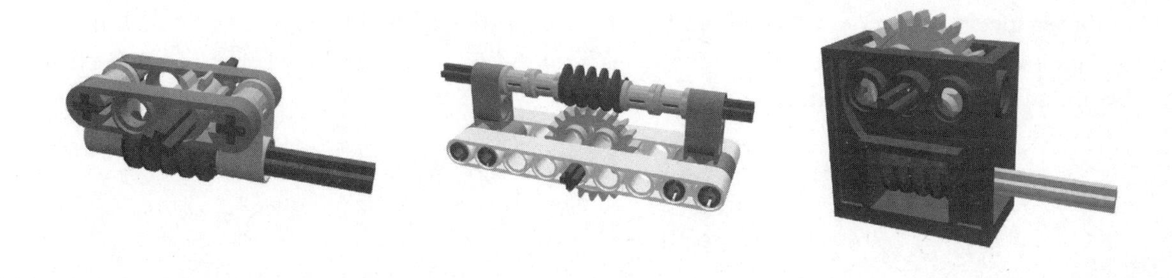

Another fact you have likely observed is that the two axles are perpendicular to each other. This change of orientation is unavoidable when using worm gears. You may also have noticed when building these assemblies that the worm gear slides easily along an axle. Sometimes this is useful, but most of the time you will have to fix it in place along the axle with bushings.

Turning to gear ratios, you're now an expert at doing the math, but you're probably wondering how to determine how many teeth this worm gear has! To figure this out, instead of discussing the theory behind it, we proceed with our experiment. Taking the middle assembly used in Figure 2.4, we turn the worm gear axle slowly by exactly one turn, at the same time watching the 24t gear. For every turn you make, the 24t rotates by exactly one tooth. This is the answer you were looking for: The worm gear is a 1t gear! So, in this assembly, we get a 1:24 ratio with a single stage. In fact, we could go up to 1:40 using a 40t instead of a 24t.

The asymmetry we talked about before makes the worm gear applicable only in reducing speed and increasing torque, because, as we explained, the friction of this particular device is too high to get it rotated by another gear. The same high friction also makes this solution very inefficient, as a lot of torque gets wasted in the process.

As we mentioned earlier, this outcome is not always a bad thing. There are common situations where this asymmetry is exactly what we want. One example would be when designing a robotic arm to lift a small load. Suppose we use a 1:25 ratio made with standard gears: What happens when we stop the motor with the arm loaded? The symmetry of the system transforms the weight of the load (potential energy) into torque, and the torque into velocity, and the motor spins back, making the arm go down. In this case, and in many others, the worm gear is the proper solution, its friction making it impossible for the arm to turn the motor back.

We can summarize all this by saying that in situations when you desire precise and stable positioning under load, the worm gear is the right choice. It's also the right choice when you need a high reduction ratio in a small space, because it allows very compact assembly solutions.

Limiting Strength with the Clutch Gear

Another special device you should get familiar with is the thick 24t white gear, which has strange markings on its face (Figure 2.5). It is a *clutch gear*, and in the next part of this section we'll discover just what it does.

Figure 2.5 The Clutch Gear

Our experiment this time requires very little work; just put one end of an axle inside the clutch gear and the other end into a standard 24t to use as a knob. Keep the latter in place with one hand and slowly turn the clutch gear with the other hand. It offers some resistance, but it turns. This is its purpose in life: to offer some resistance, then give in!

This clutch gear is an invaluable help to limit the strength you can get from a geared system, and this helps to preserve your motors and your parts, and to resolve some difficult situations. The mysterious "2.5•5 Ncm" writing stamped on it (as explained earlier, Ncm is a newton-centimeter, the unit of measurement for torque) indicates that this gear can transmit a maximum torque of about 2.5 to 5 Ncm. When exceeding this limit its internal clutch mechanism starts to slip.

What's this feature useful for? You have seen before that through some reduction stages you can multiply your torque by high factors, thus getting a system strong enough to actually damage itself if something goes wrong. This clutch gear helps you avoid this, limiting the final strength to a reasonable value.

There are other cases in which you don't gear down very much and the torque is not enough to ruin your LEGO parts, but if the mechanics jam, the motor stalls—this could be a very bad thing, because your motor draws a lot of current when stalled. The clutch gear prevents this, automatically disengaging the motor when the torque becomes too high.

In some situations, the clutch gear can even reduce the number of sensors needed in your robot. Suppose you build a motorized mechanism with a bounded range of action, meaning that you simply want your subsystem (arms, levers, actuators—anything) to be in one of two possible states: open or closed, right or left, engaged or disengaged, with no intermediate position. You need to turn on the motor for a short time to switch the mechanism from one state to the other, but unfortunately it's not easy to calculate the precise time a motor needs to be on to perform a specific action (even worse, when the load changes, the required time changes too). If the time is too short, the system will result in an intermediate state, and if it's too long, you might do damage to your motor.

You can use a sensor to detect when the desired state has been reached. If you are using NXT servo motors, you could simply use the built-in motor encoders to determine when you should start and stop the motor (we will discuss this more in Chapter 4). However, you might choose to use a different LEGO motor to power your subsystem. In this case, you will not have a rotation sensor built-in to check. Without a secondary sensor, you will have to run a motor for a specific time. If you put a clutch gear somewhere in the geartrain, you can now run the motor for the approximate time needed to reach the limit in the worst load situation, because the clutch gear slips and prevents any harm to your robot and to your motor if the latter stays on for a time longer than required.

There's one last topic about the clutch gear we have to discuss: where to put it in our geartrain. You know that it is a 24t and can transmit a maximum torque of 5 Ncm, so you can apply here the same gear math you have learned so far. If you place it before a 40t gear, the ratio will be 24:40, which is about 1:1.67. The maximum torque driven to the axle of the 40t will be 1.67 multiplied by 5 Ncm, resulting in 8.35 Ncm. In a more complex geartrain such as that in Figure 2.6, the ratio is 3:5 and then 1:3, coming to a final 1:5; thus, the maximum resulting torque is 25 Ncm. A system with an output torque of 25 Ncm will be able to produce a force five times stronger than one of 5 Ncm. In other words, it will be able to lift a weight five times heavier.

The clutch gear isn't the only way to introduce slip into a system. Later in this chapter we'll discuss pulleys and belts, another way to introduce slip into a system.

Figure 2.6 Placing the Clutch Gear in a Geartrain

From these examples, you can deduce that the maximum torque produced by a system that incorporates a clutch gear results from the maximum torque of the clutch gear multiplied by the ratio of the following stages. When you are gearing down, the more output torque you want, the closer you have to place your clutch gear to the source of power (the motor) in your geartrain. On the contrary, when you are reducing velocity, not to get torque but to get more accuracy in positioning, and you really want a soft touch, place the clutch gear as the very last component in your geartrain. This will minimize the final supplied torque.

This might sound a bit complex, but we again suggest you learn by doing, rather than by simply reading. Prototyping is a very good practice. Set up some very simple assemblies to experiment with the clutch gear in different positions, and discover what happens in each case.

Placing and Fitting Gears

The LEGO gear set includes many different types of gear wheels. Up to now, we played with the straight 8t, 24t, and 40t, but the time has come to explore other kinds of gears, and to discuss their use according to size and shape.

In studless buildings, unlike traditional studded building, the holes in TECHNIC beams stacked atop one another are the same distance apart as holes in a single beam. This means gears connected together will be the same number of holes apart, whether connected horizontally along the same beam or vertically across multiple beams (see Figure 2.7 and Figure 2.8).

Figure 2.7 Vertical Matching of Gears

Figure 2.8 Matching Gears Horizontally and Vertically

Bricks & Chips...

Idler Gears

Figure 2.7 offers us the opportunity to talk about *idler gears*. What's the ratio of the geartrain in the figure? Starting from the 8t, the first stage performs an 8:24 reduction, and the second is 24:40. Multiplying the two fractions, you get 8:40, or 1:5, the same result you'd get by meshing the 8t directly to the 40t. The intermediate 24t is an idler gear, which doesn't affect the gear ratio. Idler gears are quite common in machines, usually to help connect distant axles. Are idler gears totally lacking in effects on the system? No, they have one very important effect: They change the direction of the output!

Another common straight gear is the 16t gear (Figure 2.9).

Figure 2.9 The 16t Gear

Its radius is 1, and it combines well with a copy of itself at a distance of two. Getting it to cooperate with other straight gears, however, is very difficult.

Designing & Planning ...

Backlash

Some gears can be matched diagonally. That is, they can be matched not along a horizontal or vertical beam, but on a diagonal. Diagonal matching is often less precise than horizontal and vertical types, because it results in a slightly larger distance between gear teeth. This extra distance increases the *backlash*, the amount of oscillation a gear can endure without affecting its meshing gear. Backlash is amplified when gearing up, and reduced when gearing down. It generally has a bad effect on a system, reducing the precision with which you can control the output axle, and for this reason, it should be kept to a minimum.

When you are using a pair of 16t gears, the resulting ratio is 1:1. You don't get any effect on the angular velocity or torque (except in converting a fraction of them into friction), but indeed there are reasons to use them as a pair—for instance, when you want to transfer motion from one axle to another with no other effects. This is, in fact, another task that gears are commonly useful for. There's even a class of gears specifically designed to transfer motion from one axle to another axle perpendicular to it, called *bevel gears*.

The most common member of this class is the 12t bevel gear, which can be used *only* for this task (Figure 2.10), meaning it does not combine at all with any other LEGO gear we have examined so far. Nevertheless, it performs a very useful function, allowing you to transmit the motion toward a new direction, while using a minimum of space. There's also a 20t bevel conical gear with the same design of the common 12t (Figure 2.11). Both of these bevel gears are half a beam in thickness, whereas the other gears are one beam in thick.

Figure 2.10 Bevel Gears on Perpendicular Axles

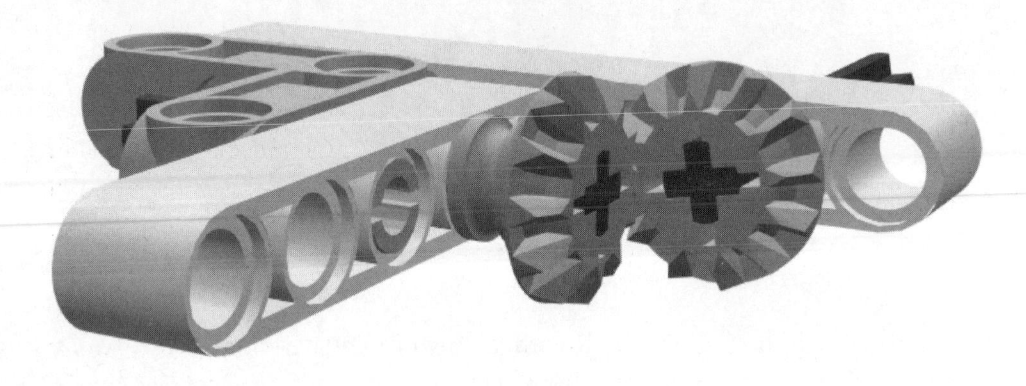

Figure 2.11 The 20t Bevel Gear

The 24t gear also exists in the form of a *crown gear*, a special gear with front teeth that can be used like an ordinary 24t, but can also combine with another straight gear to transmit motion in an orthogonal direction (that is, composed of right angles), possibly achieving at the same time a ratio different from 1:1 (Figure 2.12).

You may have noticed another group of gears in your collection. They are wider and the edges of their teeth look like the bevel gear on both sides. These are *double bevel* gears. Count the number of teeth on them. You will find they are 12t, 20t, and 36t gears (see Figure 2.13).

Figure 2.12 The Crown Gear on Perpendicular Axles

Figure 2.13 12t, 20t, and 36t Double Bevel Gears

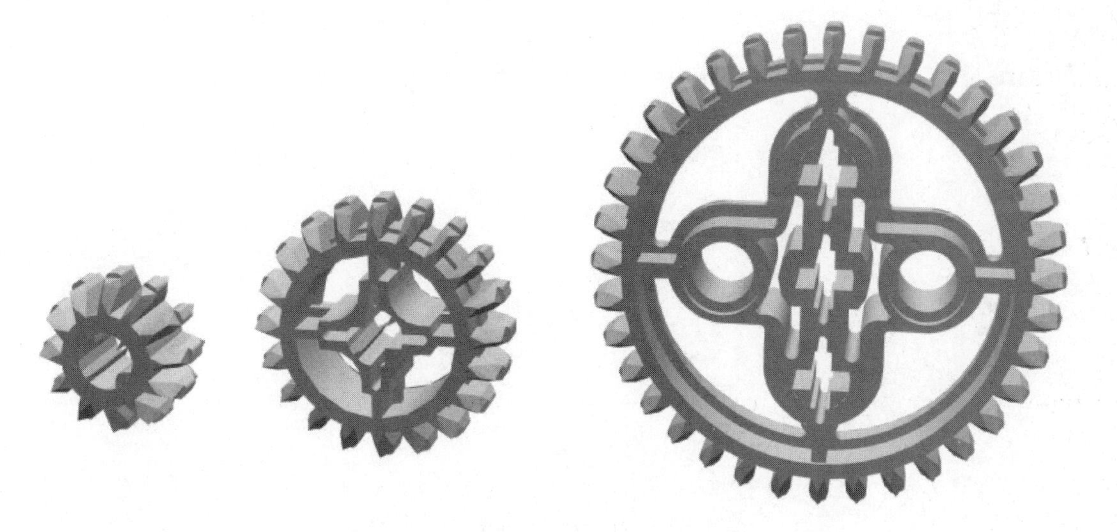

You will notice when you place these gears on a single beam that two gears of the same size will not mesh with one another. For this reason, double bevel gears are generally used in pairs of different sizes, either a 12t and a 20t, or a 12t and a 36t. These gears are designed to work well in both perpendicular and horizontal setups (Figure 2.14).

Figure 2.14 Meshing Double Bevel Gears

Mismatched pairs of double bevel gears use the same hole spacing as straight gears, so pairs of them can be used in place of a pair of straight gears, offering some new gear ratios—for example, a 12t and a 20t double bevel gear pair mesh at a distance of 2, the same as an 8t and 24t straight gear pair. Be careful when mixing and matching double bevel gears and straight gears; although they use some of the same hole spacing, double bevel gears and straight gears don't work with each other. To use them together you will have to use them in pairs (Figure 2.15).

Figure 2.15 Double Bevel Gears and Straight Gears in a Geartrain

The last gear that we'll describe doesn't look like a gear at all. In fact, it isn't known as a gear, but as a *knob wheel* (Figure 2.16)

Figure 2.16 TECHNIC Knob Wheel

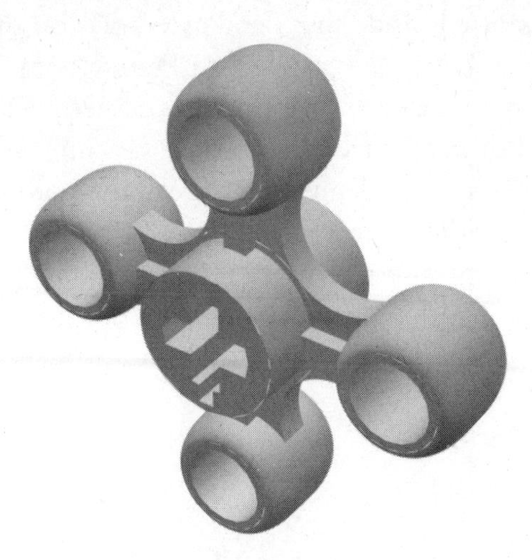

Examine the knob wheel and you will find that it is basically a 4t gear (see Figure 2.17). Connect it with another copy of itself and you will see that it works very well as a gear. Like double bevel gears, it can also work in perpendicular setups.

Figure 2.17 TECHNIC Knob Wheel on Perpendicular Axles

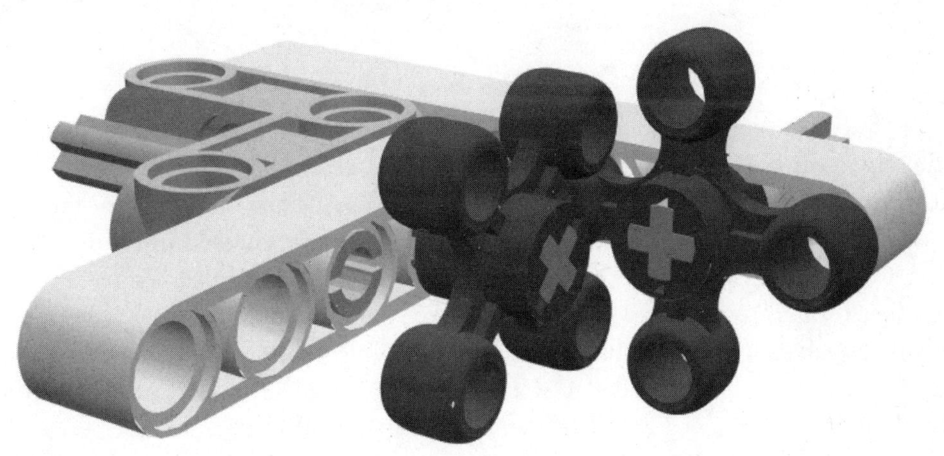

In perpendicular setups, knob wheels have one major advantage over double bevel gears. Connect two knob wheels. Examine how much area on each tooth of the knob wheel contacts a tooth on the other. Compare that to a pair of double bevel gears. The knob wheels have much more contact area. This means they can transmit much more torque from one axle to the other.

The final class of gears is actually a combination of gears and bricks. They are clear LEGO bricks with gears encased inside them and axle holes on the faces. These bricks are commonly referred to as TECHNIC gearboxes (see Figure 2.18). They come in three different types. The first is a worm gearbox that functions exactly like the worm gear block with the 24t gear discussed earlier. In fact, it even has the same gear ratio, as the gear inside is a 24t gear. The second type is a 90-degree angle gearbox. Encased within it are two bevel gears at a right angle to each other. This can be used exactly like the bevel gears previously discussed. The final gearbox type also uses bevel gears at right angles. However, instead of having just two axle holes, it has three, so it can be used as a T-connection. It can be used to split the output of a single driven axle into two outputs.

Figure 2.18 TECHNIC Gearboxes

These gearboxes are functionally no different from the bevel gears previously discussed. The same gear trains could be built with the bevel gears and worm gears already discussed. However, the gearboxes do have two major advantages. They take up very little space and are able to transmit quite a bit of torque. The gears are firmly encased in the LEGO bricks, so there is very little opportunity for the gears to slip. These gearboxes do have two drawbacks. The first is their availability. The gearboxes have been included in only a few sets and are not widely available. Second, the bricks are studded, which means they are primarily designed to be used with traditional studded TECHNIC bricks, not the studless beams included in the NXT kit and most new TECHNIC sets. This doesn't mean they cannot be used in studless constructions; as you will see in Chapter 6, using both studded and studless building techniques in a single creation is possible.

Using Pulleys, Belts, and Chains

Pulleys and *belts* are two classes of components designed to work together and perform functions similar to that of gears—similar, that is, but not identical. They have indeed some peculiarities which we shall explore in the following paragraphs. The MINDSTORMS NXT kit includes a few pulleys, but no belts. You will have to buy them separately.

Chains are also not part of the basic NXT kit. Though not essential, they allow you to create mechanical connections that share some properties with both geartrains and pulley-belt systems.

Pulleys and Belts

Pulleys are like wheels with a groove (called a *race*) along their diameter. The LEGO TECHNIC system currently includes four kinds of pulleys, shown in Figure 2.19.

Figure 2.19 Pulleys

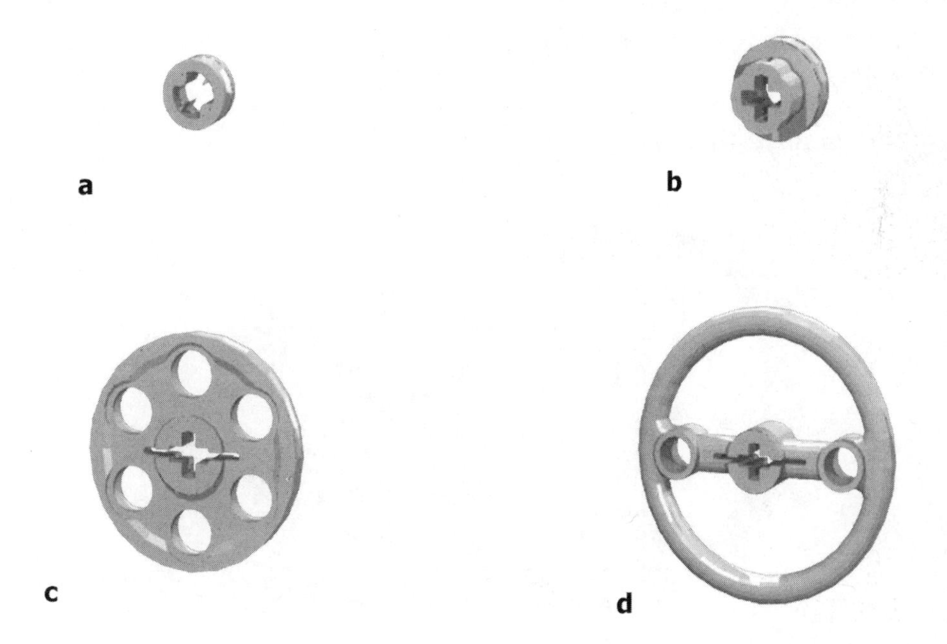

The smallest one (a) is actually the half-size bush, normally used to hold axles in place to prevent them from sliding back and forth. Because it does have a race, it can be properly termed a pulley. Its diameter is one LEGO unit, with a thickness of half a unit.

The small pulley (b) is 1 unit in thickness and about 1.5 units in width. It is asymmetrical, however, because the race is not in the exact center. One side of the axle hole fits a rubber ring that's designed to attach this pulley to the micromotor. The medium pulley (c) is again half a unit thick and 3 units in diameter. Finally, the large pulley (d) is 1 unit thick and about 4.5 units in diameter.

CLINTON-MACOMB PUBLIC LIBRARY

LEGO belts are rings of rubbery material that look similar to rubber bands. They come in four versions, with different colors corresponding to different lengths: white, blue, red, and yellow. Don't confuse them with actual rubber bands: Rubber bands have much greater elasticity, and for this reason are much less suitable to the transfer of motion between two pulleys. This is, in fact, the purpose of belts: to connect a pair of pulleys. LEGO belts are designed to perfectly match the race of LEGO pulleys.

Let's examine a system made of a pair of pulleys connected through a belt (Figure 2.20). The belt transfers motion from one pulley to the other, making them similar to a pair of gears. How do you compute the ratio of the system? You don't have any teeth to count... The rule with pulleys is that the reduction ratio is determined by finding the ratio between their diameters (this rules applies to gears too, but the fact that their circumference is covered with evenly spaced teeth provides a convenient way to avoid measurement). You actually should consider the diameter of the pulley *inside* its race, because the sides of the race are designed specifically to prevent the belt from slipping from the pulley and don't count as part of the diameter over which the belt acts.

Figure 2.20 Pulleys Connected with a Belt

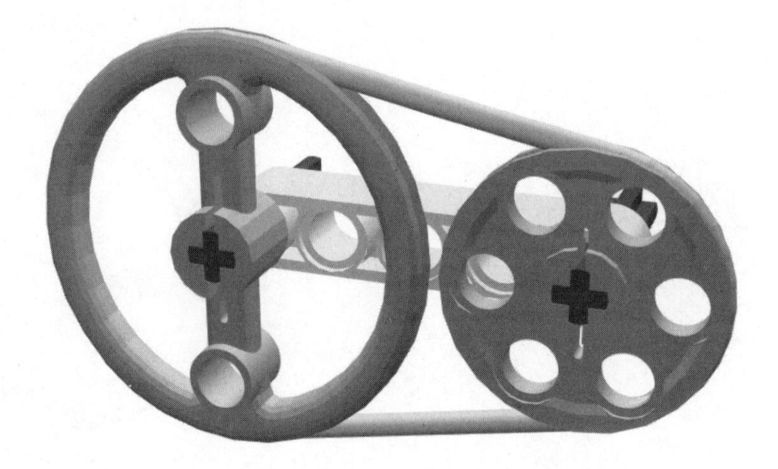

You must also consider that pulleys are not very suitable to transmitting high torque, because the belts tend to slip. The amount of slippage is not easy to estimate, as it depends on many factors, including the torque and speed, the tension of the belt, the friction between the belt and the pulley, and the elasticity of the belt.

For those reasons, we preferred an experimental approach and measured some actual ratios among the different combinations of pulleys under controlled conditions. You can find our results in Table 2.1.

CLINTON-MACOMB PUBLIC LIBRARY

Table 2.1 Ratios among Pulleys

	Half Bush	Small Pulley	Medium Pulley	Large Pulley
Half bush	1:1	1:2	1:4	1:6
Small pulley	2:1	1:1	1:2.5	1:4.1
Medium pulley	4:1	2.5:1	1:1	1:1.8
Large pulley	6:1	4.1:1	1.8:1	1:1

These values may change significantly in a real-world application, when the system is under load. Because of this, it's best to think of the figures as simply an indication of a possible ratio for systems where very low torque is applied. Generally speaking, you should use pulleys in your first stages of a reduction system, where the velocity is high and the torque is still low. You could even view the slippage problem as a positive feature in many cases, acting as a torque-limiting mechanism such as the one we discussed in the clutch gear, with the same benefits and applications. However, be careful when using belts and pulleys to allow slip in a system. Belts turning around pulleys that are not turning will cause friction and heat to build up. In time, the belt will break. Also, the belts have a tendency to jump off the pulley, causing your entire system to fail.

Another advantage of pulleys over gear wheels is that their distance is not as critical. Indeed, they help a great deal when you need to transfer motion to a distant axle (Figure 2.21). And at high speeds, they are much less noisy than gears—a facet that occasionally comes in handy.

Figure 2.21 Pulleys Allow Transmission across Long Distances

Chains

LEGO *chains* come in two flavors: *chain links* and *tread links*. LEGO produces two different types of tread links. The first shares the same hooking system with the chain links. They are freely mixable to create a chain of the required length. The second tread link is a newer introduction and uses a different hooking system, so it can be connected only with other copies of itself (see Figure 2.22).

Figure 2.22 Chain Link, Tread Link, and New-Style Tread Link

Chains are used to connect gear wheels in the same way belts connect with pulleys (see Figure 2.23). They share similar properties as well: Both systems couple parallel axles without reversing the rotation direction and both give you the chance to connect distant axles. The big difference between the two is that chain links don't allow any slippage, so they transfer *all* the torque (actually, the maximum torque a chain can transfer depends on the resistance of its individual links, which in the case of LEGO chains is not very high). On the other hand, they introduce further friction into the system, and for this reason are much less efficient than direct gear matches. You will find chains useful when you have to transfer motion to a distant axle in low-velocity situations. The ratio of two gears connected by a chain is the same as their corresponding direct connection. For example, a 16t connected to a 40t results in a 2:5 ratio.

Figure 2.23 Chain Links

TECHNIC tread links can be used to make tracked robots (see Figure 2.24). However, like most LEGO parts, they are plastic. They provide very little grip when on slippery surfaces such as wood, tile, or plastic. They work better on other surfaces such as carpet.

Figure 2.24 Tread Links

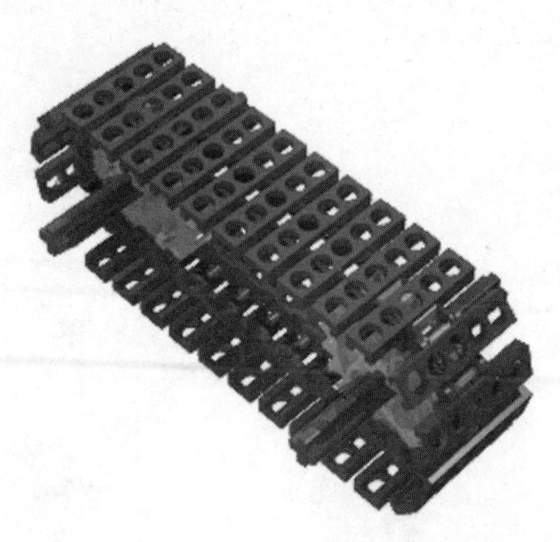

The gears used to drive chains or treads are commonly referred to as sprockets. LEGO chains and the older tread links use straight gears for their drive sprockets. The new-style tread link uses a new LEGO drive sprocket design that meshes only with the new tread link. It will not mesh with any other gear types.

Making a Difference: The Differential

We want to introduce you to a very special device now: the *differential gear.* You probably know that there's at least one differential gear in every car. What you might not know is why the differential gear is so important.

Let's do an experiment together. Take two wheels from your NXT kit and connect their hubs with the longest axle (Figure 2.25). Now put the wheels on your table and push them gently: They run smoothly and advance some feet, going straight. *Very straight.* Keep the axle in the middle with your fingers and try to make the wheels change direction while pushing them. It's not so easy, is it?

The reason is that when two parallel wheels turn, their paths must have different lengths, the outer one having a longer distance to cover (Figure 2.26). In our example, in which the wheels are rigidly connected, at any turn they cover the same distance, so there's no way to make them turn unless you let one slip a bit.

Figure 2.25 Two Connected Wheels Go Straight

Figure 2.26 During Turns the Wheels Cover Different Distances

The next phase of our experiment requires that you now build the assembly shown in Figure 2.27. You see a differential gear with its three 12t bevel gears, two 6-length axles, and four beams connected together to provide you with a way to handle this small system. Placing the wheels again on your table, you will notice that while pushing them, you can now easily turn smoothly in any direction. Please observe carefully the *body* of the differential gear and the central bevel gear: When the wheels go straight, the body itself rotates while the bevel gear is stationary. On the other hand, if you turn your system in place, the body stays put and the bevel gear rotates. In any other intermediate case, both of them rotate at some speed, adapting the system to the situation. Differentials offer a way to put power to the wheels without the restriction of a single fixed drive axle.

Figure 2.27 Connecting Wheels with the Differential Gear

To use this configuration in a vehicle, you simply have to apply power to the body of the differential gear, which incorporates a 24t on one side and a 16t on the other.

The differential gear has many other important applications. You can think of it as a mechanical adding/subtracting device. Again place the assembly from Figure 2.27 on your table. Rotate one wheel while keeping the other from turning; the body of the differential gear rotates half the angular velocity of the rotating wheel. You already discovered that when turning our system in place, the differential does not rotate at all, and then when both wheels rotate together, the differential rotates at the same speed as well. From this behavior, we can infer a simple formula:

(Iav1 + Iav2) / 2 = Oav

where *Oav* is the *output angular velocity* (the body of the differential gear), and *Iav1* and *Iav2* are the *input angular velocities* (the two wheels). When applying this equation, you must remember to use *signed* numbers for the input, meaning that if one of the input axles rotates in the opposite direction of the other, you must input its velocity as a negative number. For example, if the right axle rotates at 100 revolutions per minute (rpm) and the left one at 50 rpm, the angular velocity of the body of the differential results in this:

(100 rpm + 50 rpm) / 2 = 75 rpm

There are situations when you deliberately reverse the direction of one input, using idler gears, to make the differential sensitive to a difference in the speed of the wheels, rather than to their sum. Reversing the input means that you must make one of the inputs negative. See what happens to the differential when both wheels run at the same speed—let's say, 100 rpm:

(100 rpm – 100 rpm) / 2 = 0 rpm

It doesn't move! As soon as a difference in speed appears, the differential starts rotating with an angular velocity equal to half this difference:

(100 rpm – 98 rpm) / 2 = 1 rpm

This is a useful trick when you want to be sure your wheels run at the same speed and cover the same distance: Monitor the body of the differential and slow the left or right wheel appropriately to keep it stationary. See Chapter 8 for a practical application of this trick.

Summary

Few pieces of machinery can exist without gears, including robots, and you ought to know how to get the most benefit from them. In this chapter, you were introduced to some very important concepts: gear ratios, angular velocity, force, torque, and friction. Torque is what makes your robot able to perform tasks involving force or weight, such as lifting weights, grabbing objects, or climbing slopes. You discovered that you can trade off some velocity for some torque, and that this happens along rules similar to those that apply to levers: The larger the distance from the fulcrum, the greater the resulting force.

The output torque of a system, when not properly directed to the exertion of work, or when something goes wrong in the mechanism itself, can cause damage to your LEGO parts. You learned that the clutch gear and belts and pulleys are precious tools to limit and control the maximum torque so as to prevent any possible harm.

In addition to straight gears, you were also introduced to bevel and double bevel gears. Bevel gears are useful to transfer motion between perpendicular axles. Double bevel gears can transfer motion between both perpendicular and parallel axles, as can the knob wheel.

Gears are not the only way to transfer power; we showed that pulley-belt systems, as well as chains, may serve the same purpose and help you in connecting distant systems. Belts provide an intrinsic torque-limiting function and do well in high-speed, low-torque situations. Chains, on the other hand, don't limit torque but do increase friction, so they are more suitable for transferring power at slow speeds.

Last but not least, you explored the surprising properties of the differential gear, an amazing device that can connect two wheels so that they rotate when its body rotates, still allowing them to turn independently. The differential gear has some other applications too, because it works like an adder-subtracter that can return the algebraic sum of its inputs.

If these topics were new to you, we strongly suggest you experiment with them before designing your first robot from scratch. Take a bunch of gears and axles and play with them until you feel at ease with the main connection schemes and their properties. This will offer you the opportunity to apply some of the concepts you learned from Chapter 1 about bracing layers with vertical beams to make them more solid (when you increase torque, many designs fall apart unless properly reinforced). You won't regret the time spent learning and building on this knowledge. It will pay off, with interest, when you later face more complex projects.

Chapter 3

Controlling Motors

Solutions in this chapter:

- **Pacing, Trotting, and Galloping**
- **Mounting Motors**
- **Wiring Motors**
- **Controlling Power**
- **Coupling Motors**

Introduction

Motors will be your primary source of power. Your robots will use them to move around, lift loads, operate arms, grab objects, pump air, and perform any other task that requires power. There are different kinds of electric motors, all of them sharing the property of converting *electrical energy* into *mechanical energy*. In this chapter, we will survey different kinds of LEGO motors and will discuss how to use, mount, connect, and combine them.

Before entering the world of motors, we would like to introduce you to some basic concepts about electricity. You should be aware of a very important distinction concerning the two types of electrical current: *alternating current* (AC) and *direct current* (DC). Alternating current is the type of electricity that comes out of the wall outlets in your house, whereas batteries are the most typical source of direct current. All the electric LEGO devices, including motors, work with DC only.

To understand what DC is, imagine a stream of water going down a hill. Electricity flowing through a wire is not very different: When you connect a battery to a device such as a lamp or a motor, you enable a circuit through which electricity flows more or less like water in a stream. You know that batteries have positive (+) and negative (–) signs stamped on them: These signs indicate two poles, where the electrons flow from minus to plus, as though the minus pole were the top of the hill and, as a result, the current flows from plus to minus. You can place a water mill along a stream to convert the energy of water into mechanical energy; similarly, an electric motor converts an electric flow into motion. What would happen to the water mill if you could reverse the direction of the stream? It would change its direction of rotation. The same happens to DC motors. Every motor has two connectors, one to attach to the negative pole and the other to connect to the positive end of a DC source. You can imagine the current flowing from the positive pole of the battery into the motor, making it move and then coming out again to return to the negative pole of the battery. If you reverse the *polarity*, that is, if you swap the wires between the motor and the battery, you will change the direction of the stream and, thus, the direction of the motor.

Continuing with our hydraulics metaphor, how would you describe the *quantity* of water that's flowing in a stream? It depends on two factors: the speed of the water, and the width of the stream. Both of them have an influence on the kind of work your mill can perform. In the realm of electricity, the speed of the stream is called *voltage*, and its width (its intensity) is called *current*. They are respectively expressed in volts (V) and amperes (A), or sometimes in their submultiples, *millivolts* (mV) and *milliamperes* (mA). The amount of work that an electrical flow can perform—for example, through a motor—depends on both of these quantities. To be more precise, it depends on their product, called *power*, and is measured in watts (W).

All motors are designed to run at a specific voltage, but they are very tolerant when it comes to decreases in the supplied voltage. They simply turn more slowly. However, if you *increase* the voltage above the specific limit for a motor, you stand a good chance of burning it out.

Current has a different behavior. It's the motor that "decides" how much current to draw according to the work it's doing: The higher the load, the greater the current. The situation you should avoid at all costs when working with your NXT is to have the motor *stall* (it is connected to the power source but something prevents it from turning). What happens in this case is that the motor tries to win out against the resistance, drawing in more current so that it can convert it into power, but as it doesn't succeed in the task, all that current becomes *thermal energy* instead of mechanical energy—in other words, heat. This is the most dangerous condition for an electric motor. And here is where use of clutch gear comes into play, limiting the maximum torque and thus preventing stall situations. You will discover later in the chapter that the NXT also has an active role in protecting your motors from dangerous draws of current.

Pacing, Trotting, and Galloping

Every motor contains one or more coils and permanent magnets that convert electrical energy into mechanical energy, but you don't really need to know this level of detail. What you, as a robot builder, must remember is that every motor has a connector through which you can supply it energy, and an output shaft which draws the power. The current LEGO TECHNIC line includes several motors. All of these are 9V DC motors and have different properties and features for various applications (as shown in Figure 3.1): the ungeared motor (a), the geared motor (b), the micromotor (c), the RC motor (d), and the NXT servo motor (e). Recently, LEGO also introduced the Power Function system, which includes two more motors (f and g). There are other special motors as well: the train motor, the geared motor with battery pack, and the Micro Scout unit. These are less common, less versatile, and less useful to robotics than the ones featured here, so we won't be examining them here.

Figure 3.1 The LEGO TECHNIC Motors

a

b

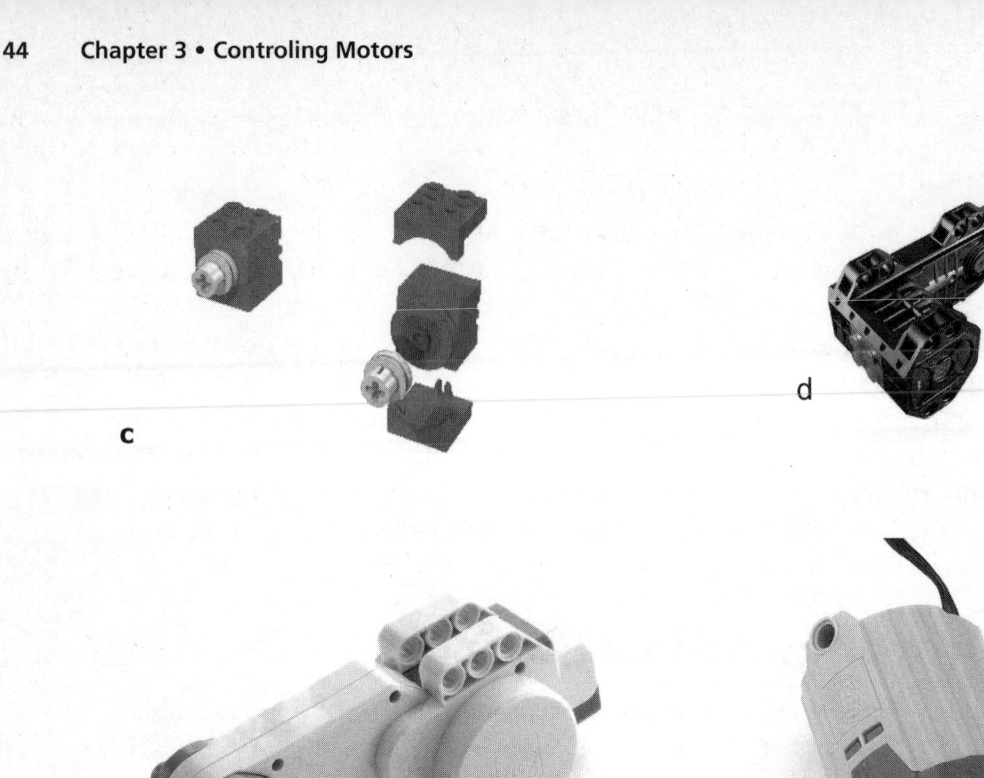

Table 3.1 summarizes the properties of these motors.

Table 3.1 Properties of the LEGO TECHNIC Motors

	Maximum Voltage	Minimum Current (No Load)	Maximum Current (Stall)	Maximum Speed (No Load)	Speed under Typical Load
Ungeared motor	9V DC	100 mA	450 mA	4,000 rpm	2,500 rpm
Geared motor	9V DC	10 mA	250 mA	350 rpm	200–250 rpm
Micromotor	9V DC	5 mA	90 mA	30 rpm	25 rpm
RC motor	9V DC	160 mA	3.2 A	1,300 rpm	900–1,200 rpm
NXT servo motor	9V DC 12 V DC*	60 mA	2 A	170 rpm	100–130 rpm
Power Function large motor	9V DC	60 mA	2 A	250 rpm	175–200 rpm
Power Function medium motor	9V DC	60 mA	2 A	450 rpm	325–375 rpm

* Handles for short periods; however, this is not recommended for extended periods.

The ungeared motor (a) has been the standard LEGO TECHNIC motor for a long time. Its axle is simply an extension of the inner electric motor shaft, and for this reason we called it *ungeared*. Electric motors usually rotate at very high speeds, and this one is no exception, turning at more than 4,000 rpm (revolutions per minute). This makes this motor a bit tricky to use, because it requires very high reduction ratios for almost any practical application, leading to very cumbersome and complex geartrains. Add the fact that it draws an amazing amount of current, and you get a pretty good picture of how difficult it can be.

This motor is still easy to find in the shops of many countries as an expansion pack (8720), but you may want to consider other types of motors for the reasons mentioned in the preceding paragraph. In this book, you won't find any example that includes the ungeared motor. Nevertheless, if you already have one, you can safely use it; it won't damage your NXT or be damaged itself. The only risk you're taking is that, under heavy loads or stall situations, it drains your batteries very quickly.

The geared motor (b) features an internal multistage reduction geartrain and turns at about 350 rpm with no load (typically 200–250rpm with medium load). It's much more efficient than the ungeared kind, and it has low current consumption. It also uses more compact geartrains. If you have the old MINDSTORMS RCX kit, you already have two of these.

The micromotor (c) is a geared motor as well. It's geared down so much that its output shaft turns at approximately 30 rpm. Nevertheless, its torque is incredibly low, well below 1

Ncm. It is also surprisingly noisy, and very easy to jam. At this point, you might wonder why you should ever consider this motor, but the answer lies in its name: because it's micro. Sometimes the size of the motor is more critical than the amount of torque and speed needed. To be used, it requires some special mounting brackets, and a small pulley to connect to its shaft (Figure 3.1c).

Bricks and Chips...

How to Release a Jammed Micromotor

A micromotor jams very easily, so you should know what to do when it occurs. The following steps should help:

1. Switch off the motor as soon as you can. Detach the cord or switch the power off; it's important not to leave a stalled motor under power for a long time because that could permanently damage it.
2. Decouple the motor from any connection (gearings, pulleys, and so on). Leave the small pulley attached to the motor shaft.
3. Holding the motor with your fingers, turn the pulley gently but firmly in the same direction the motor was turning when it jammed. At the same time, push the pulley against the motor until you hear a "click." Your motor should be okay now. If you don't know what direction the motor was rotating when it jammed, try both directions.

This procedure usually works. If it doesn't, try to power on the motor in both directions with very brief current pulses, at the same time doing what's described in step 3.

The RC motor (d) is a geared motor with approximately 1,300 rpm without load and 900 rpm under medium load. Output is delivered through a bush to join axles. It features two outputs turning at different rpms and opposite directions. The farther output is running at about 1,000 rpms without load. The higher rpm output delivers lower torque than the other.

The NXT servo motor (e) is not only geared, but also has other electronics to provide precise positioning information. This motor runs at lower rpms, but has very high torque. In the next section, you will see how to connect these motors to your NXT.

The new Power Function system motors (f and g) are also geared motors and have a special electrical connector. These motors are compact and versatile for use in small places.

Internals of NXT Servo Motor

Servo motors in industrial applications are different from regular motors because of their capability to precisely rotate the motor shaft. This is achieved by special electronics built into the motors. Similarly, the NXT servo motors are advanced in their capabilities and precision. Philippe Hurbain's Web site, NXT motor internals, is an excellent place to learn more about the internals of these motors (refer to Appendix A), and some of his material is included in the following section.

NXT servo motors have a built-in optical encoder that keeps count of rotations of the motor shaft (see Figure 3.2). This encoder is accurate up to 1 degree of motor rotation. You can use this property from your program for precise movement or positioning:

```
while (nMotorEncoder[motorA] < 1000)
// wait for motor to reach a specific location
{
. . .
}
```

This property can also keep two motors synchronized with each other and move your robot along a straight line.

Figure 3.2 Optical Encoder in NXT Servo Motor

The NXT servo motor also has built-in gears to reduce the rpms and increase the torque (see Figure 3.3). This desirable feature makes it easier to build robots without excessive geartrains, thereby reducing the complexity and size of your robot.

Figure 3.3 Internal Gears of NXT Motor

Image ©2007 The LEGO Group. Used here by special permission.

Mounting Motors

The NXT servo motor is designed for integration into the studless construction of your robot. The large rounded end is about 7 units (TECHNIC holes) high and 5 units wide, whereas the orange end of the motor is 3 units wide and 3 units high. Overall, the motor is about 14 units long, and due to its unusual shape, it requires some experience to mount on your robot. In the following paragraphs, we will discuss a few common solutions, as well as how to take advantage of this shape in your construction.

Despite its unusual shape, the NXT servo motor fits well within the standard TECHNIC grid. The elongated shape of the motor makes it easy to integrate with the primary chassis of your robot. While designing your robot, try to integrate the motors in the early stages of the design, or build the robot assembly around the motors. In Figure 3.4 (a), you can see that the large rounded end has two built-in three-hole beams on top, which you can connect to your robot's structure using TECHNIC pins. In Figure 3.4 (b), you can see the holes which can be connected using TECHNIC double pins.

Figure 3.4 Mounting a Motor with TECHNIC Beams and Pins

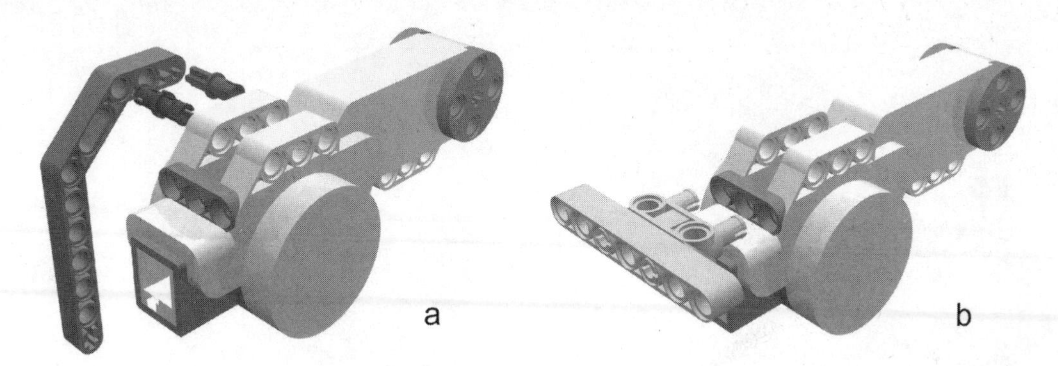

NOTE

The pictures here are mainly meant to illustrate possibilities. So, in order to let you visualize, we didn't lock the pins and the beams to the motor. In actual applications, you will complete the assembly and extend it for your needs.

In Figures 3.4a and 3.4b, the top-right end of the motor has a through hole, and the axle fits snugly into it. For simple mobile robots, you can choose to attach wheels directly onto this axle. This end also has one built-in three-hole beam to which you can connect your robot's structural beams or pins.

When it comes to transferring power along a different axis or to a different location, you have plenty of choices, but essentially you will use gears. With the high torque delivered from the NXT motor, belts are not very effective unless you can tolerate a lot of slippage. Figure 3.5 (a) shows one such assembly. Experiment with other pairs, as shown in Figure 3.5 (b), to see which best suits your needs.

As we said earlier, NXT motors are suitable for integration into your robot's structure, but when you need to reuse them in other projects, it's a challenge to keep them easily removable, while keeping the rest of the robot intact. When planning for a removable motor in your design, consider attaching motors with TECHNIC pins and a stop bush (see Figure 3.6). These bushes are easy to hold, and it's also easy to use them to pull the pins. Try to keep the connections along a single plane which effectively separates the motor from your robot's structural design, thus keeping your robot intact when the motor is removed.

Figure 3.5 Transferring Power to a Different Axis

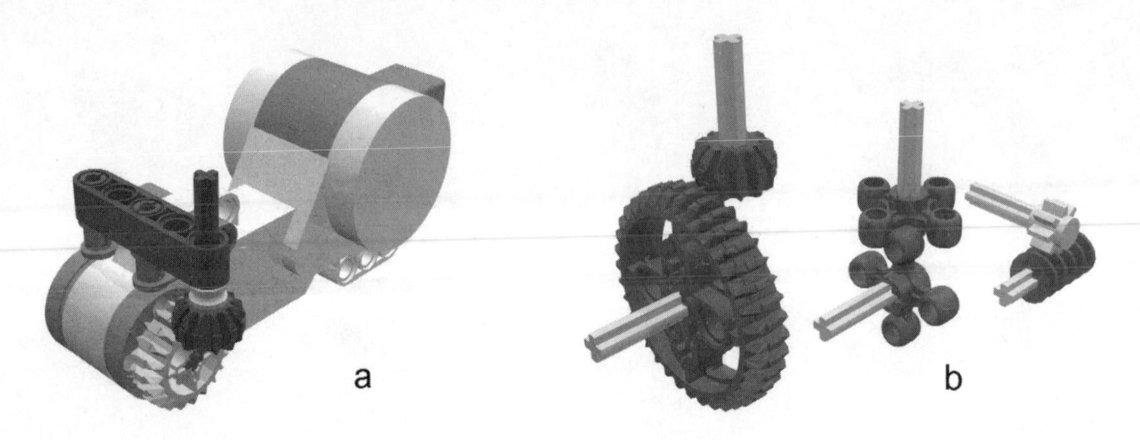

Figure 3.6 An Easily Removable Motor

Have you wondered how to connect the studded motors to studless beams? If you have RCX motors, the easiest method is to use some of your TECHNIC bricks with holes to mount the motor (Figure 3.7). Use the 1 x 2 plates with rails as brackets for the motors, and use TECHNIC bricks to hold the motors in place. Mount the studless beams over the TECHNIC bricks. Also, for additional stability at the back, you can use a TECHNIC axle joiner with four pins.

Figure 3.7 Attaching an Old Geared Motor to Studless Beams

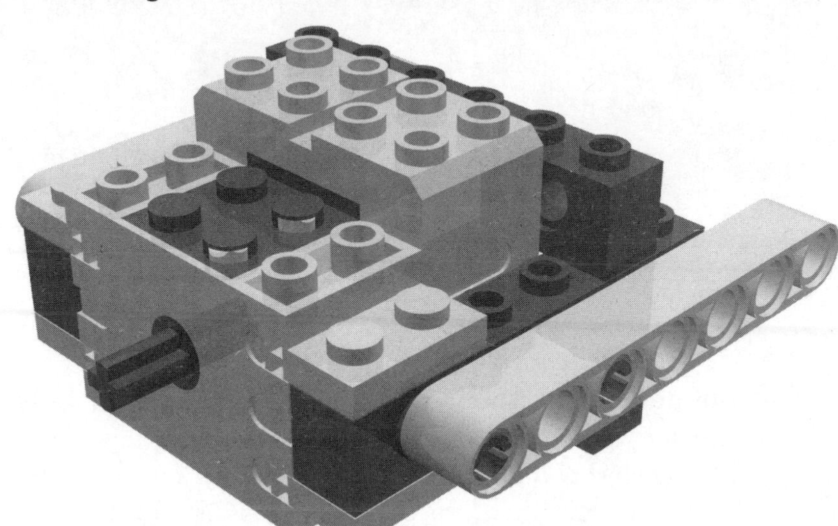

Wiring Motors

The MINDSTORMS NXT wiring system uses jacks similar to telephone jacks. Though they look similar, you cannot use regular phone wires in them. That also will keep some creative minds from plugging the NXT into the telephone network.

As we already explained, these motors are DC motors, and therefore, they are sensitive to the *polarity* with which you connect them. With NXT wiring systems, you cannot go wrong with the polarity. But if you are connecting old motors with the NXT using a compatibility cable, you will have to consider the polarity or control this property from your program.

How can you test your motors without adjusting your programming? Here are some suggestions:

- **The NXT console** Power on the NXT and press the scroll button on your NXT console until it reads **NXT Program**. Select the program using the center orange button. This is a built-in test program which allows you to create a mini program on the NXT. You can control up to two motors in this mini program.
- **The RobotC software** You can use RobotC to directly control the NXT. From the **Robot** menu, select the **NXT Brick** submenu, followed by the **Poll Brick** submenu. In the resulting window, you can control the motors directly through the **Set values into NXT** interface.
- **An external battery box** Various kinds of battery boxes are available, as shown in Figure 3.8 (a) and (b). With a box such as this and an NXT converter cable such as the one shown inFigure 3.8 (c), you can test your motor without the NXT.

Figure 3.8 LEGO Battery Boxes and Converter Cable

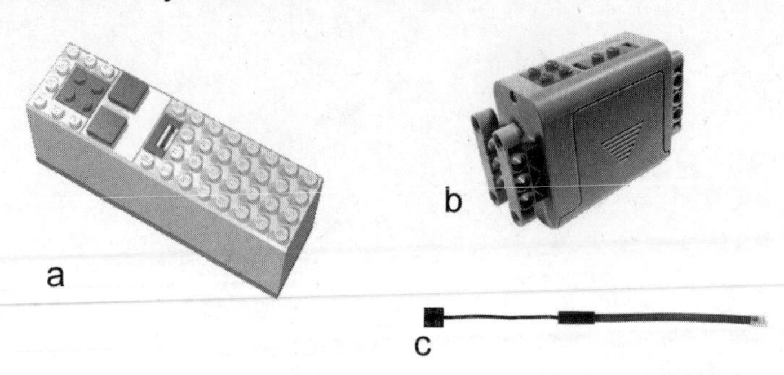

- **A Bluetooth device** You can use a Java-capable and Bluetooth-enabled cell phone to send messages to the NXT. Using a joystick or command wheel on your phone you can control two motors on the NXT. This is very useful for testing your robot during the building phase, especially when it is hard to reach the NXT console.
- **Other sources** All the components of the LEGO 9V electric system are compatible with each other. If you have a LEGO train speed regulator, or a Control Center unit, you can safely use them to run your motors using an NXT converter cable.

In some cases, you want to control more motors than the NXT ports can support. Or you may want to attach power-hungry motors to your robot, but conserve the NXT battery. For such applications, you can use a Motor Multiplexer from Mindsensors (see Figure 3.9). This multiplexer can conveniently attach up to four additional motors using RCX-style connectors. It supplies power to these additional motors from an external battery, thus conserving the NXT battery. You can also connect NXT servo motors using a converter cable.

Using Power Function Motors with the NXT

The new Power Function system is also a 9V DC system like the NXT. Because these modules use a different electrical connector, it will be a bit of a challenge to use these motors directly with NXT robots. LEGO will be developing a converter cable to connect the Power Function system with the old 9V system. And with the converter cable already developed between the NXT product and the old LEGO 9V system, it will be possible to use the LEGO Power Function together with MINDSTORMS NXT using these converter cables.

Figure 3.9 Multiplexing Motors on the NXT

The Power Function system has its own battery box and an infrared controller (see Figure 3.10). Mindsensors will be extending its NRLink to support communication with this controller. Using NRLink, the NXT can send commands to control the Power Function motors. This way, you can use the Power Function battery box on your robot, and conserve the NXT battery.

Figure 3.10 Controlling Power Functions Using an IR Interface

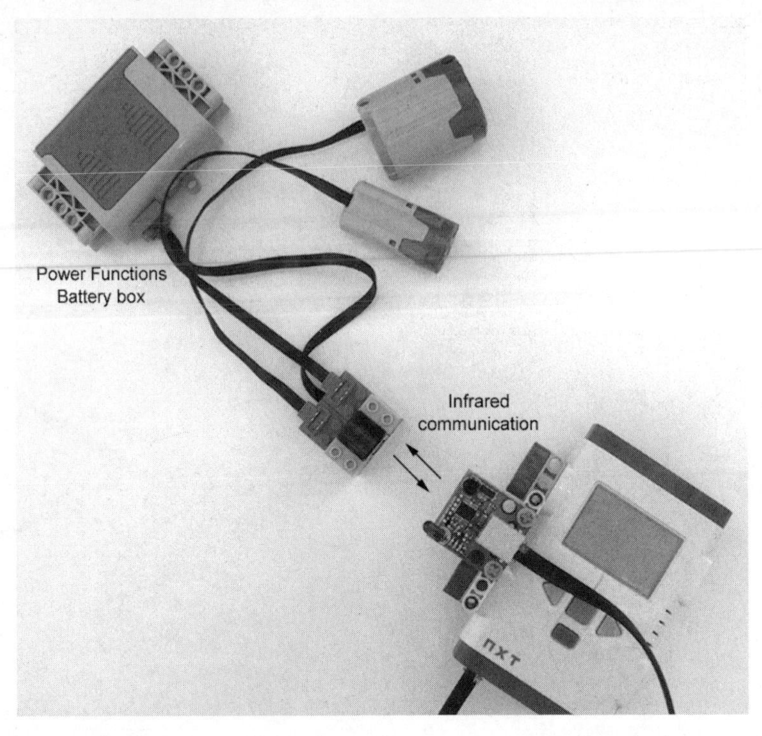

Controlling Power

You know that your program can control the power of your motors. In fact, using the RobotC *setMotorPower()* method will set the power in the range of -100 to +100, where negative values indicate reverse direction:

```
// enable motor speed regulation
nMotorPIDSpeedCtrl[motorA] = mtrSpeedReg;
. . .
// move at half speed
motor[motorA] = 50;
```

But what happens when you change this number? And why do we care? There are different ways to control the power of an electric motor. The LEGO train speed regulator controls power through voltage: The higher the voltage, the higher the power. The NXT uses the same approach as the RCX, called *pulse width modulation* (PWM).

To explain how this works, imagine that you continuously and rapidly switch your motor on and off. The power your motor produces in any given interval depends on how long it's been on in that period. Applying current for a short period of time (a *low duty cycle*) will do less work than applying it for a longer time. If you could switch it on and off hundreds of times a second, you would see the motor turning in an apparently normal way; but under load you would notice a decrease in its speed, due to a decrease in the supplied power (Figure 3.11).

Figure 3.11 Pulse Width Modulation Power Levels

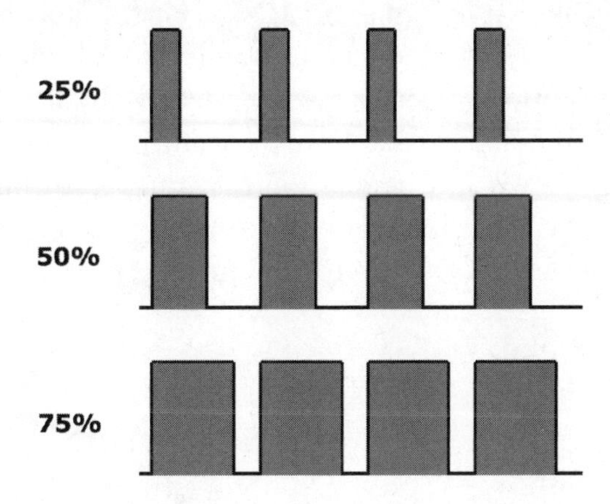

This is exactly what the NXT does. Its internal motor controller can switch the power on and off very quickly (an on/off pulse every millisecond), at the same time varying proportion between the on and off pulses. At power level 1, for every 100 pulses applied, the motor receives one on pulse; at power level 25, for every 100 pulses applied, the motor receives 25 on pulses; and so on, until you reach level 100, when all pulses are on.

Why do we care about this technical stuff? Because this explains that you aren't actually controlling speed, but power. LEGO motors are very efficient, and when the motor has no load or a very small load, lowering the power level won't decrease its speed very much. Under more load, you will see how the power level affects the resulting speed too.

Detecting Motor Overload

Often, one would like to check whether the NXT servo motors are working properly. Two common problems encountered with motors are stalling and slipping. The former happens when the load on the motor exceeds its maximal power, leading to a "frozen" motor. Checking if your motors are stalled is relatively easy by monitoring the motor encoders. If the motor should be running but its encoder doesn't increment (or it increments less than your threshold), it is stalled. Slipping occurs, for example, when your robot is stuck at a wall, but the wheels lose their grip and rotate in place. Detecting slip is more difficult than detecting stall. Guy Ziv at NXTasy.org has developed a Motor Power Meter NXT-G block that allows you to monitor the load on the motor. When the motor slips, its power is usually larger than that experienced during normal operation, which allows detection of slip conditions.

Braking the Motor

Controlling the power means also being able to brake your motor when necessary. For this purpose, the NXT features a sort of electric brake. Once again, let us explain how it works through an experiment.

Assemble the motors as shown in Figure 3.12. Note that motor (b) is locked by a beam, resulting in a stall, effectively causing motor (a) to be shorted. We know that a *short circuit* sounds like a *bad* thing, but in this particular case we mean only that the circuit is *closed*. Don't worry; your motor is not at any risk. Now try to turn the 24t with your fingers. You see? The motor offers a lot of resistance, and as soon as you stop turning, it stops too. Now disconnect motor (b) from the cable and try to turn the 24t again: It turns smoothly, and it continues to spin for a while after you have stopped turning it.

Figure 3.12 An Electric Brake

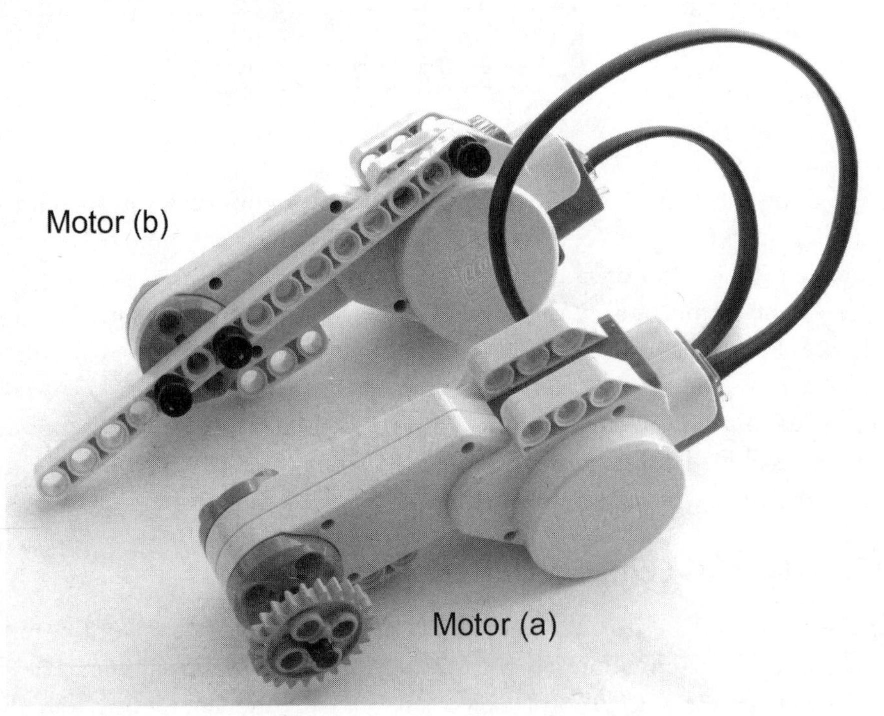

What happened? Not only is an NXT motor able to transform electricity into motion, it does the opposite too: It can be used to *generate* electricity. In our experiment, when motor (a) is shorted, the generated current is sent back into the motor, producing a force that resists the motion. This is a simple but effective technique which the NXT implements to brake the motor: When you set the motors to *brake*, the NXT not only switches the power off, but it also short-circuits the port, making the motor brake.

Bricks & Chips...

Using Motors As Generators

If you are not convinced that a motor also works as a generator, perform this simple experiment. Remove the beam that locks motor (b) and turn the gear on motor (a) while looking at the second motor. What happens? The first motor converts the mechanical energy coming from your fingers into electric current, which makes the second motor turn.

Coupling Motors

If you need more power for a task than a single motor can deliver, you will very likely need to mechanically *couple* the motors, meaning that they will work together to operate the same mechanism, sharing its load. It's like when you have to move something really heavy and you call a friend to help you: Each member of the party bears only half the total weight. Though this rule works for all electric motors in general, a specific limitation applies when attaching multiple motors to the NXT: Its current-limiting device won't allow the motors to draw as much current as they want. Consider it a constraint to the maximum power each port can pay out.

In Figure 3.13, you can see two motors connected using a single axle to a 40t gear wheel. People often wonder whether connections such as these are going to cause any problems to the motors. The answer is simply *no*. Unless you keep one of the motors stalled for more than a brief moment, they are not easy to damage. In applications such as the one in Figure 3.13, you just have to be sure the motor power of one motor doesn't oppose that of the other. The NXT wiring won't let you do it incorrectly. However, we suggest that you double-check your program to ensure that both motors are turning in the same direction.

It is true that no two motors turn at exactly the same speed, or output the same torque, but this doesn't cause any conflict. A motor doesn't *know* that another motor is cooperating on the same task. It simply reacts to the load, absorbing more current and trying to keep the speed. This works even if the motors are of different types, even if they are powered at different levels, and even if they are geared with different ratios.

Figure 3.13 Two Mechanically Coupled Motors

If you're not convinced of this, think of a simple vehicle propelled by a single motor. When the path becomes steeper, the load on the motor increases, causing it to reduce its speed. Essentially, the motor adapts itself to the load. The same happens when two motors work together; they share the load and mutually adapt themselves.

To make things easier, you can use the Synchronized Motors feature available in RobotC to run the motors together:

```
nSyncedMotors = synchNone; // No motor synchronization
. . .
nSyncedMotors = synchAC;   // Motor 'C' is slaved to motor 'A'
```

Have you ever tried riding a tandem bicycle? Your partner might be much weaker than you, but you would prefer him to pedal rather than simply ride along, watching the landscape.

Summary

NXT motors are easy and safe to use, but they require a bit of experience to get the most from them. You have seen that wiring NXT motors is very simple and you cannot go wrong with the polarity. The different mounting options require some knowledge and a bit of practice, especially if you need to keep the motors easily removable.

On the topic of coupling motors, this option is useful when you want to split a load over two or more motors to reduce their individual efforts. The only important thing to remember is that you must run them in the same direction to avoid any dangerous conflict situation in which one motor opposes the other.

As a general tip, we suggest that you make intense use of prototyping. Don't wait to finish your robot to discover that a motor is in the wrong place or has not been geared properly. Test your mechanisms while you are building them.

Chapter 4

Reading Sensors

Solutions in this chapter:

- **Digital Sensor Ports with the I^2C (Inter-Integrated) Interface**
- **The Touch Sensor**
- **The Light Sensor**
- **The Ultrasonic Sensor**
- **The Servo Motor Encoder (Rotation Sensor)**
- **Sensor Tips and Tricks**
- **Other Sensors**

Introduction

One of the most important components of a robot is the sensor. The primary purpose of a sensor is to allow the robot to interact with its environment and perform actions based on feedback from its surroundings. This process is called *autonomy*, which by definition means freedom from external law. An autonomous robot is a self-governing device that takes input from its sensors and makes decisions based on this input. This tends to be the difference between those battle-bots you see on TV which are controlled remotely and an NXT sumo-robot that has to maintain its position within a ring while trying to push its opponent out.

Since the advent of the LEGO MINDSTORMS Robotics Invention System (RIS) in 1998, a significant demand has developed for aftermarket sensors. With the initial release of the RIS, it was difficult at first to find aftermarket sensors, and many people were stuck with the touch, rotation, and light sensors included with the kit. Few were honored to have a set of the Cybermaster touch sensors that allowed for sensor multiplexing.

The introduction of the LEGO MINDSTORMS NXT has changed this. Not only is the NXT compatible with most legacy RCX sensors such as the light, rotation, temperature, and touch sensors, but also it is compatible with the numerous aftermarket sensors that started to make their debut near or at the time of release of the NXT product (more are still to come). In fact, at the time of this writing, more than 15 aftermarket sensors and communication adapters were available from third-party vendors.

Out of the box, the NXT comes with the standard light and touch sensors, and has added the new ultrasonic and sound sensors as well as built-in motor *encoders* (rotation sensors) to the three servo motors. This chapter will look at these NXT sensors with a focus on the new additions as well as a variety of third-party sensors. We will also look at the new I²C digital interface as well as some other unique aspects of the NXT system.

Digital Sensor Ports with the I²C (Inter-Integrated Circuit) Interface

The sensor interface of the NXT system has changed significantly. At first glance, you will notice that LEGO added an additional sensor port (there are now four ports). The wiring for sensors and motor connections has changed as well. It now uses a six-wire cable connector that uses an RJ-12-like connector that features an offset locking latch and is smaller than the standard RJ-12. These wires allow the NXT to support legacy analog, passive, and newer digital sensors via the I²C interface.

This custom connector is used to ensure that someone does not inadvertently try to connect his NXT to his telephone jack, or to a network hub. It is also used because the I²C interface has specific dedicated wires that must be plugged in one way only. LEGO has dedicated each wire as follows:

- Pins 1 and 2 have the same functionality as the legacy RCX sensor cables which enable the NXT to read values from the touch, light, sound, and other legacy analog sensors. Note that you can get converter cables for legacy sensors from many third-party vendors, or you can make your own, such as the one detailed on Phillip Hurbain's Web site (see Appendix A).
- Pin 3 is grounded.
- Pin 4 provides a constant voltage (~ 4.3 V).
- Pins 5 and 6 are used to communicate with the digital sensors via the I^{2}C serial bus. The ultrasonic, color, and digital compass sensors (to name a few) are accessed using this method.

The I^{2}C interface is a multimaster serial computer bus developed in the 1980s by Philips Semiconductor. It was initially used to connect low-speed peripherals to motherboards, embedded systems, and so on. It's likely that the computer you are using right now has some form of I^{2}C implemented, such as hardware to read its CPU temperature.

The NXT brick has an I^{2}C communication channel for each of the four input ports. The I^{2}C serial bus uses pins 5 and 6—with pin 5 carrying the clock signal and pin 6 carrying the data signal. The NXT acts as the "master," always invoking communications, and the sensor acts as the slave. This system supports up to 128 different slaves (or addresses). Can you imagine the possibilities? How about a combination of 128 motors and sensors, all multiplexed on the same port? Don't sweat it; these ideas are already being addressed with third-party sensor suppliers such as HiTechnic and Mindsensors. Both have created sensor and motor multiplexers.

The general communication flow is as follows:

- The master (NXT) initiates communications with the slave (sensor).
- The master sends a start message with an address, and then sends a command.
- The master waits for a reply.
- The slave responds.
- The master reads the response and code provides further actions for the robot.

LEGO used I^{2}C for the ultrasonic programming block in NXT-G (LEGO's NXT graphical programming software). At the time of this writing, new custom I^{2}C NXT-G sensor blocks have been appearing weekly on fan-related blogs (e.g., www.nxtasy.org) that are opening up NXT-G to allow for custom sensor integration into this graphical development environment. RobotC has I^{2}C functionality built in, with many samples already available; you simply need to know the memory address locations of the registers from which to read the sensor results, which are typically provided by the vendor that is selling the sensor. In most cases, vendors also have sample code that you can use to get started.

NOTE

NXT-G is the programming environment LEGO provides with the NXT set. It is developed by National Instruments and provides you with a graphical interface for programming your NXT robot. It is easy for people of all ages to use to program their robots. Chapter 7 provides more detail on NXT-G as well as other third-party development platforms.

I^2C integration on the NXT allows you to add digital devices to your robots. Digital devices have the added advantage of being able to include parameters such as device names, configuration parameters, calibration information, and so on. This makes these devices smart and allows for additional devices to be developed to the same standard. It also allows them to be chained much in the same way that USB devices can be chained with USB hubs.

LEGO chose to implement I^2C on the NXT, which has opened the door for a variety of sensors that will work with the NXT. I^2C is one of the most significant improvements to the MINDSTORMS NXT and will open the door for a plethora of sensors, devices, and communications adapters for years to come. Now you will be able to make that 28-motor walking robot that you always dreamed of!

Bricks & Chips...

High-Speed Communications Port

One of the lesser-known new features of the NXT brick is the high-speed communications port that was implemented on port #4. First, don't worry; you can still use this port the same way you use the other three for your sensors. It just has a little extra added touch!

LEGO had an RS-485 communications chip implemented behind the normal input circuit on port #4. Through this port, the NXT intelligent brick supports high-speed bidirectional communication on a multipoint data line over wide distances out-of-the-box. At the time of this writing, no known devices require this feature, but the door is open to future devices that require higher-speed communications. Who knows; maybe it may even serve as a port for memory expansion.

It is thought that LEGO may use the P-NET communications protocol (www.p-net.org) for any future devices that it creates. The P-NET Fieldbus enables components such as computers, intelligent sensors, actuators, and so on to be connected via a common two-wire cable. Communication is bidirectional and data is transmitted digitally. The advantage is that linked components can connect inline via the two-

Continued

wire cable and operate independently. From a robotics standpoint, this would allow you to create robots with numerous sensors, all strung together via a single cable, and it would allow for easy expansion by just plugging new sensors into the line. The concept is similar to USB (Plug and Play), but with a flavor of the older BNC (10Base2) computer networking approach (series connected components).

The Touch Sensor

The touch sensor (see Figure 4.1) is probably the simplest and most intuitive member of the LEGO sensor family. Other than having the NXT "look," not much has changed with this sensor. The most notable difference is a migration to the three-hole studless connection and the axle slot on the front. It works more or less like the pushbutton portion of your doorbell: When you press it, a circuit is completed and electricity flows through it. The NXT is able to detect this flow, and your program can read the state of the touch sensor

Figure 4.1 The Touch Sensor

One of the sensor's most common applications is to act as a *bumper*. Bumpers are a simple way of interacting with the environment; they allow your robot to detect obstacles when hit, and to change its behavior accordingly.

In NXT-G the *sensor block* has three states: pressed, released, and bumped. These are important, as they enable you to determine how your robot will react in different situations. For example, envision a robot that uses the sensor to detect the passing of an object on a conveyor. If that object gets stuck, the sensor state remains "pressed"; if it passes, the state is "bumped." You could program your robot such that it reacts to "pressed" by reversing the conveyor temporarily to get the object unstuck. Alternatively, in RobotC, you would build a function that watches the sensor. Once pressed, it would monitor the time that it remains pressed. If this time exceeds a predefined limit, you could trigger your robot to react accordingly. This method works particularly well in sumo-robot competitions. You configure the sensor to detect when you are in contact with your opponent. When the sensor state is pressed, you continue in the direction you were going, and maybe even power up the motors, assuming you are pushing your opponent out of the ring.

A bumper typically is a lightweight mobile structure that actually hits the obstacles and transmits this impact to a sensor, closing it. You can invent many types of bumpers, but their structure should reflect both the shape of your robot and the shape of the obstacles it will meet in its environment. A very simple bumper, such as the one in Figure 4.2, could be perfectly fine for detecting walls, but might not work as expected in a room with complex obstacles, such as chairs. In such cases, try experimenting. Design a tentative bumper for your robot and move it around your room at the proper height from the floor, checking to see whether it can detect all the possible collisions. If your bumper has a large structure, don't take it for granted that it will impact the obstacle in its optimal position to press the sensor. Our example in Figure 4.2 is actually a bad bumper, because when contact occurs, it hardly closes the sensors at the very end of the traverse axle. It's also a bad bumper because it transmits the entire force of the collision straight to the switch, meaning an extremely solid bracing would be necessary to keep the sensor mounted on the robot.

Figure 4.2 A Simple Bumper

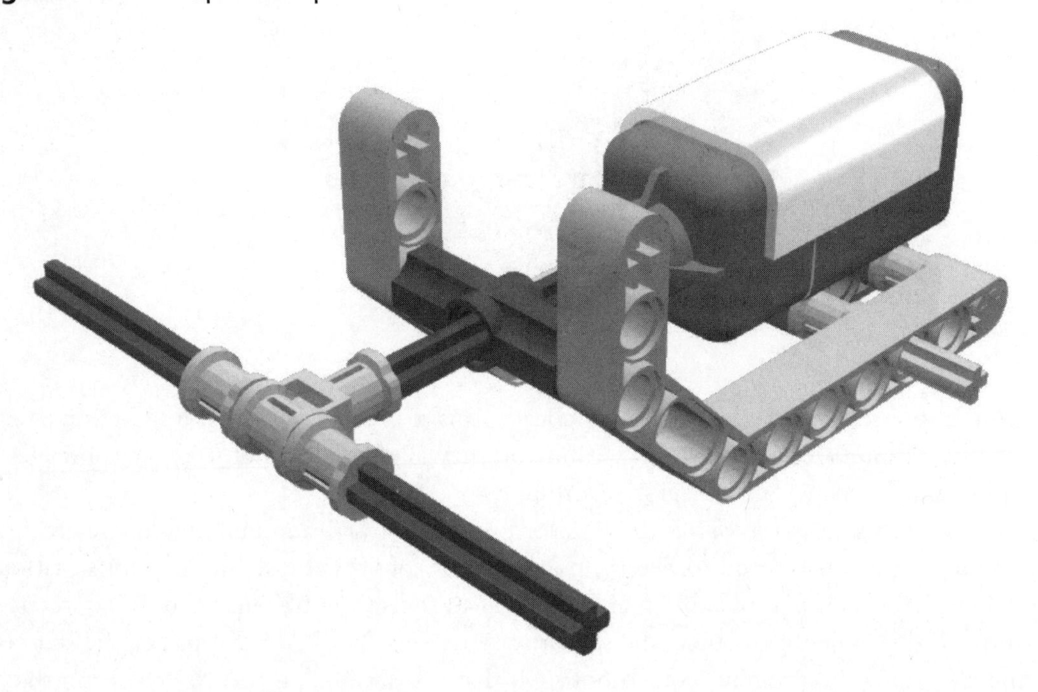

Be empirical. Try different possible collisions to see whether your bumper works properly in any situation. You can use the built-in touch sensor test program on the NXT brick which will display the status of the sensor on the LCD, or write a very short program that loops forever, producing a beep when the sensor closes, and use it to test your bumper.

When talking of bumpers, people tend to think they should *press* the switch when an obstacle gets hit. But this is not necessarily true. They could also *release* the switch during a collision. Look at Figure 4.3. The rubber bands keep the bumper gently pressed against the sensor; when the front part of the bumper touches something, the switch gets released.

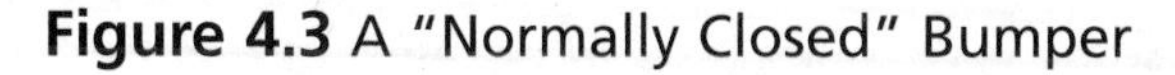

Figure 4.3 A "Normally Closed" Bumper

Actually, there are some important reasons to prefer this kind of bumper:

- Because of its ability to be connected via studless beams, the new NXT touch sensor is sturdier than the legacy RCX one. However, you still want to minimize the impact force on parts. With this setup, the force doesn't transfer to the sensor itself.
- The rubber bands absorbing the force of the impact preserve not only your sensor, but also the whole body of your robot. This is especially important when your robot is very fast, is very heavy, is very slow to react, or possesses a combination of these factors.

Bumpers are a very important topic, but touch sensors have an incredible range of other applications. You can use them like buttons to be pushed manually when you want to inform your NXT of a particular event. Can you think of a possible case? Actually, there are many. For example, you could press a button to order your NXT to "take a reading of the ultrasonic sensor now," and thus test distance readings. Or you could use two buttons to give feedback to a learning robot about its behavior, good or bad. The list could be long.

Another very common task you'll demand from your sensor is *position control*. You see an example of this in Figure 4.4. The rotating head of our robot mounts a touch sensor that closes when the head looks straight ahead. Your software can rely on timing to rotate the head at some level (right or left), but it can always drive back the head precisely in the center simply waiting for the sensor to close.

Figure 4.4 Position Control with a Touch Sensor

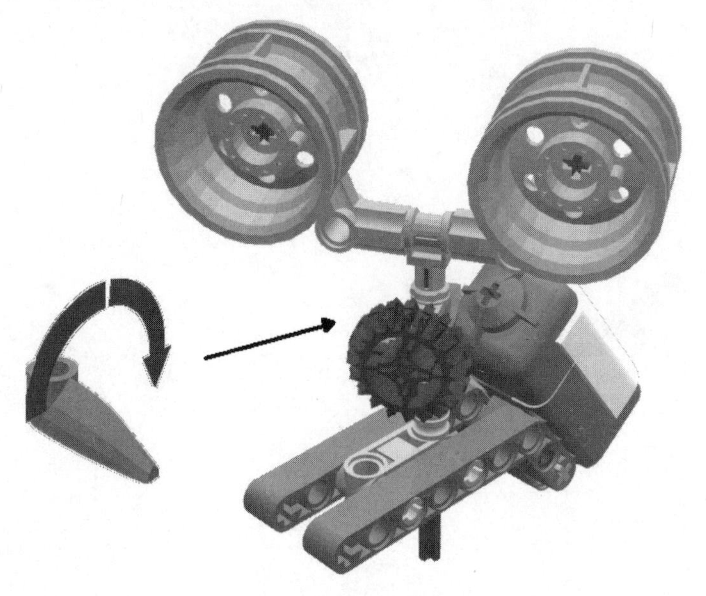

There would be many other possible applications in regard to position control. What matters here that you explore many different approaches before actually building your robot. Let's create another example to clarify a bit. Suppose you're going to build an elevator. You obviously want your elevator to stop at any floor. You may think of having a switch at every level, so when one of them closes, you know that the cab has reached that level. Okay, nice approach. There's one small problem, however; you have just one sensor, and an elevator with only two floors doesn't seem like such an interesting project to you. You could buy more sensors, but this simply pushes your problem one floor up, without solving the general case. Meanwhile, you have used up most of the input ports of your NXT. Suddenly, an idea occurs to you: Why not put the sensor on the booth instead of on the structure? With a single sensor on the booth, and pegs that close it at any floor, you can provide your elevator with as many floors as you like. You see, by reversing our original approach, you found a much better solution. Are the two systems absolutely equivalent? No, they aren't. In the first, you could determine the absolute position of the booth, and in the second, you are able to know only its relative position. That is, you do need a known starting point, so you can deduce the position of the cab counting the floors from there. Either requires that the cab be at a specific level when the program starts, or that it use a second sensor to detect a specific floor. For example, place a sensor at the ground level so that the very first thing your program has to do when started is to lower the elevator until it detects the ground level. From then on, it can rely on the cab sensor to detect its position.

Now your elevator is able to properly navigate up and down. You have one last problem to solve: How do you inform your elevator which floor to go to? Placing a touch sensor at every floor to call the elevator there is impractical. What could you do with a single sensor? Can you apply the previous approach here too?

Yes. You can count the pushes on a single touch sensor. For example, three clicks means third floor, and so on. Now you are ready to actually build your elevator! Of course, with the advent of the integrated encoders in the NXT servo motors, this approach has limited appeal. Later we will show you how you can address the same challenge with a single servo motor to both move the elevator and detect which floor it is on.

Bricks & Chips...

Counting Clicks

The following examples are written in RobotC. Counting how many times a touch sensor is pressed requires some tricks. Suppose you write some simple code, such as this:

```
short nCount = 0;
   while(true)
    {
            if(SensorValue(sensor1) == 1)
           {
                ++nCount;
               nxtDisplayClearTextLine(4);
               nxtDisplayString(4, "Count: %d", nCount);
     }
    }
```

Continued

Your code executes so fast on your NXT that during the short instant you keep the sensor pressed, it counts too many clicks. Thus, you need to have it wait for the button to be released before counting a new click:

```
 short nCount = 0;
while(true)
{
       if(SensorValue(sensor1) == 1)
        {
              ++nCount;

              while (SensorValue(sensor1) == 1)
// The NXT will wait until the sensor is released before continuing.

             nxtDisplayClearTextLine(4);
      nxtDisplayString(4, "Count: %d", nCount);
  }
}
```

Now, your code properly counts the transitions from off to on. There's one last feature you must introduce in your code: You want the counting procedure to end when it doesn't receive a click for a while. To do this, you employ a timer that measures the elapsed time from the last click:

```
short nCount = 0;
short nInterval = 200; //milliseconds. You could use whatever you want here.

   ClearTimer(T1);

    do
    {
          if(SensorValue(sensor1) == 1)
           {
                 ++nCount;

                 if (time10[T1] < nInterval)
                 {
      ClearTimer(T1);
      }
```

Continued

```
                //wait till the sensor is released or the timer hits the
interval time
                //while (SensorValue(sensor1) == 1 || time10[T1] >
nInterval)
                while (SensorValue(sensor1) == 1)

                 nxtDisplayClearTextLine(4);
            nxtDisplayString(4, "Count: %d", nCount);
        }

    }
    while (time10[T1] < nInterval);

   nxtDisplayClearTextLine(4);
   nxtDisplayString(4, "Finished");
  nxtDisplayString(5, "Closing...");
   wait10Msec(100);
```

Let's say your interval is two seconds. When the counting procedure begins, it resets the timer and the counter to 0, and then starts checking the sensor. If nothing happens in two seconds, it exits the loop and is finished. If a click occurs, it counts it, waits for the user to release the button, and resets the timer so that the user again has two seconds for another click before the procedure ends.

The Light Sensor

Saying that the light sensor (Figure 4.5) "sees" is definitely too strong a statement. What it actually does is detect ambient (surrounding) light and measure its intensity. But in spite of its limitations, you can use it for a broad range of applications.

Figure 4.5 The Light Sensor

The most important difference between the touch sensor and the light sensor is that the latter returns many possible values instead of a simple on/off state. These values depend on the intensity of the light that hits the sensor at the time you read its value. In both NXT-G and RobotC, default settings return values in the form of percentages ranging from 0 to 100. However, in RobotC you can also read the sensor in Raw mode. This allows for a higher degree of granularity, returning values ranging from 0 to 1,023.

When reading the sensor, the more light there is, the higher the percentage will be. What can you do with such a device? A possible application is to build a light-driven robot, a *light follower* as it's called, that looks around to find a strong (or the strongest) light source and directs itself toward it. Provided that the room is dark enough not to produce interference, you could then control your robot using a flashlight.

Bricks and Chips...

Sensor Readings

A common mistake that some people make is to assume that this sensor can "see" color. It can only see varying intensities in light conditions, essentially a wide range of black-and-white variations. For example, if you take a reading of a black-and-white LEGO brick, it will return two distinct values. This is because the amount of light reflected off the black brick is less intense than the white brick. The sensor is reading that intensity and translating it into a reading.

This ability to trace an external light source is interesting, but it's probably not the most amazing thing you can do with this sensor. There is another feature of this device: Not only does it detect light, but it *emits* some light as well. A small red LED provides a constant source of light, thus allowing you to measure the reflected light that comes back to the sensor. Both NXT-G and RobotC allow you to set the sensor to generate light (or not).

Designing & Planning...

Reading Ambient Light

The original RCX light sensor is not a great device to measure external sources, as its sensitivity is too low. The emitting red LED is so close to the detector that it strongly influences the readings. The newer NXT version has seen some improvements with the LED being moved below the sensor, with a plastic shelf separating the two.

When you want to measure reflected light, you must be careful to avoid any possible interference from other sources. The amount of light reflected by a surface

Continued

depends on many factors—mainly its color, texture, and distance from the source. A black object reflects less light than a white one, whereas a black matte surface reflects less light than a black shiny surface. Plus, the greater the distance of the objects from the sensor, the less light returned to the detector. These factors are interdependent, meaning that with a simple reading from your light sensor, you cannot tell anything about them. But if you keep all the factors constant except for one, you are now able to deduce many things from the readings. For example, if your sensor always faces the same object, or objects with the same texture and color, you can use it to measure its *relative distance*. On the other hand, you can place different objects in front of the sensor, at a constant distance, to recognize their color (or, more accurately, their *reflection*).

Measuring Reflected Light

To illustrate the concept of measuring reflected light, let's prepare an experiment. Take your NXT, turn it on, attach a light sensor to any input port, and configure the port properly using the built-in test program on the NXT. Configure the sensor as *Reflected Light* and set the port. Do a quick test. The red LED should illuminate. Prepare the environment. You need a dark room, not necessarily completely dark, but there should be as little light as possible. The NXT sensor test program allows you to view the value of a sensor in real time on the LCD screen. Run the test program to begin viewing the sensor value. Now you can proceed. Put the sensor on the table. Take some LEGO bricks of different colors and place them one by one at short distances from the sensor (about 0.5 inches, or 1 to 1.5 cm). Keep all of them separated from each other at the same distance, and look at the readings. You will notice how different colors reflect a different amount of light (you might want to write down the values on a sheet).

For the second part of the experiment, take the white brick and move it slowly toward the sensor and then away from it, always looking at the values in the display. You see how the values decrease when you increase the distance. You can find a distance where the white brick reads the same value you have read for the black one at a shorter distance. You cannot tell the distance *and* the color at the same time, but if you know that one of the properties doesn't change, you can calculate the other. It's important to stress again that in both cases, you must do your best to shield your system from ambient light.

Bricks & Chips...

Understanding Raw Values

Understanding raw values is an advanced topic, and not strictly necessary to successfully using the NXT system. With that said, it does help to understand how sensors process readings and convert these to results.

The NXT converts the electrical signals coming from sensors into whole numbers in the range of 0 to 1,023, called *raw values*. When you configure a port to host a specific kind of sensor in your program, the NXT automatically scales raw values to a different range, suitable for that particular kind of sensor. For example, readings from touch sensors become a simple 1 or 0 digit, meaning on or off, whereas readings from the ultrasonic sensor convert into distance values ranging from 0 to 100 (centimeters or inches). Similarly, light sensor readings are converted into percentages.

Why should you need to know about this conversion? Well, for most applications, the percentage light value returned by the NXT works well, but sometimes you need all the possible resolution your sensor can provide, and this conversion into percentages masks some of the resolution your light sensor is capable of. Let's explain this with an example. Suppose that in two different conditions, your sensor returns raw values of 707 and 713. Convert these numbers into percentages. Considering that the NXT uses whole numbers only, and the numbers are being converted down to a smaller range, there will be some loss of detail. So, 707 and 713 are likely to return the same value as a percentage. In most situations this granularity of readings is not very important, but sometimes even such a small interval matters.

If you program your NXT using NXT-G, the graphic LEGO environment, the built-in sensor blocks provide scaled values. However, fans are producing new custom blocks daily via the LabView Toolkit, which allows for these sensors to be read in raw mode. If you use alternative programming tools (such as RobotC), you can choose to receive the unprocessed raw values directly, taking advantage, when necessary, of their finer resolution.

Line Following

Probably the most widespread usage of the light sensor is to make the robot read lines or marks on the floor where it moves. This is a way to provide artificial landmarks your robot can rely on to navigate its environment. The simplest case is *line following*. The setup for this project is very simple, which is one of the reasons it's so popular. Despite its apparent simplicity, this task deserves a lot of attention and requires careful design and programming. Pay attention to what happens when the sensor "reads" a black line on a light floor.

When the sensor is on the floor, it returns, say, 70 percent, whereas on the black line, it returns 30 percent. If you move it slowly from the floor to the line or vice versa, you notice that the readings don't leap all of a sudden from one to the other. They go through a series of intermediate values. This happens because the sensor doesn't read a single point, but rather a small area in front of it. So when the sensor is exactly over the borderline, it reads half the floor and half the black strip, returning an intermediate result.

Is this feature useful? Well, sometimes it is, and sometimes it's not. When dealing with line following in particular, it is *very* useful. In fact, you can (and should) program your robot to follow the "gray" area along the borderline rather than the actual black line. This way, when the robot needs to correct its course, it knows which direction to turn: If it reads too "dark," it should turn toward the "light" region, and vice versa.

When you need to navigate a more complex area—one, for example, that includes regions of three different colors—things get more difficult. Imagine a pad divided into three fields: white, black, and gray. How can you tell the gray area from the borderline between the white and the black areas? You can't, not from a single reading, anyway. You must take into consideration other factors, such as previous readings, or you can make your robot turn in place to make it gather more information and understand where it is. To handle a situation such as this, your software is required to become much more sophisticated.

Of course, with the advent of aftermarket sensors such as the HiTechnic color sensor, variable colored line following may become more feasible. More on this later.

Designing & Planning…

Calibrating Readings

Sometimes you can't know in advance what actual values your sensor is going to read. Suppose you're going to attend a line-following contest: You cannot be sure of the values your sensor will return for the floor and the black line. In this case, and as a good general practice, it is better not to write the expected values as constants in your program, but rather to allow your robot to read them by itself through a simple calibration procedure. Staying with the line-following example, you can dedicate a free input port to a touch sensor to be manually pressed when you put your robot on the floor and then on the black line so that it can store the maximum and minimum readings. Or you can program the robot to perform a short exploration tour to uncover those limits itself.

The light sensor is such a versatile device that you can imagine many other ways to employ it. You can build a form of proportional control by placing a multicolored movable block of LEGO parts in front of it. Figure 4.6 shows an example of this kind. When you push or pull the upper side of the beam, the sensor reads different light intensities.

Figure 4.6 An Analog Control with a Light Sensor

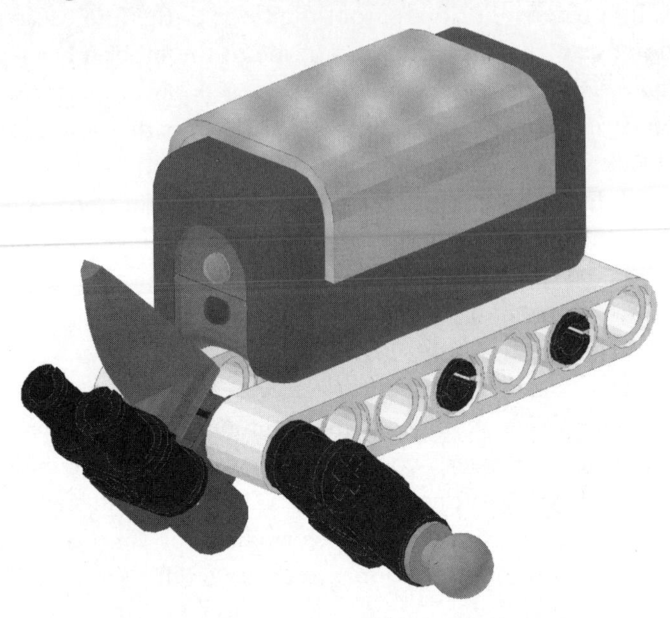

Combining the light sensor with a lamp brick (not included in the NXT kit) you get a photoelectric cell (Figure 4.7); your robot can detect when something interrupts the beam from the lamp to the sensor. Notice the double-split TECHNIC axle joiner in front of the sensor to reduce the possible interference from ambient light.

Figure 4.7 A Photoelectric Cell

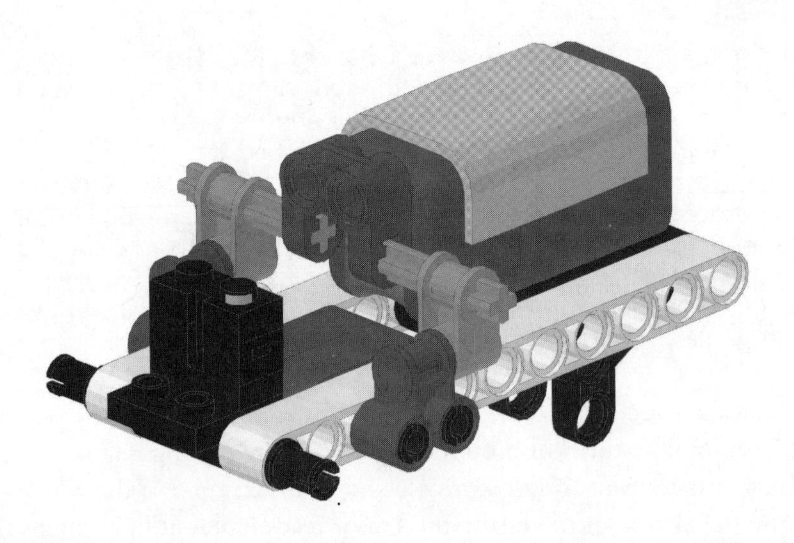

The Ultrasonic Sensor

One of the most anticipated additions to the NXT is the *ultrasonic* sensor (sometimes referred to as the sonar sensor). This sensor had its origins as a third-party add-on sensor developed by HiTechnic for the RCX. The idea of using sound to detect obstacles was popular enough that LEGO decided to make this sensor a standard part of the NXT kit. Figure 4.8 shows the sensor.

Figure 4.8 The NXT Ultrasonic Sensor

The principle behind ultrasonic detection is that the sensor emits high-frequency ultrasound (beyond the limit of human hearing), which bounces off objects and is read back by the sensor. The time that each pulse takes to bounce back to the sensor determines the distance from it. The longer the interval, the further the object is, and vice versa. Bats use the same principle as a means of navigating and of locating their prey. The technique is called echolocation and allows bats to distinguish between an insect and a falling leaf. Now there is a challenge to try with your NXT!

A common way to program the sensor is to set it up to constantly poll the environment, and then use the value returned to enable your robot to react (e.g., steer away, or back up). In both NXT-G and RobotC, readings are normalized, returning values between 0 and 100. NXT-G allows these to be set in inches or centimeters. RobotC also supports raw, un-normalized readings; however, most just use the normalized values.

Here is a sample RobotC ultrasonic test program:

```
while(true)
{
   nDist = SensorValue(sonarSensor);
   nxtDisplayTextLine(1,"Distance:%d",nDist);
   wait10Msec(100);
```

```
        eraseDisplay();
    }
```

Try to build yourself a robot that can navigate a room without bumping into objects. Figure 4.9 shows a sample robot built to test the ultrasonic sensor using a simple four-wheel drive skid-steer loader platform.

Figure 4.9 CT1 Sample Robot Using the Ultrasonic Sensor

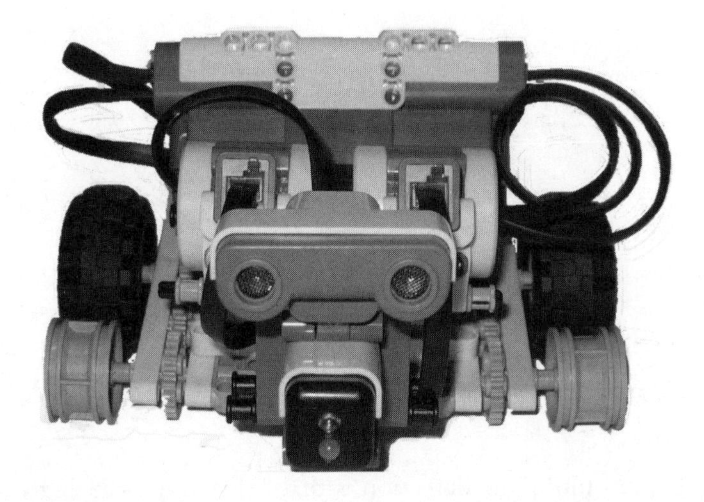

This is the first sensor from LEGO to implement the I^2C interface described at the beginning of this chapter. Through NXT-G, LEGO developed the *sensor block* to read the I^2C messaging and interpolate the values to provide a distance reading in inches or centimeters.

When choosing which sensor to use, you should remember the advantages and limitations of each sensor type. In particular, the ultrasonic sensor has a wide range and is capable of working with all lighting conditions (because it relies on sound only). On the other hand, its beam is relatively wide (about 30 degrees), so even point objects appear broad when you scan around with it. The IR sensors based on SHARP sensor technology (such as the new Mindsensors.com IR sensor) have a narrow beam, giving better spatial resolution, but are more sensitive to light conditions. Furthermore, the two sensor types have different "tastes" with respect to target materials—ultrasound is absorbed by soft materials, making them invisible to the ultrasonic sensor, whereas a mirror would reflect ultrasound back to the sensor but prevent IR sensor detection.

Proximity Detection

Previously with the RCX, users who wanted to perform proximity detection were stuck with limited options. One option was to purchase a third-party proximity sensor such as the Techno-Stuff Dual Infrared Proximity Detection (DIRPD) sensor. This sensor, combined with a language such as NQC, could be programmed to detect left, right, and center obstacles and proved very effective at navigation and obstacle avoidance. To demonstrate its effectiveness, look at the WallFollower robot at www.plastibots.com. Another option was to use the standard light sensor with the RCX IR communication feature. With some NQC code, one could program the RCX to send pulses of IR and have the light sensor read these. The timing of response for them would yield a distance reading.

The introduction of the ultrasonic sensor to the NXT provides similar functionality out of the box. Through NXT-G, you can quickly write a simple program that can be used to detect obstacles in front of your robot and have it react accordingly. You can try doing this yourself by building TriBot from the NXT-G Robo Center and configuring it such that the sensor is used to avoid obstacles as it drives around.

Fans such as Guy Ziv (at www.nxtasy.org) have gone a step further and taken advantage of the I^2C interface to use the sensor in "ping" mode. What does this mean? Well, out of the box, you really can use one of these sensors on only one robot at a time. If you try to configure two ultrasonic sensors to run at the same time, both sensors get initialized when the program starts, each will emit ultrasound at the same time and cause confusion between readings due to bouncing signals between them.

To solve this, Guy used a custom block in NXT-G by configuring two sensors on the same robot and having the software alternate (at a rapid pace) the sending and reading of ultrasound for each sensor. The general process is as follows:

- Ultrasonic Sensor 1: Send signal (ping), read result, store value.
- Ultrasonic Sensor 2: Send signal (ping), read result, store value.
- Compare both values and react accordingly.

If you have two ultrasonic sensors, you can get into developing some sophisticated proximity detection with your robot. You could use them to better decide which side of your robot an obstacle is on, or use the values to produce a "radar" type of map on the NXT display to do some environment mapping.

The Servo Motor Encoder (Rotation Sensor)

The legacy RCX rotation sensor was always known for its lack of reliability with readings when turning at both low and high speeds. Robot makers had to play with code to provide stability to readings returned from this sensor.

LEGO decided to integrate an encoder (rotation sensor) directly within its new NXT servo motors (Figure 4.10). There are two benefits to this: The encoder functionality was improved, and the NXT received three rotation sensors built right into the motors that don't require additional sensor ports! The interactive servo motor (as it's also referred to) allows you to measure both speed and distance in a variety of formats, including degrees, rotations, and seconds. It acts as both a motor and a rotation sensor, and has a dedicated block for each of these in NXT-G. In RobotC, you would simply set the parameters for driving the motor as you normally would while using other commands to read the encoder values to measure rotation.

Figure 4.10 The NXT Servo Motor with Built-in Encoder

Figure 4.11 shows an internal view of the servo motor with the encoder (in blue) located to the left of the larger orange drum (the motor). In reality, the encoder is actually a black wheel that contains 12 holes which allow the optical sensor to read 24 on/off states with each full rotation. This provides the NXT with a great deal of resolution to detect position down to the nearest degree. From the image, you can also see how the NXT motor is internally geared. There is enough torque to drive wheels/tracks directly. Even though the RIS motors are also internally geared, they have limited torque that usually required an additional geartrain—especially in sumo competitions!

Figure 4.11 The NXT Servo Motor—Internals

Having an encoder built directly into the servo motor allows robot designers to develop more sophisticated drive mechanisms that enable your robot to do things such as drive straight, even over rough terrain. This functionality works out of the box with NXT-G. When you program your robot, the *move block* pairs two motors together, enabling the NXT to monitor the encoders of both motors while correcting them on the fly to ensure that the robot is tracking straight. The general idea is that the program monitors rotations on both motors. If one falls behind, it adjusts the speed of one motor to compensate for the lag, which keeps the robot driving straight.

You can try this yourself by creating the TriBot from within Robo Center (sample robots in NXT-G). Following the programming guide, you will use a *move block* to allow for both drive motors to be synchronized. Once built, run the robot and follow along beside it. Press your finger to one of the wheels and then let go. Note how at first you slow down one side of the robot, but then it speeds that side up to bring the robot back to driving in a straight line.

As mentioned earlier, you can use a single servo motor to both move an elevator as well as determine which floor it is on. With the new level of accuracy in these motors, you can determine the position of the elevator by performing some simple tests to find which angle values represent each floor. To do this, create the elevator unit and manually rotate the motor while viewing the encoder rotation values in NXT-G (or the RobotC *poll brick* window). Jot down the rotation angle for each floor. Then, simply identify in your program these angles as stop points for the elevator unit.

The encoder functionality is very powerful for the future of NXT robots, as it opens the door by enabling your robots to be "location-aware" by performing tasks such as room mapping. The sky is the limit here.

Bricks & Chips...

How the Servo Motor Encoder Works

The NXT motor encoder detects movement similar to the way an older computer mouse (with the ball) works. One of the first things Philippe Hurbain (Philo) did when he got his NXT set was to dismantle the servo motors to have a look under the hood. His site (see Appendix A) provides detail on this. Figure 4.12 shows the motor and encoder components cut away from the rest of the motor. The encoder wheel (the black wheel to the right) is driven directly off the motor. The wheel has a number of holes in it which allow the optical sensor to detect on/off states as it spins. A beam of light is generated from the optical sensor (the gray square box covering the encoder wheel) on one side of the encoder wheel and shines through to the other, which falls upon a photocell. As the motor spins, the encoder rotates and causes light to alternate through a series of on/off states. The sensor picks this up and passes the information to the NXT for processing. Unlike the older rotation sensor, because the motor is directly coupled to the encoder wheel, direction is automatically handled, as the NXT always knows which way it is driving the motor based on the programming done in the software.

Through RobotC, it is possible to detect motor stall conditions by monitoring the motor encoder rotation as your motor turns. Knowing the power and expected output of the motor, you can match this with the speed at which the encoder is actually turning and detect when the motor has stopped turning. This has an added bonus, as you could conceivably create a robot that does not need a touch sensor to detect when it has hit a wall. You can simply monitor the motor rotations and judge it this way—when you hit a wall, you can detect that the motors have either slowed or stopped, and after a period of, say, one or two seconds, force a decision to back up or turn.

NOTE

The NXT supports three types of sensors: passive, active, and digital. The main difference is that passive sensors do not require a current generator to supply power to the sensor, whereas active sensors do. Digital sensors use I^2C communication and typically have a microcontroller to handle sampling of the environment.

Passive sensors include the touch sensor (NXT and RCX), light and sound sensors (NXT), and the RCX temperature sensor. Active sensors include the RCX light and rotation sensors. Digital sensors include the NXT ultrasonic sensor and numerous third-party sensors such as color, compass, pressure, gyro, and acceleration sensors.

Figure 4.12 The NXT Servo Motor Encoder

Sensor Tips and Tricks

Sooner or later, you will probably find yourself without the proper sensor for a particular project. For instance, you need three touch sensors, but you have only one. Is there anything you can do? There's no way to turn any sensor into a light sensor or a temperature sensor, but touch and rotation sensors are at some level replaceable.

Even with the addition of an extra sensor port on the NXT, there are still times when builders need additional ports. There are aftermarket solutions, such as sensor multiplexers, but these can be expensive, especially if you want to do this for only one robot or to test an idea. In the following sections, you'll find some common and well-tested tips that can help.

Emulating a Rotation Sensor

There's a long list of possible alternatives to the rotation sensor. All the suggested methods are based upon counting single impulses generated by a rotating part. They all work well, but usually they don't detect the direction of rotation. In many cases, this is not a problem, because when coupled with a motor you *know* which direction your sensor is moving. Of course, with the addition of three rotation sensors with the NXT, the need for something such as this is more limited now than it was for the RCX. However, it is still important to show ideas on how, with some simple parts, you can emulate other sensors.

The assembly pictured in Figure 4.13 shows an axle with a TECHNIC knob wheel that closes a touch sensor. Each complete rotation would account for four "ticks" on the sensor. If geared properly, this could provide a decent resolution for determining the state of a component of a robot. For example, you could use it to determine which floor an elevator is at by counting the number of ticks per floor. This is the principle: Use either a cam or any other suitable part that, while rotating, periodically pushes the sensor.

Figure 4.13 Emulating a Rotation Sensor with a Touch Sensor

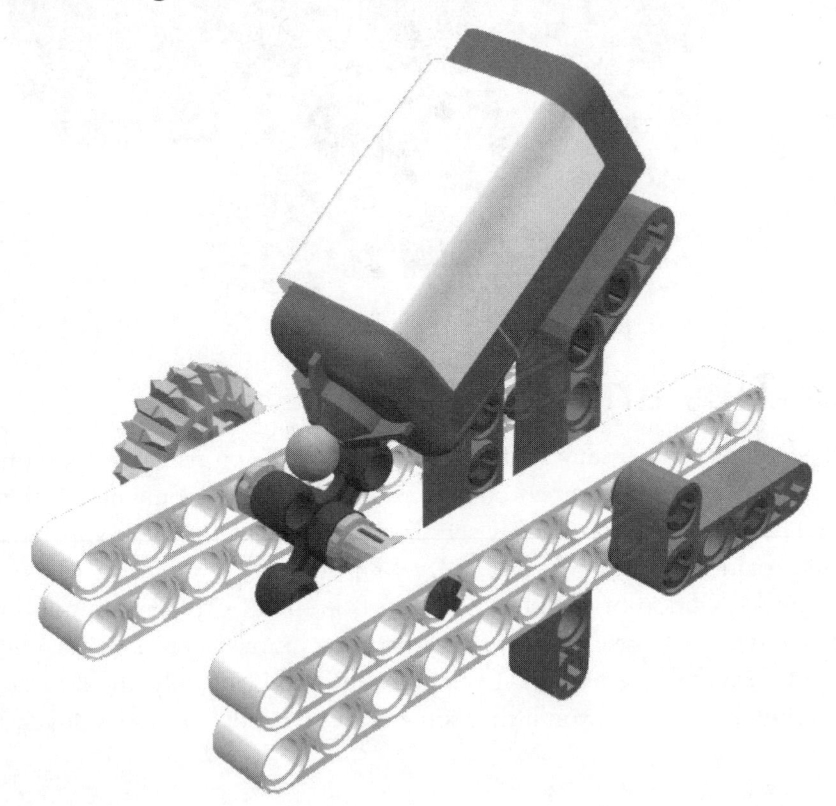

Making a rotation sensor out of a light sensor is not very different: Build some kind of rotating disk with sectors of different colors, and count the transitions from one color to the other (Figure 4.14). The general tip applies to this case too: Try to insulate the sensor from external light sources as much as possible.

Figure 4.14 Emulating a Rotation Sensor with a Light Sensor

Other LEGO electric devices, though not actual sensors, can be successfully employed to emulate a rotation sensor. These are typically available from sources such as BrickLink and are not hard to find.

One such device is the polarity device. Connect it as shown in Figure 4.15, and configure it as a touch sensor. With every turn, it closes the circuit twice. Note that you would need an NXT converter cable, as this uses the older RCX connection.

Figure 4.15 Emulating a Rotation Sensor with a Polarity Switch

Connecting Multiple Sensors to the Same Port—Multiplexing

After the RIS came out, some people found that they needed more than the three sensor ports provided on the RCX. Expansion of sensor ports on the RCX was limited to touch sensor multiplexers such as the one developed by Techno-Stuff (Figure 4.16).

Figure 4.16 Techno-Stuff Switch Multiplexer

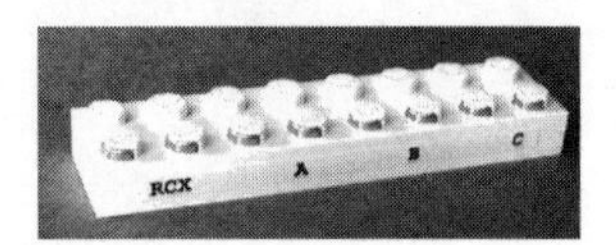

The principle behind its construction is based on wiring the sensors in a series and with each connection having a resistor wired between each sensor connector and the port labeled "RCX." All that would be needed is an RCX-to-NXT converter cable to enable your robot to work with it. For each connector, the resistor would vary in rating. In the case of the Techno-Stuff multiplexer, the resistors are 62K, 30K, and 15K. The RCX sensor port would be set to read raw data (light sensor) and the values returned would vary depending on which sensor or sensor combination was clicked. This was quite powerful, as you could sense eight combination sensor states for three sensors on one port.

If you were lucky enough to own a LEGO CyberMaster 8482 set, you had three touch sensors that already had variable resistance built in. All you needed was a unique cable configuration (Figure 4.17) that would allow the NXT to see each sensor separately. Reading these sensors as raw (light sensor) would allow you to read varying resistance values depending on which sensor combination was pressed.

In Figure 4.17, you can see a number of RCX cables connected to each other, as well as running off to each sensor. The purpose of this is to have the wiring such that the sensors are connected in parallel to each other. This wiring allows the NXT to detect all the combinations sensors being pressed by the electrical resistance being returned to the NXT port. If you want to know more about how this works, look on the Internet for Ohm's Law.

Figure 4.17 RCX Touch Sensor Multiplexer

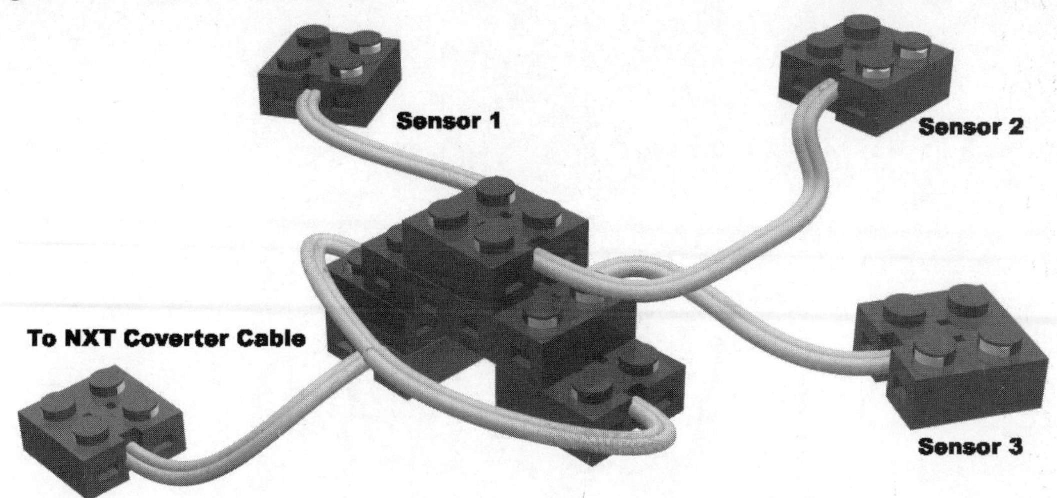

A bit has changed with the NXT. If you have legacy sensors and multiplexers, you could probably connect them to your NXT, but you will need a converter cable to go from the NXT to the RCX connector. However, a number of new sensor multiplexers are on the horizon. Both HiTechnic and Mindsensors are in the running and have products in the works.

With the advent of the I^{2}C interface, the line between motor and sensor multiplexing has been blurred a bit. Both HiTechnic and Mindsensors have something in the works here. Because this chapter is focused on sensors, we won't go into detail on the motor multiplexers. Generally speaking, they are similar to what will be described here, with the difference being that they are used to drive multiple motors off one motor port on the NXT.

HiTechnic has a sensor multiplexer (Figure 4.18) that connects to a single NXT sensor port, allowing four additional sensors to be connected. This allows seven sensors to be connected at once. Imagine what you can do with four of these! The I^{2}C interface is the foundation for all of this and enables the device to handle any combination of any sensor type, including light, ultrasonic, compass, color, pressure, and other I^{2}C sensors. It also supports legacy sensors using the converter cable.

Figure 4.18 HiTechnic Sensor Multiplexer

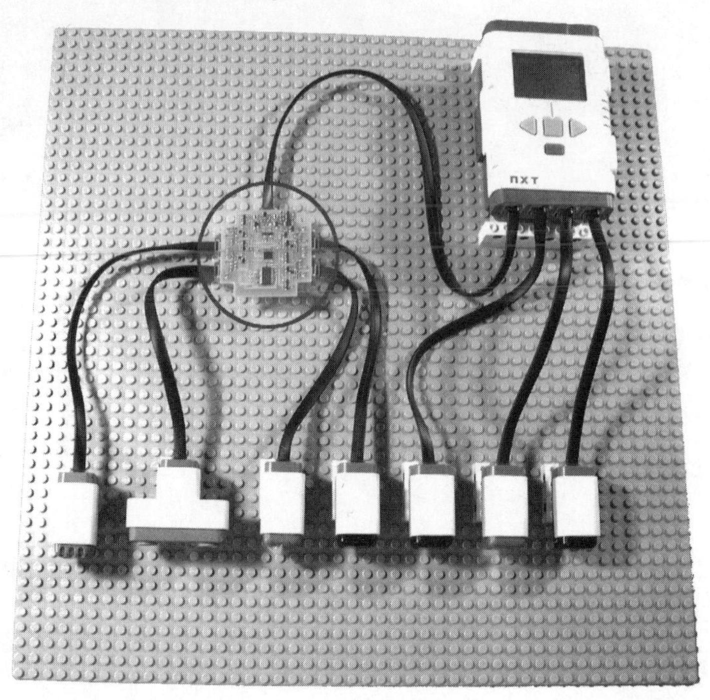

Fans of the original MINDSTORMS line will be the first to tell you that the inclusion of a fourth sensor port on the NXT is a welcome addition. Some believe that four is still not enough, but with the introduction of the rotation sensors in the servo motors and sensor multiplexers, fans have far more options now to build complex robots with single NXT units. Of course, one could always buy another NXT and use Bluetooth communications to enable them to think together. But that is a discussion for another chapter.

Although this chapter is focused on sensors, you may be interested in having a look at the Mindsensors motor multiplexer in Chapter 3. It acts in a similar way to the aforementioned sensor multiplexer, but it allows you to control four additional motors from one NXT sensor port and it uses the I^{2}C interface.

Other Sensors

In the months after the release of the MINDSTORMS RIS, a whole new market for add-on sensors opened up. The introduction of the NXT in 2006 has propelled this market to new highs. Currently a handful of aftermarket companies, such as HiTechnic, Mindsensors, and Techno-Stuff, are actively marketing sensors for the NXT (and RCX).

The NXT is backward-compatible with most legacy RCX sensors, which provides users with a wide array of choices for additional sensors. Currently a number of sensors are available for both the NXT and the RCX, ranging from pressure, tilt, gyro, color, passive infrared, motor/sensor multiplexers, compass, IR communication, acceleration, and sound sensors (to name a few). There has even been discussion about GPS integration! There is not enough room to cover all these sensors in this chapter, or it would become a book itself! However, we will look at some of these to give you a taste of what is out there. If you don't see something here, a quick search of the Internet for LEGO sensors or NXT sensors will give you a night's full of reading.

The Passive Infrared Sensor

The passive infrared (PIR) sensor (Figure 4.19) by Techno-Stuff was originally intended for use with the RCX. However, you can also use it with the NXT using a converter cable. The PIR sensor detects passive infrared radiation (heat sources). Remember the movie *Predator*? Remember how the alien "saw" the soldiers it was hunting? The PIR sensor "sees" heat much in the same way. You can use this sensor to build a robot that can seek out heat sources not evident to the human eye.

Figure 4.19 PIR Sensor

You can configure the PIR sensor as a light sensor in either NXT-G or RobotC. Readings will indicate detection of an object in front of it. The sensor technology has a 20-foot range and a 180-degree swath. The infrared technology used in this sensor is different from TV remote controls, so they will not interfere with it.

If you have this sensor, you can try setting up a small program that plays a sound when you wave objects in front of it. First try moving something cool to the touch, such as an unfilled glass or cup—you should not get any results, as it is not giving off much (if any) infrared radiation. Now try moving your hand in front of the sensor—you should hear a sound. Your results may vary depending on the level of infrared being given off by your hand. During tests, the sensor was quite effective at detecting people walking in front of it from a few feet away, but it did nothing when it was right in front of a wall or door (as expected).

Can you think of an idea for using this sensor on your robot? How about making a robot that can follow your pet around? Or how about a robot that can find a small light source (such as a light bulb) in a maze? Note that you should not use a candle or open flame for such a test, as that is a fire hazard. Stick to things that are safe!

The Pressure Sensor

Pneumatics plays an important role in many robot builders' designs and creations. For more detail on this, look at Chapter 11, as it discusses the use of pneumatics with the NXT. Integrating pneumatics into NXT robots can be both exciting and challenging at the same time, as you are combining motor, gear, and drive systems with pneumatic cylinders to perform varying tasks.

In some cases, you may find the need to use pneumatics for things such as hand claws or providing motion. In other cases, it is necessary to maintain a state of constant pressure and to monitor it so that compressed air is available when needed. The same principle is used in air compressors that you can buy at your local hardware store. A compressor usually has a motor, sensor, piston, and tank to store compressed air. When started, the compressor turns on and begins filling the tank, until at some point, the sensor detects that there is enough air pressure in the tank, which tells the motor to stop pumping. In these compressors, the sensors can be simple mechanical triggers, or electronic sensors.

Figures 4.20 and 4.21 show two pressure sensors developed by Mindsensors and Techno-Stuff. They are both very similar in operation and specifications. The main difference is how the product is presented and its mounting approach. Mindsensors tends to go with the raw PCB approach, but also provides standard TECHNIC pin holes for mounting to your robot. Techno-Stuff provides the sensor electronics stuffed into a traditional LEGO brick, and you can mount the sensor the same as you would any LEGO 2 x 4 brick.

Uses for this sensor vary in different designs. One such idea would be to connect it inline to a pneumatic circuit via a T-junction air tube. Your robot design would likely include a pump, storage tank(s), and pneumatic piston(s), with the sensor being linked to the system to monitor the air pressure. While the robot is operational, some of the air is lost through switches and junctions (which are not perfectly sealed). This sensor can be used to monitor the pressure and trigger a motor pump to switch on to refill the storage tanks.

Programming it as a light sensor would allow the NXT to monitor the pressure and trigger the compressor motor on or off based on a preset threshold, thus ensuring that enough air is in storage and ready for the robot to perform its tasks.

Figure 4.20 Mindsensors Air Pressure Sensor

Figure 4.21 Techno-Stuff Air Pressure Sensor

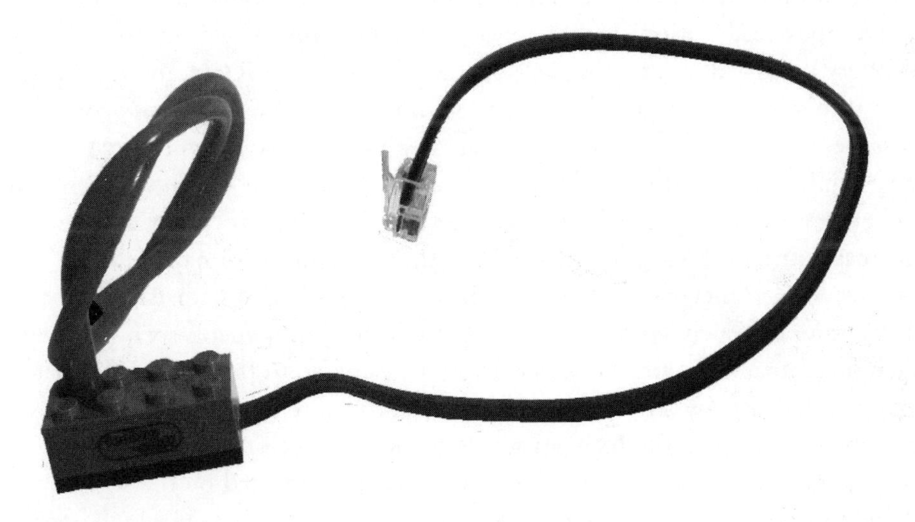

Think of some other ways you can use a pressure sensor. How about using it inline on a robotic claw? Here's the idea:

1. Build a claw hand to grab objects and use the sensor to measure pressure feedback as the claw grabs the object.
2. Use a motor to open/close the claw and to mount a pneumatic piston inline on one side—a sort of "spring-loaded" finger.

3. Mount the pressure sensor inline with the pneumatic piston so that movement in the piston directly causes a change in the readings of the pressure sensor.
4. As the claw closes on the object, the pneumatic piston will compress as the grip gets tighter.
5. Use the values returned to provide feedback. For example, you could have the NXT play sound tones depending on how hard it is gripping. Or you could code the grabbing motor to close the claw only so much, depending on the amount of desired pressure.

The Acceleration Sensor

Measurement of the change in velocity of an object is called *acceleration*. When an object goes from a standstill to moving, its velocity changes and it begins to accelerate. When that object slows, its velocity is reduced and therefore it is decelerating. Acceleration sensors are useful for measuring changes in movement of a robot, detecting when it is tilting (for balance), determining whether it is climbing a hill, or going down, and so on.

Numerous acceleration sensors are on the market for use with the NXT system. Techno-Stuff offers a two-axis sensor that you can use for both the RCX and the NXT. HiTechnic offers a three-axis sensor that allows acceleration measurement in the x-, y-, and z-axes. Mindsensors is also in the game here, offering a variety of two- and three-axis sensors with varying sensitivities ranging to +/- 5 G. Figure 4.22 shows a variety of these sensors.

There are varying types of accelerometers, including piezo electric, surface micromachined capacitive (MEMS), thermal, and electromechanical (to name a few). The technology behind an accelerometer is quite simple. In the case of MEMS, for each axis a tiny polysilicon structure is suspended above a silicon wafer via a polysilicon spring. Together, they form a differential capacitor with the wafer below. Gravity keeps the suspended structure hovering at a specific location; when the sensor is moved, the acceleration causes the suspended structure to move relative to the fixed structure below it, which causes a change in capacitance. This change is measured and converted to values that are used by the software.

Specific programming approaches for these sensors depend on which version you have. Each supplier offers samples for you to work with in both NXT-G and RobotC. HiTechnic has an NXT-G acceleration sensor block that allows you to read x-, y-, and z-axis values. Mindsensors provides RobotC sample code using the I^2C bus.

Figure 4.22 Acceleration/Tilt Sensor

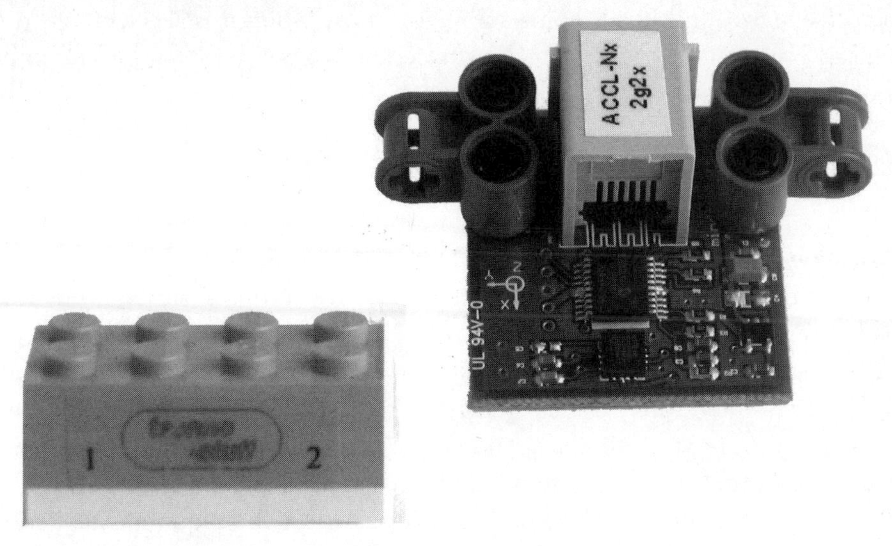

When using these sensors on your robots, there are some considerations you have to be aware of to ensure reliable readings. Some robot designs cause the robot to "jolt" when moving (especially when starting and stopping), which you will have to account for when programming your sensor. This "jolting" effect may be enough to cause the sensor to return erroneous readings that, if not programmed correctly, could trigger a reaction when you don't desire one. To address this, you can write your sensor reading functions in such a way as to smooth out the values.

For example, let's say you are building a sumo robot that can detect when it is about to be flipped over and you are using the acceleration sensor to detect tilt and initiate a reaction. If you simply take each reading (say, every 200 ms) and use that value to react, you may find your robot reacting unnecessarily due to its own driving actions. Instead, you can write a routine that will take (for example) 20 readings, every 10 ms to 20 ms apart, and then average the values to return a more reliable result. This way, for the brief moment that the sensor received a few erroneous readings, they will be averaged with the 18 correct readings which will allow your robot to interpret and ignore this. When your robot does get hit by its competitor, most (if not all) of the readings will be suggesting a hit and then it can react appropriately.

It is also important to consider mounting locations. If you want to have a highly sensitive response, mount the sensor on a long antenna-like contraption. Each small movement at the base will be translated into a larger movement at the sensor. If you desire a less sensitive response, mount the sensor close to your robot's center of gravity (the base), away from motors and other mechanisms that will cause vibration.

So, what can you build with an acceleration sensor? How about a weigh scale? Or a robot that has self-leveling, separately tracked wheel drive units where the base keeps itself parallel to the ground? Because the sensor uses gravity to measure tilt, you can do things such as detect when your robot is going uphill, or know when it is accelerating too fast back down. You can even build a robot that shifts its center of gravity forward as it climbs to give it more holding power and better balance.

The Compass Sensor

As technology advances, robots are becoming more advanced than they have ever been before. Providing the capability for a robot to become "location-aware" takes us one step closer to robots that will be able to know where they are and navigate their environment based on an internal "map" database of its surroundings.

There has been some discussion about integrating GPS receivers into NXT robots, but although they are accurate, they are not (currently) accurate enough to provide the resolution to navigate, say, a room with chairs and tables. This also does not take into consideration GPS signal quality issues indoors. One would have to spend tens of thousands of dollars to get a GPS device that could return something even close to the requirements here.

One solution for providing location awareness is *dead reckoning*. This process begins with a known position, or "fix." Then the robot will proceed to navigate by recording its movement. The problem is you can only really record time, distance, and speed. Although the NXT servo motors allow for precise tracking of the robot's direction, they can't tell you when the robot has actually slipped on a slick floor or loose terrain. This is where the heading comes in. A compass sensor will provide a means to do this (see Figure 4.23).

Figure 4.23 Mindsensors Compass Sensor

A compass sensor is not affected by the interaction between the robot wheels and the ground. It uses the Earth's magnetic field to determine the position (heading) of the sensor. These values are converted into measurements in one-degree resolution for use by the NXT.

Adding a compass sensor into the mix while performing dead reckoning allows you to track heading along with other parameters. Together, these values would provide you the means to record your robot's actions and allow for them to be played back. Conceivably, you could build a robot that records time, distance readings, speed, and heading values to a file, and then reads the file to have it navigate back along the track it just took. There is a challenge that will keep you busy for a night.

The NXT-to-RCX Communication Bridge

Mindsensors has developed an NXT-to-RCX bidirectional communication adapter (Figure 4.24) to allow users with RCX programmable bricks to control them using the NXT. You can also use the adapter to control some LEGO IR-based sets such as the 7897 Passenger Train set. Although this device is not actually a sensor itself, it connects to a sensor port and provides a means to link the NXT to other devices that use infrared for communications and control.

Figure 4.24 NrLink: NXT-to-RCX Communication Bridge Sensor

The NrLink adapter uses the I²C interface and protocols for communication on the NXT. The device has a number of configuration parameters, such as user-selectable short/long-range IR signal communication with the RCX, and because it is developed using I²C, it can also coexist with other LEGO or third-party digital sensors on the same port.

Using supplied sample code, you can easily connect to an RCX within range, and issue commands to run programs, motors, start/stop tasks, and so on. So, if you have an older RCX brick and want to expand the functionality of your NXT kit, this device may provide you with that extra needed expansion. Have a look at Chapter 3, as it shows a sample using this sensor to communicate to the LEGO Power Function IR interface, which will allow you to expand motor control via an unwired connection. This is quite useful when creating robots where you can't afford to have cables tangle up.

The Color Sensor

The HiTechnic Color Sensor (Figure 4.25) is one of the more welcome additions to the line of aftermarket sensors for the NXT. For both the RCX and the NXT, robot builders were pretty much limited to receiving variations of black-and-white values using the provided light sensor. This sensor changes all that by allowing your robot to see a variety of colors.

Figure 4.25 Color Sensor

The color sensor uses three different colored LEDs to illuminate the surface of objects. It reads the intensity of each color reflected back, and the variation of this intensity for each color enables it to determine the color of the object. The sensor uses the I²C interface to communicate with the NXT and returns the color number (0–17). It also returns the raw *and* normalized red, blue, and green values so that you can customize the response based on exact readings (custom code is required for this).

With these options, you may think the sensor offers limitless possibilities. Although it does offer enough possibilities for most uses, it is important to note that this sensor is affected by ambient light similar to the way the stock light sensor is. More important, in order to get an accurate reading, you should place the sensor directly above (or in front of) the object it is reading, and it must be close to the object (approximately 1 cm). One reason for this is that reading the object at close range reduces the chances of ambient light entering the equation; in addition, this requires less power.

What can you do with a color sensor? How about building a brick sorter? Surely every LEGO robot builder wishes he could have a robot help him sort his LEGO bricks. This sensor will help you on your way to achieving that goal. Figure 4.26 shows Brick Sorter (www.plastibots.com). It demonstrates how you can use the sensor to read the color of a LEGO brick and provide a color to allow the NXT to sort it into the correct container. With some additional tweaking, it would be fairly easy to enhance this robot to also sort gears, pins, and so on.

Figure 4.26 Brick Sorter

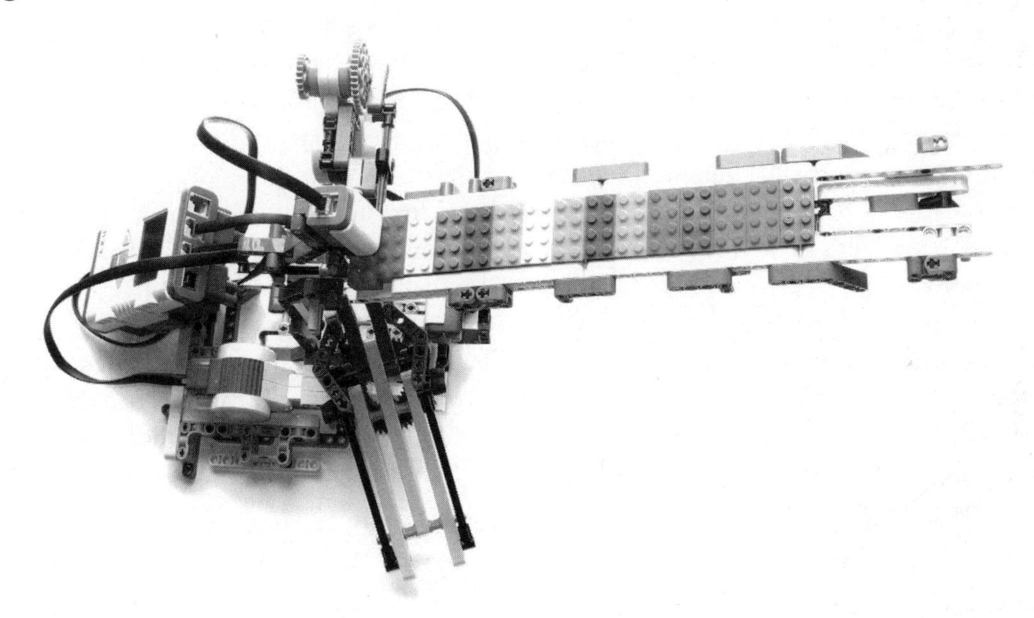

This sensor offers many more possibilities than just allowing you to sort LEGO bricks. Can you think of an idea for how to use this sensor? How about using it for line following in a multicolored line-follower competition? You haven't heard of such a competition? Why not try to create one yourself? Just get some colored tape and lay it down on the floor. To make it more challenging, have some colored tape crisscross and see whether your robot can determine the correct path to travel. Then, once you figure it out, try to see how fast you can make your robot do the circuit.

Summary

In this chapter, we've introduced you to a number of sensors, including the standard NXT sensors (touch, sound, light, ultrasonic, and servo motor encoder) as well as a number of sensors provided by third-party suppliers. Their basic behavior is easy to understand, but here you've discovered that if you want to get the very most out of them, you must understand their requirements and limitations. The touch sensor, for example, seems to be a simple device, but with some clever work on your part, it can become an important tool for counting clicks, or it can make a good bumper.

You were also introduced to the implementation of the I^{2}C communications interface to the NXT. This feature has opened the doors for endless sensor/motor control possibilities with the NXT and external devices. People have been interfacing all sorts of sensors and controllers to the NXT, and more are yet to come. There has even been discussion of connecting the NXT to GPS receivers or similar devices to make NXT robots location-aware.

Another notable addition to the NXT system is the sound sensor, which allows your robot to interact with the environment by "listening." There is also the addition of the ultrasonic sensor, which has opened the door to building robots that are able to react before they hit an object. The integration of a rotation sensor to each of the three NXT servo motors provides your robot with precise motor control for things such as driving and positioning. More important, the NXT effectively has the capability to run with seven sensors out of the box (four sensor ports and three servo motor encoders).

Sensors are an important part of robot design and functionality. Remember to think outside the box with each challenge you are faced with. If you require multiple touch sensors for a robot, but you have only one, think of another way you can meet the requirement. As discussed in this chapter, you can use a light sensor to act as a touch sensor.

Numerous aftermarket sensors are available for the NXT, and more are coming. If, for example, you are looking to expand motors or sensors on your NXT, you don't have to buy another NXT. Instead, have a look at the various multiplexers available out there. Stay tuned, as there are more cool sensors to come that we could not talk about here due to nondisclosure restrictions.

Chapter 5

What's New with the NXT?

Solutions in this chapter:

- Notable Enhancements
- The NXT File System
- The LCD Screen
- Digital Interfaces and Bluetooth
- Future Possibilities

Introduction

Several changes are introduced in the NXT compared with its predecessor, the RCX. This chapter covers the major new technologies in detail, explains how they work, and provides examples of how people have used them in their robots. Specifically, it covers the new LCD, file system, Bluetooth, and digital interfaces. (We intentionally omitted several improvements, because we cover them elsewhere in the book.) The chapter concludes with a discussion of some novel ways in which you can use the NXT thanks to all of these new technologies.

Notable Enhancements

In this section we'll discuss some enhancements to the MINDSTORMS NXT, including studless construction, ethical connectors, rechargeable battery packs, and flash memory.

Studless Construction

If you are accustomed to LEGO TECHNIC construction, you will get the hang of NXT robot construction, as the NXT set is almost entirely studless. However, if you primarily did studded construction before, this is a paradigm shift in robot design, and it takes a while to get used to. The MINDSTORMS NXT kit introduces a few new pieces that were specially designed for this kit. For example, as of this writing, the TECHNIC Beam 3 x 3 Bent with Pins (often referred to as a *Hassenpin* after Steve Hassenplug) is available only with the set. As a matter of fact, this part, which is useful for 90-degree connections, was spurred on by members of the original group of adult fans (MINDSTORMS User Panel) who were part of the NXT project when it was just getting started. Early on, Hassenplug and a few others recognized the need for such a piece to aid in the construction process that was mostly all studless building. This piece is one of the most popular pieces in this set today.

Electrical Connectors

The electrical system of the MINDSTORMS NXT kit introduces a new connector for the sensor and motor cables. The connector is similar to an RJ12 phone jack, but if you look carefully, you will notice that the latch is not in the center. It is offset to the left edge, preventing use of regular phone cables, whether accidental or intentional. You can still connect the new electrical system with the old motors and sensors using a legacy compatibility cable.

Rechargeable Battery Pack

Though the rechargeable battery pack doesn't come with the retail set, it is a thoughtful addition as an NXT accessory. The battery pack has a socket on the side to attach the charger, which can conveniently recharge the batteries without you having to remove the pack from the NXT. Compared to the original battery cover, the depth of the battery pack is slightly greater, and when connected to the NXT, it protrudes at the bottom. This is something worth noting and you need to account for it in your robot design. One nice addition on the software side in terms of the NXT-G and RobotC is that it is possible for you to detect whether the robot is using a rechargeable battery pack, as well as evaluate the battery's power level (see Figure 5.1).

Figure 5.1 Rechargeable Battery Pack

Flash Memory

Another convenience that is worth mentioning is the flash memory. The flash memory of the NXT brick retains firmware and all your files, even when you remove the batteries for an extended period. This is a significant change from the days of the RCX, when you would have to reflash the unit after you replaced the batteries. The NXT flash can store up to 256KB of data. This is a lot more space compared to the RCX. However, as we will discuss later in this chapter, you can create several new types of files, and those files can occupy space quickly.

Multiple Types of Sensors

The NXT system supports two types of sensors: new sensors that use the Inter-Integrated Circuit (I^2C) interface, and older analog sensors such as those that came with the RCX kit. There is no specific rotation sensor, as it is now built into the NXT servo motors. The Ultrasonic sensor is a new addition to the sensor family. This sensor communicates with the NXT brick using the new digital interface. The redesigned light sensor is also a major improvement over its predecessor. Chapter 4 provides additional details on these and other third-party sensors.

The NXT File System

The NXT has introduced a file system that stores up to 64 files. The firmware allows users to create and delete files, rename them, and modify their contents. Filenames have a three-character extension, separated from the filename by a period. The name itself can be up to 15 characters long. The extensions start with *r*, and by this convention, the executable files on the NXT have an extension of .rxe. Table 5.1 lists the recognized file types with their extensions.

Table 5.1 NXT File Types

File Type	Extension(s)
Data files	.rdt
Executable files and Try Me programs	.rxe, .rtm
Icon files	.ric
Hidden menu files	.rms
Program files	.rpg
Sound files	.rso
Hidden system files	.sys
Temporary hidden files	.tmp

The NXT brick user interface shows these files in the form of a menu structure starting with "My Files," and then branching into "Software files," "NXT files," "Sound files," and so on (see Figure 5.2). Each menu then shows the files stored in NXT memory. The screen sample in Figure 5.2 is from NeXTScreen, a utility developed by John Hansen (see Appendix A for details).

Figure 5.2 NXT File Menu Navigation

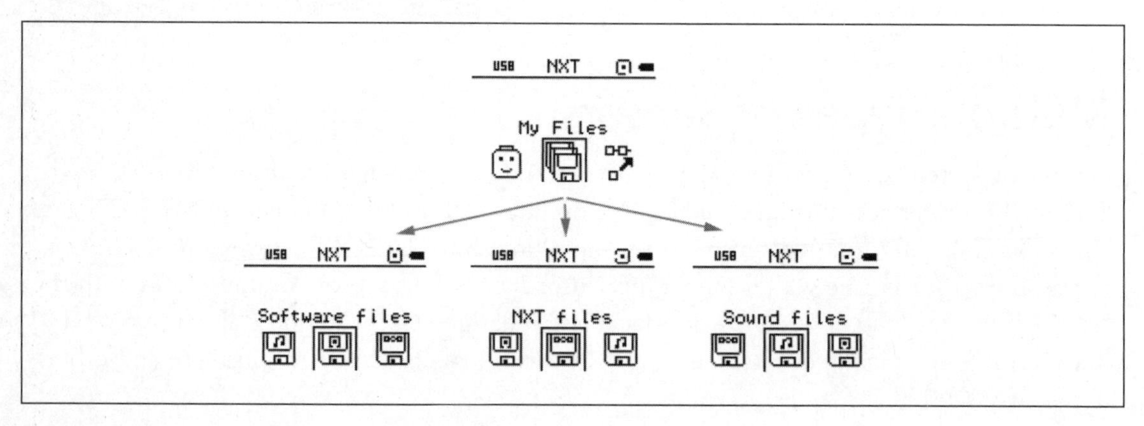

File-Handling Functions

The NXT-G and the RobotC software offer functions to work with files. The NXT-G software provides a block to manipulate files on the NXT. This block supports read, write, close, and delete operations on files. For the read and write operations, it supports reading and writing of the data type as "text" or "number."

The RobotC software has several advanced functions for file handling and manipulation. The read/write functions support data types such as byte, short, float, long, and string. RobotC also supports functions to search files that may be on the NXT.

Using File Space Efficiently

The file space on the NXT is limited, and it's wise to use that space efficiently, as you never know when you will need to add that extra logic in your program.

You can delete a few files that come installed on the NXT, if you want to reclaim some space in lieu of functionality. For example, you can delete the Try Me programs, sound files, and so on. If you need that functionality at a later date, you can always download those files onto the NXT.

The key to efficient use of the file system is to write small programs. You can do so by reusing code wherever possible. The first time you use a block or function in your program, all the code required for that function needs to be linked into your program, and that increases the program's size. Any subsequent use of the same block or function does not increase the program's size much; rather, there is only a minuscule increase to account for the subsequent calls to that function. Lately, LEGO has been publishing mini versions of some popular blocks that are optimized to make smaller programs (for details, see Appendix A).

The LCD Screen

The NXT brick has an LCD screen that is much larger than its predecessor. The screen is a monochrome matrix of 100-by-64 pixels. When you are interacting with the NXT, the user interface on this LCD is your window to the NXT's internals.

The firmware provides text as well as pixel-level support for drawing on this screen. You can also specify the name of an icon file that is residing on the NXT to draw as a picture. RobotC has several functions for displaying text on the screen. Using these functions, you can choose a large or small font and display text at any position on the screen. RobotC also has several functions for drawing lines, rectangles, circles, ellipses, and images on this screen. Figure 5.3 shows an example of text sizes and lines.

Figure 5.3 The LCD

Here is the RobotC program that was used to create the screen image shown in Figure 5.3:

```
task main()
{
  // Display small font text lines
  nxtDisplayTextLine(1, "Building Robots");
  nxtDisplayTextLine(2, "with LEGO");
  nxtDisplayTextLine(3, "Mindstorms NXT");
  // Display large font text
  nxtDisplayBigStringAt(1, 20, "SYNGRESS");
  // Draw the lines
  nxtDrawLine(10, 28, 90, 28);
  nxtDrawLine(5, 25, 95, 25);
  // wait for a minute before ending the program
  wait10Msec(6000);
}
```

Games

One interesting feature the NXT offers for drawing images is its icon file format. The icon files have an .ric extension on the NXT file system. The file generally contains instructions which define the image. Unlike a bitmap image, this file can contain instructions with parameters for drawing lines, circles, rectangles, and so forth. Of course, the bitmap instruction is also available.

The format for these files is well defined, and the images the NXT software comes with use the same format. You can also create your own files and upload them on the NXT. Moreover, in theory, you could create such files from within your NXT program, and use them. That opens up numerous possibilities for game designers! There is, however, limited support for testing these files outside of the NXT. The NXT-G software can render only the bitmap instruction from this file format. The other instructions must be tested on the NXT itself.

Ross Crawford wrote a Tic-Tac-Toe game to run on the NXT screen (see Figure 5.4). The program uses the buttons on the NXT for choosing location and placing circles, and it uses the NXT features of drawing lines, circles, and text to draw the game board and messages (for details, see Appendix A).

Figure 5.4 Tic-Tac-Toe on the NXT

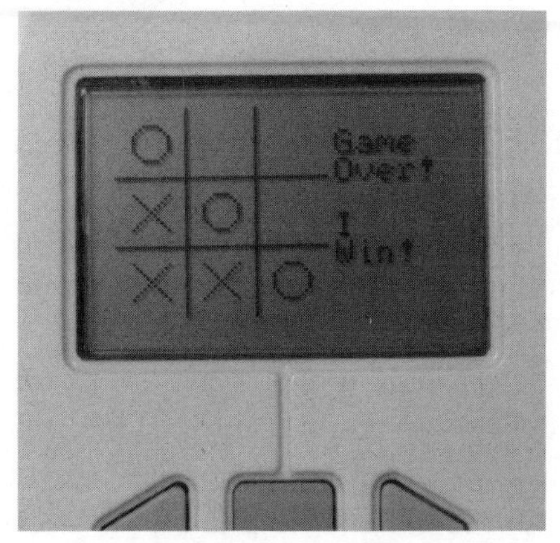

The NXT display functions are also useful for making charts and graphs on the screen. With the NXT file system, as you will see in the next section, data gathering and logging are easy. You can write programs to read these files and, using the display functions, render this data as charts or graphs.

Digital Interfaces and Bluetooth

To communicate with sensors, motors, and other peripheral devices physically attached to the NXT, the new system introduces I^2C and USB interfaces, whereas for wireless communication, the NXT has a Bluetooth interface. The USB interface is used to communicate with a Windows PC or a Macintosh, and I^2C is used to communicate with sensors and motors.

The NXT software comes with required USB drivers, making the NXT a Plug and Play device. The USB system consists of a host device connecting multiple peripheral devices, with peripherals being able to connect to only one host. The USB interface on the NXT is a peripheral USB port, and thus it can connect to any computer with a USB host port. At the same time, you can't attach another USB peripheral device, such as a memory stick, to the NXT and expect it to work.

The I^2C interface, also known as IIC, is not a popular household name, but it is used in a wide array of gadgets and devices (and even in some household items). For instance, it is commonly used in electronics devices where space efficiency is key, such as mobile phones and PDAs. For the most part, you won't have to work with the I^2C interface directly, unless you are developing specialized hardware. In any case, this is a very versatile interface in the NXT, and it holds promise for a wide variety of next-generation attachments and extensions to the NXT system. Already a few third-party sensors are available that use the I^2C interface, and more are being developed at a rapid pace. For more information on this, refer to Chapter 4, which covers the I^2C interface in greater detail.

Bluetooth Communication

In addition to the wired interfaces mentioned previously, the NXT supports a wireless Bluetooth interface. This is a major improvement over the former infrared interface, which the RCX had, in terms of speed, power consumption, and standardization. The NXT can selectively turn this interface on and off through programming functions or on-screen menus, thus further controlling power consumption.

Bluetooth on the NXT can operate in either master or slave mode. While connecting to a PC, it is always in slave mode; however, you can configure it to operate in master mode while connecting to other slave NXTs. You can connect up to four NXTs to each other, with one being the master and the others being slaves.

Each NXT has a unique Bluetooth name, and this name is displayed on the top line of the screen. While referring to each other by this unique name, you can make a convoy of NXT robots that can work as a team!

A Surveillance Robot Using NXT and Bluetooth

The standardized Bluetooth interface also lets the NXT communicate with other Bluetooth-enabled devices. What does this mean? Well, as an example, sending commands to the NXT from a mobile phone is a breeze. You can attach a mobile phone to your NXT robot, send the robot on a surveillance mission, take photos or video, and have the mobile phone transmit the data to a computer!

For instance, Martyn Boogaarts developed a robot (see Appendix A) that has a camera attached for taking pictures (see Figure 5.5). The robot's head swings around slowly in increments of 5 degrees. Using an ultrasonic sensor, it looks for any objects in its vicinity, and as soon as it spots an object, it sends Bluetooth commands to the camera to take a picture. Then it continues on its mission.

Figure 5.5 Coco5 by Martyn Boogaarts

A Bluetooth-Based Remote Controller

Using the Bluetooth interface, the NXT motor, and a Mindsensors acceleration sensor, Philippe "Philo" Hurbain created a remote controller (see Figure 5.6) that transmits spatial information to an NXT robot (see Appendix A). Using the acceleration sensor, the remote controller transmits forward or backward tilt and left or right tilt information to the robot. In addition to this information, you can use your thumb to press the touch sensor on the left side to transmit on/off commands. There is a double-bevel gear on the motor that you also can turn with your thumb. Using this gear, the remote controller can transmit information such as rich motor encoder values, which the robot can use for navigating.

Figure 5.6 NXTiiMote Remote Controller

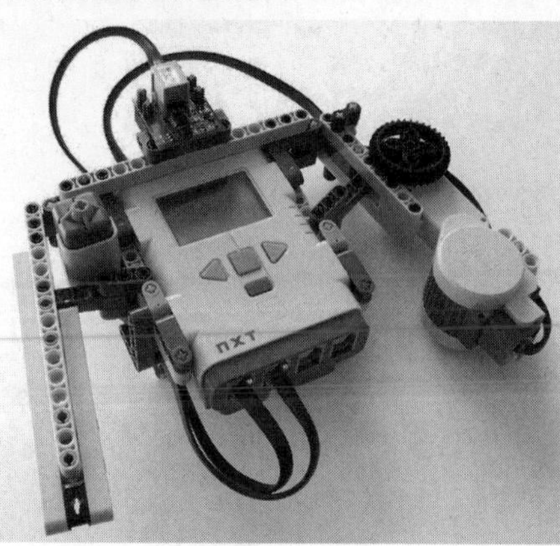

Spatial Motion Controllers

The Wii remote controller by Nintendo, popularly known as the Wiimote, is another amusing device that you can integrate with the NXT (see Figure 5.7). In fact, a few folks have tried to integrate the Wiimote with the NXT directly, with little success. The Wiimote uses the Bluetooth HID profile technology, whereas the NXT supports serial port profile technology. Unfortunately, these two don't work together, unless you are good at hacking the Wiimote and you know what you are doing. However, it is possible to use a programmable mediation Bluetooth device which communicates on both profiles. A mobile phone running a special-purpose NXT Mobile Java application is a perfect example (see Appendix A for more information on NXT Mobile applications). The application would interpret Wiimote commands on the HID profile, and transmit them as serial commands to the NXT. Jose Bolaños created a robot (see Appendix A) for which he used a computer running a Bluetooth program as a mediation device.

Figure 5.7 Wiimote Integration

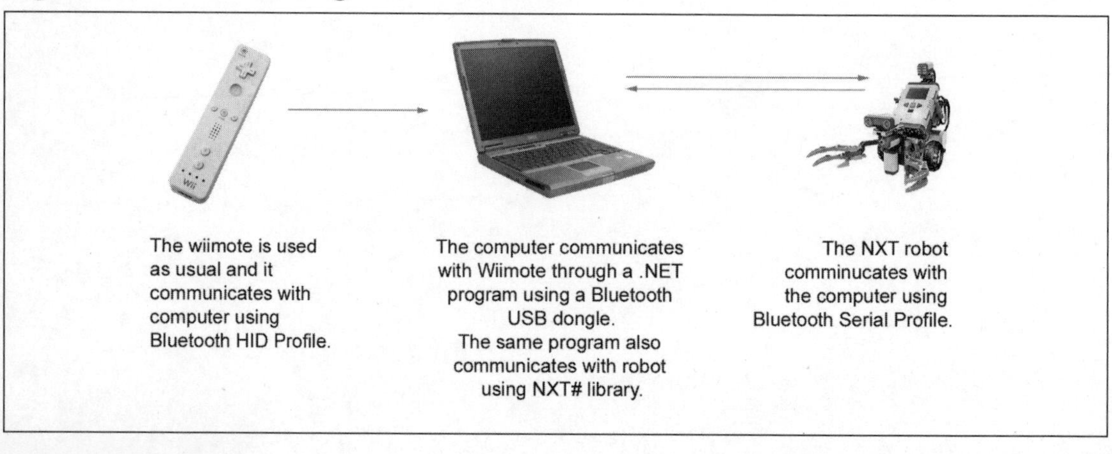

I²C for Spatial Motion Controllers

Spatial motion interpretation on the NXT is not an unattainable ambition, though. Paul Tingey created a controller to steer and control the RobotArm-56 (see Appendix A). He used the Mindsensors compass and acceleration sensors for the motion sensing, and wired them to the NXT with a smart algorithm (see Figure 5.8).

Figure 5.8 Wii-like Controller

Future Possibilities

If you could peer into a crystal ball to see future scenarios for using the NXT, what would you see? In this section we'll describe some of them.

An NXT Robot Controlled from a Web Server

Imagine a robot to be sent on a mission to explore uncharted territory—say, a dark closet. You are the mission controller and you have all the computing power to design the control station. You want to see where the robot is going, and control and steer him in the right direction. At the same time, you want to let the robot have sufficient autonomy to do things on its own, such as negotiating its path to advance farther. And you may not be anywhere near the robot.

You can do this in several ways, and one of them is to create a Web-based application, which you can access from anywhere and anyplace, using any browser. The application would communicate with the NXT robot using Bluetooth and gather pictures from the robot, which would then be displayed on your browser. The steering controls that you design for the Web page would transmit the commands to the NXT in real time as you operate the controls.

While designing such a robot, you have choices for the Web application architecture, too. For example, if you were to go with the Internet Information Server (IIS), you would have plenty of support in .NET. If you chose to go with the Apache server, you would have adequate scripting support for Bluetooth communications. In fact, NXT-specific Bluetooth functions are available now for Perl, too.

NXT Puppet Show

Now imagine a stage designed for puppet shows; one similar in size to a LEGO theme—say, a meadow. In that meadow, a boy, *made out of an NXT*, is guarding his sheep, *again made out of an NXT*. The boy plays his flute for a while and falls asleep ... *a feat which is now possible thanks to the sound capability on the NXT*. The sheep is grazing around happily. From the edge of the table, the show performer places an NXT wolf in the meadow, and turns it on. As soon as the wolf comes to life, the sheep is scared and starts to scurry around. The wolf pounces at the sheep and knocks her down. The sheep is screaming, and all that sound wakes up the boy. The boy looks around, and finds the wolf and chases him away.

Okay, back to the drawing board. What do you need to make this kind of puppet show? The Bluetooth identifiers of NXTs can clearly differentiate the boy from the sheep or the wolf. As soon as the NXT is turned on, the Bluetooth identifier of the wolf can broadcast itself and the programs in the sheep can choose to respond to scurry around, whereas the programs in the boy ignore it. One hurdle in a story like this is locating other puppet, if the puppets are not predictably placed on the stage, or if they have moved around to a random location. To see things in the area, therefore, a third-party camera for the NXT would be great! As you can see, the possibilities for stories are limitless.

GPS and the NXT

Wouldn't it be nice to let your robot wander around the neighborhood and let it know where it is? A distinct possibility: a robot that could deliver flowers at the other end of town, if it only knew the coordinates of the house! A GPS would come in handy here. A Bluetooth-enabled one.

Several Bluetooth GPS modules are available today. These modules are designed to connect to mobile devices, such as PDAs or laptops, and they provide location information. The actual mapping is done in the software running on the PDA or laptop. You can use the same concept in the NXT robot. There are a few hurdles, however. First, the GPS unit needs to be facing a clear sky, so your robot arena needs to be outdoors. Second, the GPS resolution, at best, is half a meter. Therefore, a small robot would be left on its own to position itself within that zone.

Summary

In this chapter, we covered several new introductions in the MINDSTORMS NXT system as compared with its predecessor, the RCX. Most of these introductions represent the latest technologies made available for your robot to play with.

For instance, the LCD screen provides a better user interface for the NXT, and allows users to create interactive games and other screen applications. The NXT file system is simple and provides several functions for manipulating files and their contents. And I^{2}C and Bluetooth have opened up possibilities for integrating your robot with several other standardized gadgets and sensors.

As you can see, plenty of creative possibilities are within your reach. You just need to put your mind and your NXT to work!

Chapter 6

Building Strategies

Solutions in this chapter:

- Studless Building Techniques
- Maximizing Modularity
- Loading the Structure
- Putting It All Together: Chassis, Modularity, and Load
- Hybrid Robots: Using Studless and Studded LEGO Pieces

Introduction

Having discussed motors, sensors, geometry, and gearing, we'll now put all these elements together and start building something more complex. LEGO robotics should involve your own creativity. You won't find any rules or style guides in this chapter, simply because there aren't any. What you will find are some tips meant to make your life easier if you want to design robust and modular robots.

Studless Building Techniques

Building with studless LEGO pieces is different from traditional studded building. Simply stacking pieces together to make strong robots and structures is not an option with studless building. As you build with studless pieces, you may notice that studless structures tend to be more flexible than similar studded structures. Don't confuse *flexible* with *weak*. Properly built studless structures can be just as strong as those built with studded pieces. The two building systems are different, and they require different building techniques.

The main structural component of any robot built with studless pieces will be the studless beam. Let's start with two white studless beams oriented as shown in Figure 6.1. There are many ways to connect these two beams, and all of them have advantages and disadvantages. Figure 6.2 shows four components that you can use to connect the two beams.

Figure 6.1 Two Parallel Beams

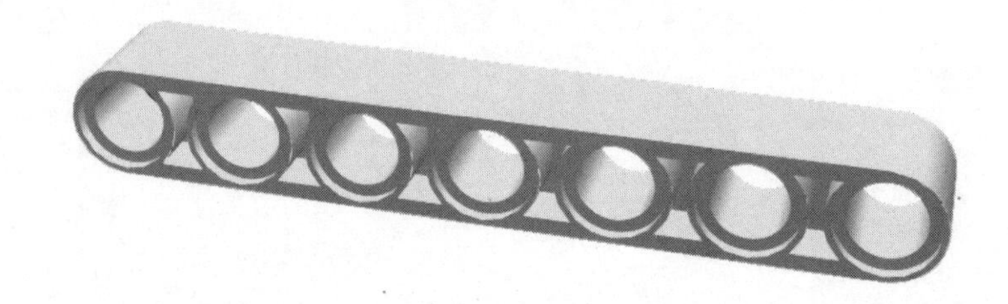

Figure 6.2 Methods to Connect Parallel Beams

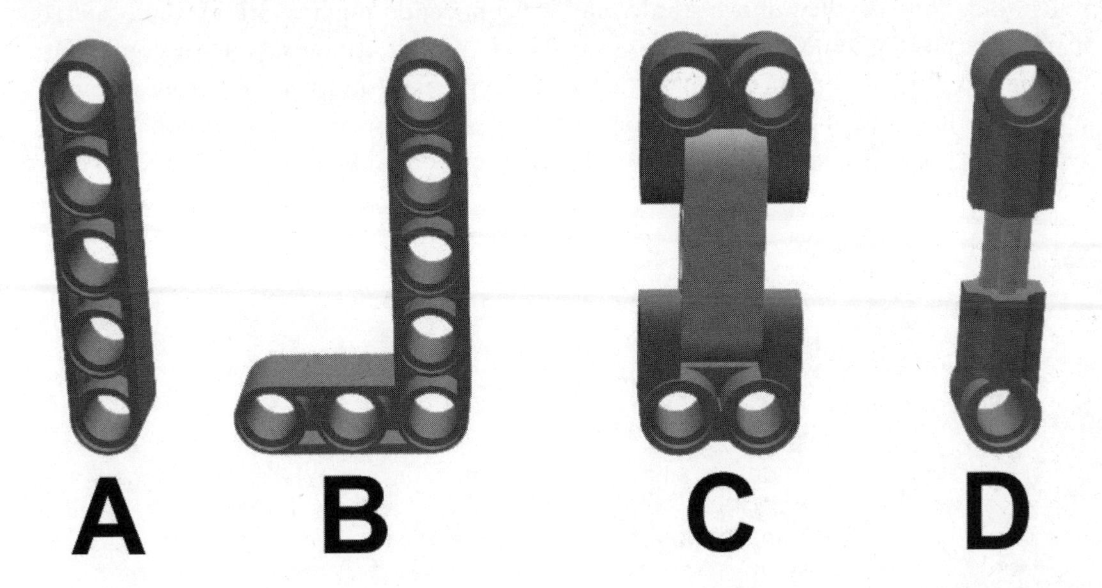

Option A is a simple beam that you can pin to each parallel beam. A single beam won't be very useful, but as a pair, they will connect the two beams and keep them parallel to one another. This type of connection is called a **parallel linkage**. In it, the two white beams will always be parallel to each other, as will the two gray beams. This will work well in some situations, but most of the time you'll want a more rigid assembly.

NOTE

Remember to use the *black and blue pegs* (or **pins**) when connecting beams. They fit in the holes with much more friction than the gray or tan ones, because they are meant for building rigid components and structures. The *gray and tan pegs*, on the other hand, were designed for building movable connections, such as levers and arms.

You can use option B to make a more rigid assembly. Replace one of the gray beams with the L-shaped piece. Be sure to use three pins to connect it to the white beams. You'll discover that the assembly is both strong and fairly rigid.

You can use option C on its own, to connect the two white beams. It will keep them parallel to one another and not allow them to move. However, this option is less rigid than a combination of options A and B. Use option C to connect the two white beams. Now gently press the two white beams together with your thumb and forefinger. It doesn't take much force to make them move a little bit. This connection might work in some cases, but if you start putting some weight (such as our NXT) on this, it might bend or come apart.

Option D is another method. You can use it much like you used the two beams in option A. This method has one major drawback. It will not work under tension. In other words, you can pull it apart easily. Option D will work only in compression when the two white beams are being forced together.

Bricks & Chips...

Tension and Compression

Tension and compression are two forces that can act upon objects. *Tension* is a force that attempts to stretch or lengthen an object. *Compression* is the opposite; it will try to compact or shorten an object. Option D in the preceding section is a good example. In compression, when the two ends are being forced together, this option can work well. However, in tension, when the two ends are pulled apart, the object will come apart. All the other options work well in both tension *and* compression. They will not fail easily in either situation, at least within the reasonable limits of plastic LEGO pieces.

Now let's change the orientation of the two white beams and try to connect them in a different way. Figure 6.3 shows several ways to connect beams in this new orientation. The first two options, A and B, use pieces designed specifically for this purpose. These two connectors will keep the two white beams together and parallel to each other in all directions. Both work very well in many situations. However, both fail when placed under tension. In other words, when the two white beams are pulled apart, the pieces easily separate.

Options C and D are also good ways to connect the two beams. However, as constructed, both of them fail in compression, when the two white beams are pressed together. You can construct options C and D so that they will not fail in compression by placing the two white beams outside of the two black connectors.

Figure 6.3 Connecting Parallel Beams

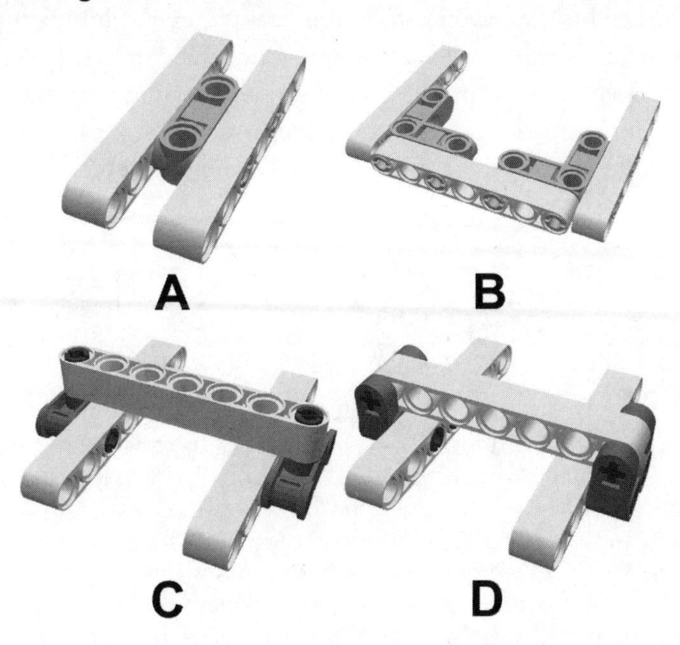

So, what can you do to keep the assembly together in both tension and compression? Starting with option C or D, you can simply add two more connectors, as shown in Figure 6.4. This prevents failure in both tension and compression. This type of connection also has another major advantage. Unlike options C and D, where the two white beams can still move in one direction or another when connected, these two beams are fixed in position and cannot pivot, rotate, or slide. This results in a strong and rigid assembly that will not easily fall apart.

Figure 6.4 Assembly to Resist Tension and Compression

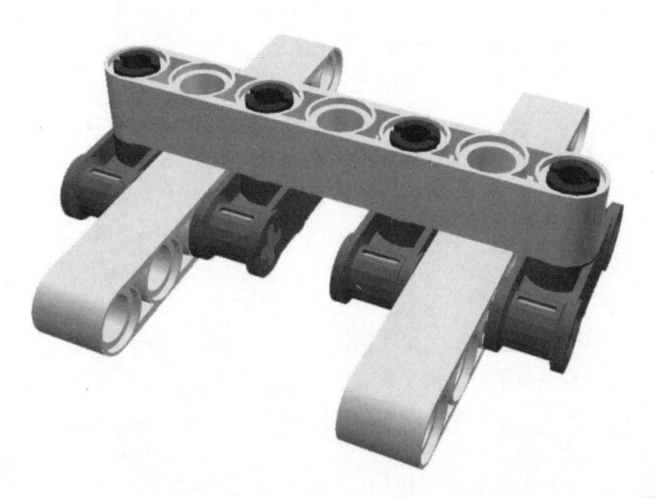

Despite our insistence on the importance of connecting beams for a strong and rigid assembly, there's no need to go beyond the minimum required. Eliminating unnecessary parts can result in a smaller, more compact, and lighter weight assembly. Weight is, actually, a very important factor to keep under control, especially when dealing with mobile robots. The greater the weight, the lower the performance, due to the inertia caused by the mass and because of the resulting friction the main wheel axles must endure.

Bricks & Chips...

What Is Inertia?

In physics, **inertia** is the tendency that objects have to resist changes when in a state of motion or rest. Objects at rest tend to stay at rest, whereas objects in motion tend to stay in motion, moving with the same direction and speed. All objects have this tendency, but some more so than others: chiefly because inertia depends on **mass** (quantity of matter). A good example of how mass affects inertia comes from something with which most people have direct experience: shopping carts! When the cart is empty, you can easily start and stop it, or change its direction, with minimum effort. The more stuff you put inside it, however, the more strength you need to maneuver it. Why? Because its objective mass, and thus its inertia, has increased. Similarly, the greater the mass of your robot, the more force is required from its motors when accelerating or braking.

Maximizing Modularity

While building your robot, you will likely have to dismantle and rebuild it, or parts of it, at least, many times. This isn't like following someone's detailed instructions; it's more of a trial-and-error process. Unless you're a very experienced builder and are blessed with clear ideas, your design will develop in both your mind and your hands at the same time.

For this reason, it's best to make your model as easy to take apart as possible, or, to term it more appropriately, your robot should be **modular** in construct. Building in a modular fashion also gives you the opportunity to reuse components in other projects, without having to rebuild common subsystems that already work. This is not always possible, because when you want something really compact, you have to trade some modularity in favor of tighter integration. Nevertheless, it's a good general building practice, especially when constructing very large robots.

Building with the NXT motors and sensors requires a little more planning than building with legacy LEGO motors and sensors. Their unique shape and unique location of mounting points make them difficult to simply attach anywhere with little planning. For this reason, you'll probably want to plan where they will be in the finished design and build them into the structure of your robot from the very beginning. Doing so will probably make your finished design more compact as well as stronger. Unfortunately, it will probably be nearly impossible to easily remove your motors without destroying a large part of the surrounding structure.

So, how can you maximize modularity while at the same time tightly integrate the NXT, motors, and sensors into a robot? By creating subassemblies that, when combined, make up the complete robot. A subassembly would have the motors, sensors, and/or NXT tightly integrated within it. Assembling the multiple subassemblies should be easy. This way, you can remove, rebuild or improve, and reinstall a single subassembly without altering the other subassemblies. The bucket robot in Figure 6.5 illustrates these ideas. You can break down this robot into three subassemblies.

Figure 6.5 Bucket Robot

Remove the two pins on each side that connect the top motor to the turntable (see the arrows in Figure 6.5) and the wire connecting the motor to the NXT. This separates the bucket subassembly from the rest of the robot. The NXT is a subassembly of its own. You can remove it by simply pulling up on it. Figure 6.6 shows the three separate subassemblies. You can modify or change each one entirely without affecting the others as long as the mounting points where they connect to each other do not change.

Figure 6.6 Bucket Robot Subassemblies

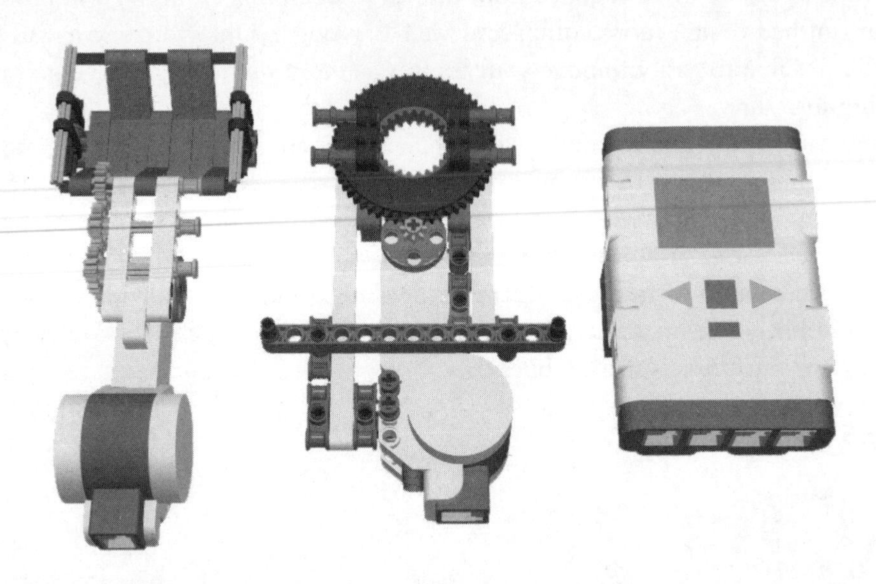

NOTE

One good reason to make your NXT easily detachable is that you must be able to change batteries when necessary. The most common solution is to keep the NXT at the very top of your robot—this way, you can also easily access the push buttons and read the display. It is also a good idea to have easy access to the sensor and motor ports, as well as the USB port on the NXT while it is installed in your robot. That way, you can attach motor and sensor cables as well as the USB cable, all without having to remove the NXT from your robot.

Modular building has another advantage. At some point, you may build a very large LEGO robotic creation and wish to share it with friends at school, a local club meeting, or a LEGO event. If it is modular, you could take it apart into its individual subassemblies and reassemble it at your destination. Transporting multiple subassemblies is usually easier than transporting one very large structure.

Loading the Structure

Even the most minimal configuration of a mobile robot has to carry a load of about 450 g (16 oz): the weight of one NXT (with batteries) and two motors. Adding cables, sensors, and other structural parts can easily push you up to 600 g (21 oz) or beyond. Should you worry about this mass? Is its position relevant?

The first factor you need to consider is friction. You should take all possible precautions to minimize it. This is especially true where the structure attaches to the wheels because it is there that you transfer all the weight to the wheels by way of the axles. The wheel acts as a lever: The greater the distance from its support, the greater the resulting force on the axle. Such forces tend to bend axles, twist beams, and produce plenty of friction between the axle and the beam itself. For this reason, it's important that you keep your wheel as close as possible to its supporting beam. Figure 6.7 shows three examples: "a" being the worst case and "c" the best.

Figure 6.7 Keep a Wheel As Close As Possible to Its Supporting Beam

It is also a good idea to support the load-bearing axles with more than a single beam whenever possible. The three examples shown in Figure 6.8 are better than those in Figure 6.7, with 6.8c being the best among all the solutions shown so far. The use of two supports—one on either side of the wheel, as on a bicycle—avoids any lever effect created by the axle on the support, thus reducing the friction to a minimum.

Figure 6.8 Two Supporting Beams Are Better Than One

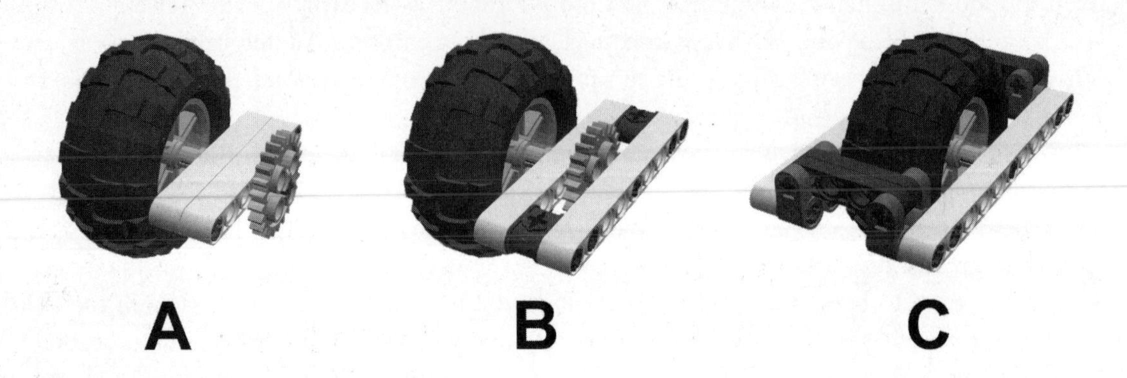

Having your gearing as close as possible to your supporting beams is just as important (see Figure 6.9). Positioning your gears next to supports will help to reduce or eliminate any gear slippage. If you do not place your gears near supports, the axles they are attached to may bend a little, which might allow the gears to slip when they are placed under load.

Figure 6.9 Place Gears Close to Supporting Beams

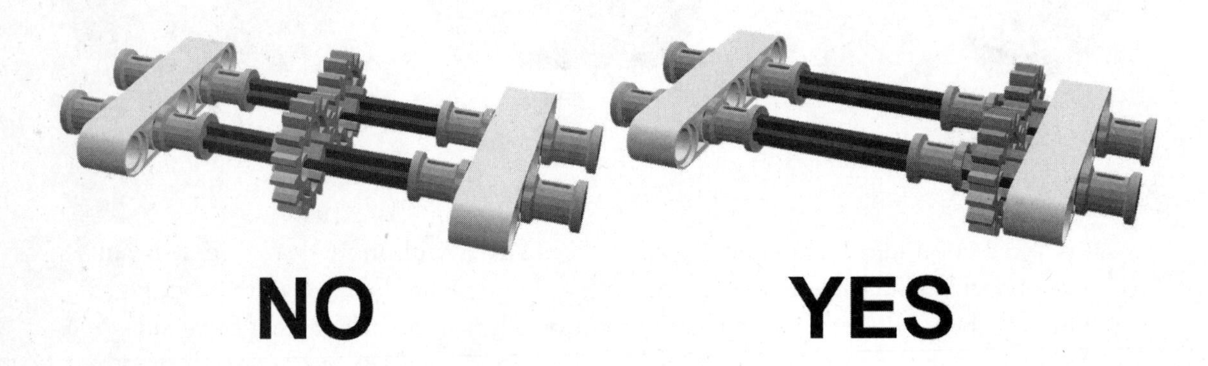

It is also important to place supports in the same orientation as the gears. If your gears are oriented vertically, your main support should be in the vertical direction. When placed under load, the majority of the force being placed on the gears and axles will tend to push them away from one another, so your support needs to be oriented to resist that force. Likewise, if your gears are in a horizontal orientation, your supports should be horizontal as well. Figure 6.10 illustrates this idea. In Figure 6.10, the main supports for the gears are placed vertically, while the gears are lined up horizontally with each other. When placed under load, the gears are going to try to push away from one another, primarily in the horizontal direction. Because there is little support in the structure to resist this, there is a good possibility that the gears will slip. Simply placing a short beam between the two gears connecting the two axles together will reduce the possibility of gear slip.

Figure 6.10 Gears In-Line with Supporting Beams

Ideally, the supports would be exactly in-line with the axles so that the axles would run through the holes in the beam being used for support. However, that is not always possible. In these cases, placing the supports as close as possible to the gears is advisable. Figure 6.11 shows three possible options, with "a" being the worst and "c" the best.

Figure 6.11 Gears Placed Near Supports

The position of the NXT has a strong influence on the behavior of mobile robots. It's actually the shape and weight of the whole robot that determines how it reacts to motion, but the NXT (with batteries) is by far the heaviest element and thus the most relevant to balancing load. Recalling the concept of inertia will explain *why* balancing load is important. As explained earlier in this chapter, any mass will tend to resist a change in motion—in some cases, to resist *acceleration*. The greater the mass, the greater the force needed to achieve a given variation in speed.

Putting It All Together: Chassis, Modularity, and Load

The following example summarizes all the concepts discussed so far in this chapter. You can build the chassis shown in Figure 6.12 using only parts from the NXT kit. Its apparent simplicity actually conceals some trickiness. Let's explore this together.

Figure 6.12 A Complete Platform

The motors are integrated into the chassis for this robot. They provide the power for movement and act as structural components. The upper beams are attached to the motors and lower beams using two L-shaped beams for improved rigidity. In front, the two sides of the robot are connected using two horizontal beams. The upper is designed to hold up under compression, and the lower resists tension forces. Two separate supports also help to increase the rigidity of the entire structure. Figure 6.13 shows the bottom view of the robot.

Figure 6.13 Bottom View

Wheels are placed as close as possible to the supports to reduce any lever effects that could introduce more friction into the system. The axles are also supported on both sides of the motors. The motor hubs tend to bend and move a little bit, so the axles are supported on both sides of the motor to further reduce any lever effects.

This platform is also very modular. Even though the motors are tightly integrated into the chassis, this platform comes apart easily so that each individual subassembly can be worked on and modified. The caster wheel assembly simply pulls off the back of the platform, as shown in Figure 6.14.

Once you have removed the rear caster assembly, remove the two axles and bushings from the 7-length beam on the front, and then remove the beam. With a gentle pull, the left and right drive assemblies will separate from each other. You now have the three separate subassemblies that make up the robot platform (see Figure 6.15). Even the wheels can be easily removed for use elsewhere.

Figure 6.14 Removing the Rear Caster Assembly

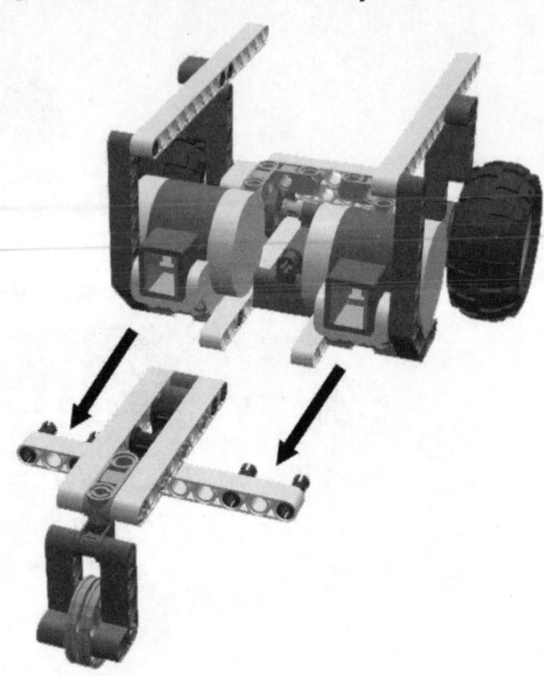

Figure 6.15 Separating the Left and Right Drive Assemblies

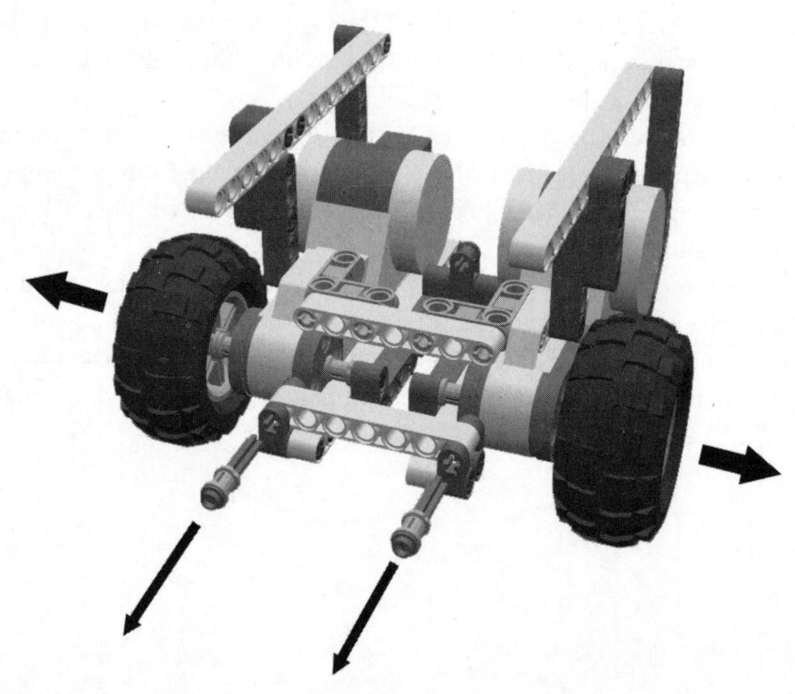

The truth is that if you own only the NXT kit, you probably won't have enough parts to build another robot unless you dismantle at least most, if not all, of your existing robot. If you have more LEGO TECHNIC parts, however, you might be able to leave your platform or a select subassembly intact and reuse other parts in a new project.

This robot is a good platform to use to experiment with load and inertia. Write a very short program that moves and turns the robot. You don't need anything more complex than the following pseudocode example, which will drive your robot briefly forward and backward, and make it turn in place:

```
start left & right motors reverse
wait 2 seconds
stop left & right motors
wait 2 seconds
start left & right motors forward
wait 2 seconds
start left & right motors reverse
wait 2 seconds
stop left & right motors
wait 2 seconds
start left motor forward
start right motor reverse
wait 2 seconds
stop left & right motors
```

Place your NXT in different locations and test what happens. When it is slightly forward of the main wheel axles (see Figure 6.16), the caster wheel tends to lift off the ground when quickly switching from forward to reverse. This creates an unstable condition that could result in the robot failing to work properly.

As you move the NXT rearward toward the caster wheel, the robot becomes more stable (see Figure 6.17). However, too much weight on the rear caster wheel is not the best position either. The more weight placed on a caster wheel, the poorer it will perform. Too much weight and it may not work at all.

Figure 6.16 Poor Positioning of the Load (NXT) Makes This Robot Unstable

Figure 6.17 Too Much Weight on the Rear Caster

Proper positioning of the load would be somewhere in between. For this platform, ideal placement of the NXT would be just above or a little behind the front drive wheels (see Figure 6.18). The drive wheels, not the caster, will carry most of the robot's weight. The NXT is also not so far forward that the whole robot becomes unstable.

Figure 6.18 Proper Positioning of Load

Hybrid Robots: Using Studless and Studded LEGO Pieces

Some of you may have a substantial collection of LEGO pieces besides those that came with your NXT kit. Many of them may be studded LEGO TECHNIC pieces. Because the NXT kit comprises almost exclusively studless pieces, what should you do with all your studded LEGO TECHNIC bricks? Should you throw them away? Absolutely not! Like all LEGO pieces, studded and studless TECHNIC pieces are specially designed to fit well together, and using both in robotic creations can allow for some very interesting innovations. Sometimes creations using both studded bricks and studless beams are called **hybrid** creations, because they are a mixture of studded pieces and building techniques and studless pieces and building techniques.

We already discussed connecting legacy motors and sensors to the NXT with converter cables. Figure 6.19 shows some ways that you can attach legacy motors and sensors to studless beams so that you can include them in your robotic creations.

Figure 6.19 Legacy Sensors and Motors on Studless Beams

As discussed earlier in this chapter, pure studless assemblies often have more flexibility than similar studded assemblies. Oftentimes this is okay, even desirable, but sometimes it isn't. Figure 6.20 shows a differential assembly using a special studded TECHNIC brick to house the differential gearing. Studless beams attached to the sides of the TECHNIC brick show how you can integrate studded and studless pieces in this assembly. As discussed in Chapter 2, the spacing between gears is very important, especially when using a crown gear. A structure using studless pieces to house the differential and gear assembly could have been built, but it would probably have used more pieces and been more flexible than the single TECHNIC brick. This is an example where using both studded and studless LEGO elements have resulted in a better assembly.

Another reason to combine studded and studless elements may be to attach studded LEGO plates to your creation to hide mechanisms and give it a clean outer appearance. Simply pinning studded LEGO TECHNIC bricks to your studless beams and attaching plates to those bricks can accomplish this. You may even want to add your RCX to your NXT robot. Adding studded TECHNIC bricks to your studless chassis will give you studded mounting points for your RCX. You can use the same method to attach a stationary studless creation to a LEGO base plate.

Figure 6.20 Hybrid LEGO Differential Assembly

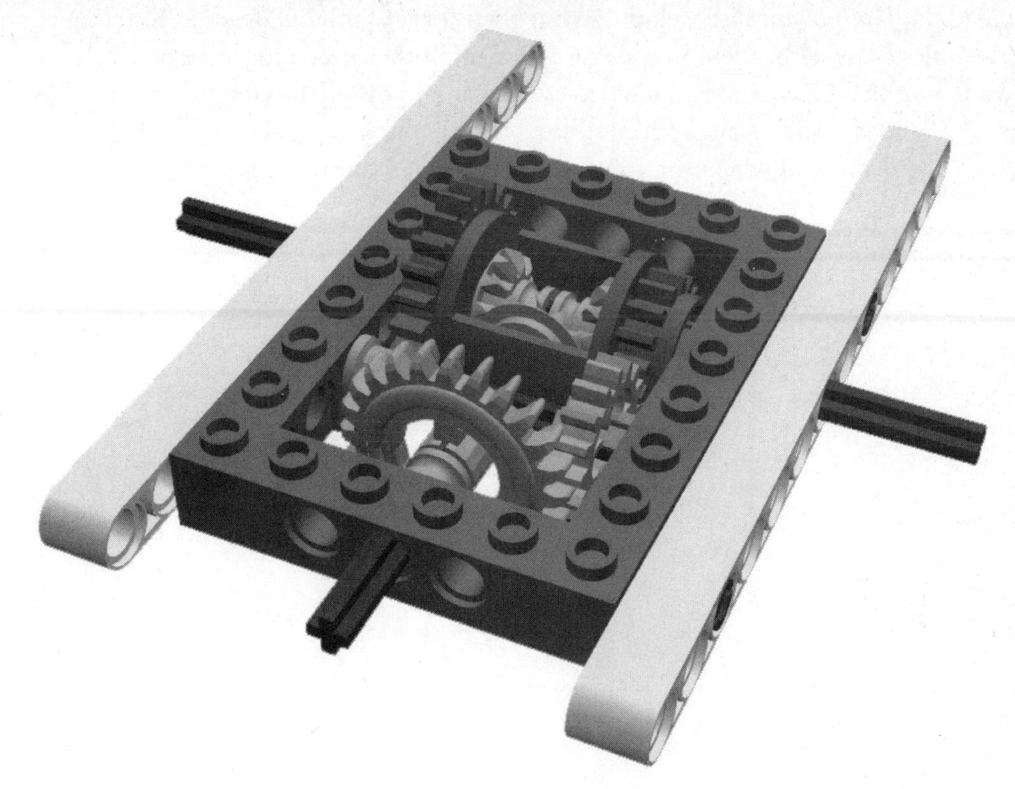

Summary

Recalling the key ideas we've presented in this chapter will serve you well in building LEGO robots with your NXT.

Using the studless building techniques—such as how to connect and brace parallel beams—will help to make your robot chassis strong and reliable. Remember that the proper placement and attachment of beams will do more to create a strong chassis than simply adding more parts. Keep in mind that a solid yet lightweight structure is the goal.

Modularity will allow you to reuse some components in other projects. It will also allow you to move entire subassemblies from one robot to another. You can also modify or rebuild individual subassemblies without having to change the entire robot. Modular building will also help if you need to take your robot apart for transport or storage.

Balance is the key to stable vehicles. Keep the overall mass of your mobile robots as low as possible to reduce inertia and its poor effects on stability. Experiment with different placements of the load, mainly in regard to the NXT, to optimize your robot's response to both acceleration and deceleration and to improve stability.

Keep in mind that the ideas presented in this chapter are suggestions only. They are intended only to aid you in developing your own successful building style. Sometimes you may be able to use all of them in a single robot, and other times you may be able to apply only a few of them. These ideas aren't rules. It may be possible to violate one or more of them at a time and still create a successful robot. Use them as a guide, but feel free to abandon the main road whenever your imagination tells you to do so.

Chapter 7

Programming the NXT

Solutions in this chapter:

- What Is the NXT Programmable Brick?
- Introduction to Programming the NXT Brick
- Using RobotC
- Using Other Programming Languages
- Code Samples
- Running Independent Tasks

Introduction

This book is not about programming. There are already many good resources about programming languages and techniques, and about programming the NXT in particular. However, the nature of robotics (often called *mechatronics*) is such that it combines the disciplines of mechanics, electronics, and software, meaning you cannot discuss a robot's mechanics without getting into the software that controls the electronics that drives the machine. Similarly, you cannot write the program without having a general blueprint of the robot itself in your mind. This applies to the robots of this book as well. Even though we are going to talk mainly about building techniques, some projects have such a strong relationship between hardware and software that explaining the first while ignoring the latter will result in a relatively poor description. For these reasons, we cannot simply skip the topic. We need to lay the foundation that will allow you to understand the few code examples contained in the book.

In the previous chapters, we mentioned the NXT many times; this chapter assumes that you are familiar with the documentation included in the MINDSTORMS kit and that you know what the NXT is. The time has come to have a closer look at its features and discover how to get the most from it. We will describe its architecture and then give you a taste of the broad range of languages and programming environments available, from which you can choose your favorite. Our focus will be on two of them in particular: NXT-G Code, the graphics programming system supplied with the kit; and RobotC, the text-based C programming language available from LEGO Education.

The last sections of the chapter provide two complete code examples, both of which are meant to help explain how to write well-organized code that is easy to understand and maintain, and are designed to familiarize you with the programming structures you'll find later in the book.

What Is the NXT Programmable Brick?

The NXT is a powerful computer. You may think of a computer as being merely a PC with a keyboard, mouse, and monitor; devices created to allow human users to interface with the computer, none of which are available on the NXT. However, many commonplace items—Apple's iPod, cell phones, portable game consoles from Nintendo and Sony, VCRs—have "embedded" computers that provide their functionality. The NXT is more similar to these embedded computer devices.

Instead of the keyboard, mouse, and monitor found on a PC, the NXT has a small liquid crystal display (LCD) and four push buttons. The NXT's LCD is 100 pixels (i.e., dots) wide by 64 pixels high—large enough to display eight lines of text with 16 characters per line. The LCD can also be used to display icons or graphics—in other words, black-and-white pictures. You could even write a program to plot a graph on the display showing how the value of one of its sensors changes over time.

The NXT has four input ports. These are used to connect the four types of sensors (touch, ultrasonic distance, sound, and light) developed by LEGO for the NXT. Many additional NXT-compatible sensors have subsequently been developed by third-party companies (see Appendix A), including accelerometers, compasses, higher-performance light sensors and distance sensors, pressure detection, and so forth.

There are three output ports on the NXT. Variable-speed motors are connected to these ports. The motors have an integrated position detection capability or "encoder" that is used to keep track of how far the motor has traveled. One complete revolution of a motor will change the motor encoder value by 360 counts. The position detection is similar in function to the odometer on a car that measures how far the car has traveled.

The NXT has both a wired USB port and a wireless communications link. You can use either of these links to connect your NXT to a PC. The wireless link uses the industry-standard Bluetooth protocol and the NXT can connect to other compatible Bluetooth devices. For example, some cell phones come with a Bluetooth link and by loading a special software program in your phone you can then use your phone as a remote control for the NXT. You also can use the Bluetooth link to connect one NXT—the "master"—with up to three other NXTs—the "slaves"—for some interesting multi-NXT applications.

The NXT has 256 KB—in other words, 256,000 bytes—of flash memory, and 64 KB of random access memory (RAM). Flash memory retains its value even when the NXT is powered off; this is the same type of memory used to store songs in an iPod or to store telephone directories in a cell phone. The contents of RAM memory are lost when the NXT is powered off; RAM is used to store transient values created while a program is running. The flash memory is used to implement a small file system of up to 64 files that can be stored on the NXT. One type of file that you want to store on the NXT is the program file you've written. Another type of useful file is a sound file that can be played over the NXT's speaker.

The NXT is just like a PC in that it has an "executive" control program or operating system that provides overall control over the other programs on the NXT. Operating systems for PCs include Windows and Linux. The NXT's operating system was written specially for the NXT and does not have a name. About half of the 256 KB of flash memory is used to store the NXT's operating system and the other half is used for the file system.

A total of 256 KB of flash memory may not seem like a lot when you compare it to the disk drive on your PC, which can have 100,000 times this capacity. But the programs written for the NXT are far smaller than the programs on a PC. Typically there's room on the NXT to store around 20 or more user programs.

MINDSTORMS: A Family of Programmable Bricks

The NXT is the second generation of LEGO's MINDSTORMS product line.

The first generation was introduced in the mid-1990s and centered on the RCX programmable brick. There were several derivatives of the RCX, with slightly different features. They all shared the same Hitachi H8 8-bit computer and had a similar programming language. The *Cybermaster* brick had two built-in motors and used a wireless link instead of an infrared link. The *Scout* brick was a lower-functionality subset of the RCX. The NXT is far more capable than the RCX. Some of the improvements include in the following:

- The NXT's computer is 10 to 50 times faster than the computer in an RCX. It has five times as much memory. Internally the computer uses a 32-bit native format instead of 8 bits.
- The NXT has four sensor ports versus three on the RCX. Many more types of sensors are available for the NXT than for the RCX.
- The NXT motors have integrated odometers/position sensors. The RCX does not.
- NXT motors are "smart" with their built-in encoders. On the RCX, you only were able to "apply a specified percentage of the maximum available power" to a motor. If the batteries were weak or there was a lot of resistance to movement—driving up an incline, more friction/binding in the mechanical construction—the motor would operate at different speeds.
- Motors on the NXT can be directed to "move at a specified percentage of the maximum speed using feedback from the motor odometers to adjust the applied power up or down to achieve this speed." The NXT motors will have consistent speed even if the batteries are weak or there is higher friction!
- The NXT has both USB and Bluetooth communications links. The RCX had a single infrared communications link that was slower than and not nearly as robust as the NXT's links.
- A large part of the RCX's program memory was read-only; once programmed in the factory during manufacture, it could not be upgraded. All of the NXT software is stored in flash memory that can be upgraded.
- On the RCX, user programs and a significant chunk of firmware were stored in volatile RAM memory whose contents were lost when batteries were exhausted or removed. On the NXT, nonvolatile flash memory is used.

Introduction to Programming the NXT Brick

The NXT is not useful without a user-written "program" that describes its behavior on how the outputs—in other words, the motors—should react to changes in the inputs (i.e., the sensor values). Without a program, the NXT is an expensive paperweight! Fortunately, you don't have to have a college degree in computer science to program the NXT; LEGO has developed three different environments for programming the NXT and they all make it quite easy to write a program for the NXT.

You may already be a computer "programmer" and don't realize it. For example, when you enter numbers into your cell phone directory you are "programming" your cell phone. When you enter formulas into a spreadsheet, such as "SUM" the values in a column of your spreadsheet, you use a form of computer programming. When you define formatting styles in a word processor—for example, indent paragraphs 1 inch with a line of space before and after the paragraph—you are creating a "program" for your word processor that describes the "behavior" for text entered into a file.

A common theme in these examples is that they are applications written by computer experts to make it very easy for the end user to customize or program how the application behaves and deals with its data/environment. The end user does not have to be an expert.

The NXT is similar to the aforementioned examples. LEGO partnered with industry experts to develop three different applications that allow you to easily enter data that creates a program for the NXT. The data that you enter is a "program" for the NXT. You program the NXT by running one of these three applications on your PC to create the program; once the program is created, you then use the application to download (or transfer) the program to the NXT, where it can be "run" or "executed." When you run the program on the NXT, the NXT is behaving autonomously—independent of human control—according to the behavior described in your program.

There are three programming environments for the NXT because they are optimized for the needs of different types of users. Table 7.1 summarizes the three programming solutions.

Table 7.1 Programming Environments for the NXT

Programming Environment	Description
NXT-G	Developed by National Instruments for LEGO. Ships with the retail version of the NXT kit. A graphical programming environment. You drag code blocks (represented by icons) that describe different behaviors—turn motor A on at 50 percent of full power—and connect them with lines to describe program behavior.

Continued

Table 7.1 continued Programming Environments for the NXT

Programming Environment	Description
	Works only with the NXT. Best for writing short, simple, uncomplicated, basic programs, and for younger (preteen) owners.
ROBOLAB	Developed by Tufts University for LEGO's Education division. Available from LEGO's Education division. Another graphical programming environment. Not quite so intuitive as the NXT-G programming language, but a bit more feature-rich. Originally developed for the RCX. Supports both the NXT and RCX. Best for users (e.g., schools) that have a mixed environment of NXTs and RCXs and want to use a graphical programming solution. If you already have RCXs and are adding some NXTs, you may want to use ROBOLAB.
RobotC	Developed by the Robotics Academy at Carnegie Mellon University for LEGO's Education division. CD version is available from LEGO's Education division or can be downloaded directly via the Internet. A text-based programming language. Uses the popular industry-standard C programming language. Supports both the NXT and RCX brick. Suitable for novice and experienced programmers.

When you purchase an NXT from a retail supplier, it comes with a CD containing the NXT-G programming environment. An NXT purchased from LEGO's Education division does not come with programming software; you need to purchase one of the three available environments.

How Does a Program Run?

The NXT has its own operating system, just like a PC has a Windows or Linux operating system. The operating system is a powerful "executive" program that manages all of the devices (sensors, motors, buttons, LCD display, timers, memory, etc.) that are part of the NXT.

In an embedded system such as the NXT, the operating system is often called *firmware*—in other words, software for an embedded system stored in nonvolatile flash memory—and it provides control over the hardware of the embedded system.

An operating system also provides other base capabilities that are found on the NXT. On the NXT, it implements a file system that can store up to 64 files. It provides a menu system for managing files—in other words, execute, run, or delete the files—similar to, but naturally not as powerful as, the functionality found in the Windows desktop and Explorer menus!

You write your program for the NXT using one of the programming environments. Then the PC "compiles" the program into a compact format suitable for execution by a computer and transfers it to the NXT. Once it's on the NXT, you use the buttons to navigate to the operating system's "run program" menu item. Pushing the Enter button (the orange button) will "run" the program contained in the selected file.

This compact format of a program is often called *bytecode*. A byte contains a single character of data. A user program is composed of a series of instructions that describe the desired or programmed behavior. It takes several bytes of data to describe one instruction—hence, the term *bytecode*. "Set motor power level" is a typical instruction; the first byte is the opcode specifying the instruction type, the second byte indicates which of the three motors to manipulate, and the third byte is the power level to apply.

Bytecode is not instructions that the NXT's CPU can execute directly. Instead, the operating system contains an "interpreter" application that looks at each bytecode and converts it to the appropriate functionality. Interpretative programming languages are very common—Java is a popular example, where a user's program is converted into Java bytecode which is then interpreted when the program is run.

Using NXT-G

NXT-G (NXT Graphical) is the graphical integrated development environment (IDE) provided with the NXT set. NXT-G is targeted at children and adults with no programming experience, and for this reason it is very easy to use.

You write a program simply by dragging and connecting *code blocks* into a sequence of instructions, more or less like using actual LEGO bricks. Different kinds of code blocks correspond to different functions: You can control motors, watch sensors, introduce delays, play sounds, and direct the flow of your code according to the state of sensors, timers, and so forth. NXT-G also provides a simple way to organize your code into *MyBlocks*, or groups of code blocks that you can call from your main program as though they were a single code block.

When you think your code is ready to be tested, the NXT-G IDE will compile (translate) your code into bytecode and download it to the NXT through the USB or Bluetooth link as a file on the NXT.

The intuitiveness of NXT code makes it the ideal companion for inexperienced users, but as you become more expert, you may notice some drawbacks:

- The NXT-G graphical programming interface is not suitable for large programs. Graphical programming is great for small programs, but it becomes hard to follow when the complete program will not fit on a single screen (say, more than 20 blocks).

- User-defined variables are a useful and common programming concept. NXT-G supports variables but they are awkward to use.
- NXT-G provides access to the vast majority of the NXT's operating system functionality. But some useful features can't be accessed via NXT-G. For example:

  Arrays—a list of related variables—are a very useful programming construct that is not available in standard NXT-G programming. (Some advanced NXT users have enhanced NXT-G with additional blocks that provide primitive array support, but using arrays remains awkward and difficult.)

  You can use NXT-G to display the value of a variable on the NXT LCD as long as it is a whole number. But other IDEs provide stronger formatting capabilities such as "display a number with two decimal points in the fractional part," or "center a text line on a screen" instead of always left-aligned. Or you might want to draw a column of nicely aligned numbers on a screen; other IDEs would allow you to "display a numerical value padding with blanks as required to take precisely five characters."

  NXT-G supports only "integer" or whole number variables. It doesn't provide access to "floating-point" or fractional numbers. There are many times when you'll want to use fractional values; for example, you might want to keep track of the distance your robot has traveled in inches—the calculation is easy; just multiply the wheel diameter (say, 2.67 inches) by the encoder count, and divide by 360 because there are 360 counts per revolution. Or you might want to use the sine function which ranges from 0.0 to 1.0.

- Some other IDEs have much better capabilities for debugging errors in your program. All but the simplest programs are extremely likely to have errors.
- Trigonometric functions (sine, cosine, arctangent) are useful in calculating a robot's position.
- The interface is very intuitive but can become somewhat tedious as you become expert and know exactly what you're doing. It tends to take a lot more keystrokes and mouse clicks to write the same logical code in a graphical environment than it does using a text-based language.

Sooner or later, you may start to want a more powerful language. Fortunately, alternatives such as the RobotC programming environment are still easy to use but provide more power and flexibility.

Using RobotC

The NXT-G IDE was optimized for the target market of young children—say, 8 to 14 years of age. It's a great solution for this market segment. It also is a great fit for any inexperienced user with no previous computer programming knowledge.

But many users will want more. They may want an IDE that uses a standard programming language. Or they may want a solution that offers more power and flexibility. RobotC is a great alternative for this type of user.

- The RobotC IDE uses the industry-standard C programming language. The C language, and its "big brother," C++, is the most popular and most used programming language over the past 20 years. Linux and Windows are both predominantly written in C. When you learn to program in RobotC, you're really learning how to write programs in C; a useful skill!
- The RobotC IDE has been optimized to make it easy to learn. So far, more than 2,000 students have been taught RobotC in the classroom; at the end of the first 90-minute class, they were programming and running their first RobotC programs for the NXT!
- RobotC has a "basic" and an "expert" mode; in the "basic" mode a lot of the advanced functionality is hidden, and it is ideal for use by novices. Many other C development environments are targeted at the experienced professional programmer and the inexperienced user is easily overwhelmed.
- RobotC's basic mode is great for the novice user. The expert mode is suitable for the intermediate and advanced user; this mode currently has the most features and programming flexibility of any programming environment for the NXT.
- RobotC has a number of extensions that have been added to simplify and facilitate control of robotic controllers such as the NXT.

RobotC was developed by the Robotics Academy (RA) at Carnegie Mellon University (CMU). CMU is one of the top three U.S. colleges for robotics research. The RA is part of the National Robotics Engineering Consortium. The RA is a worldwide leader in developing robotics education and training material for the precollege student. A wealth of training material has been developed for RobotC. A lot of this is Web-based and targeted for individual, self-based instruction and learning (see Appendix A).

Programming in RobotC is conceptually the same as programming in NXT-G. You write your program on the PC. When you think it is complete, you used the RobotC IDE to compile the program into bytecode and transfer (download) it to the NXT file system where it can be run or executed. The difference is that NXT-G is a graphical programming language—you drag and connect blocks on a diagram—whereas RobotC is a text-based programming language—you write the program as a text file.

A neat enhancement in RobotC is the ability to automatically trace, on your PC, the execution of your program. All the internal values (sensors, motors, user-defined variables) are automatically polled by the PC for display on its monitor. You can even temporarily suspend your program and walk through the execution of your program logic one line of code at a time.

Usually only the simplest programs work the first time you try to run them. RobotC has a very powerful interactive debugger which is really terrific in terms of helping you find and correct errors in your programming.

Some of the projects discussed in this book actually require that you go beyond the limits imposed by NXT-G code. This is the main reason we chose RobotC to illustrate the few programming examples we've included. RobotC also has the advantage that, being a textual language, it allows for a very compact representation that better suits the format of a book.

Later in this chapter, we'll discuss a few simple programs and how you can program them in both NXT-G and RobotC.

Using Other Programming Languages

LEGO made a bold move in releasing the source code for the NXT firmware and providing a detailed software development kit (SDK) for advanced computer programmers to modify or write their own firmware. The vast majority of NXT users will never need to look at or use the SDK. However, a small handful of users are using this SDK to create their own firmware and IDEs for the NXT. Since the release of the SDK in August 2006, several new programming environments for the NXT have been developed and are in various states of readiness, as outlined in Table 7.2.

Table 7.2 Programming Environments for the NXT

IDE/Programming Language	Description
NBC/NXC/BricxCC	NBC is an assembler for the NXT-G bytecode. NXC is a high-level text-based language for the NXT that generates NXT-G bytecode instructions. BricxCC is an IDE that supports both NBC and NXC. All three of these platforms are relatively mature and stable. John Hansen is the primary developer of the NXT IDE. For many years, he has been the developer of an earlier version for the RCX.

Continued

Table 7.2 continued Programming Environments for the NXT

IDE/Programming Language	Description
pbLUA	pbLUA is an implementation of the LUA language for the NXT. Ralph Hempel is the sole developer of the LUA application. Ralph is well experienced with the MINDSTORMS products, having developed a version of the FORTH language (pbFORTH) for the RCX.
LeJOS NXJ	LeJOS NXJ is a subset of the Java language for the NXT. It is currently in alpha release. A core team of about four developers are working on leJOS for the NXT. Most of the team members were already developers of the leJOS version for the RCX.

All of the programming solutions in Table 7.2 are open source and, unlike the solutions available from LEGO, are free and noncommercial (see Appendix A for references).

All of the alternative and the LEGO-provided solutions are interpretative systems that execute bytecode created from a compiler. NXC and NBC use the NXT-G bytecode. LUA and Java have bytecode that is unique to these languages; LUA and Java are standard programming languages that run on a variety of platforms. NXC is a language unique to the NXT.

The LUA and Java implementations are currently in an early stage of development and are not ready for widespread deployment, although this will change over time.

Using NBC/NXC

The NXT Byte Codes (NBC) and NXC (Not Exactly C (NXC) languages use the same firmware as NXT-G. The advantage of this approach is that you can program in either the NXT-G graphical language or in NBC/NXC without having to reload the NXT firmware. The other solutions utilize different firmware interpreters that do not support NXT-G programs. (A dual-version firmware solution is currently in development for RobotC that also supports NXT-G-based programming; it should be available by the time this book is published.)

NBC was developed first; it is a very low-level language requiring detailed knowledge of the individual NXT-G opcodes. It has subsequently been superseded by the higher-level NXC language which is similar to, but different from, C.

NBC is generally similar to the C programming language but has many differences due to the limits and capabilities found in the NXT-G bytecode. For example, it does not support floating-point variables. In addition, it implements arrays, but in a proprietary fashion that is not common to standard programming languages.

NXC and Bricx Command Center (BricxCC) are the most advanced of the alternative programming environments for the NXT. NXC + BricxCC is quite a powerful combination. The drawbacks are that the programming language is unique to the NXT—it is not C—and it is constrained by the limitations and performance of the NXT-G interpreter. For example, it doesn't support floating-point variables, and array support is different from that found in most other programming languages.

BricxCC is quite a powerful IDE. It was originally developed for the RCX and has been subsequently enhanced to add support for the NXT and NBC/NXC.

Using pbLUA

The pbLUA language (the name stands for *programmable brick LUA*) is the result of Ralph Hempel's experience in designing and programming embedded systems. According to Wikipedia.org, "Lua is a dynamically typed language intended for use as an extension or scripting language, and is compact enough to fit on a variety of host platforms. It supports only a small number of atomic data structures such as boolean values, numbers (double-precision floating point by default), and strings. Typical data structures such as arrays, sets, hash tables, lists, and records can be represented using Lua's single native data structure, the table, which is essentially a heterogeneous map."

Using LeJOS NXJ

LeJOS NXJ (the name stands for Lego Java Operating System NXJ) is a subset of the Java language. It was originally developed for the MINDSTORMS RCX and many of the original developers have been involved in a similar port to the NXT. LeJOS is an open source project and is under continuous development. It became available in early 2007 as a limited-feature alpha release, and it is still incomplete for widespread use.

Significant effort is still required to get the full NXT feature set enabled in leJOS. For example, a menu system is not available, so after you have downloaded your program, you can run it only once. If you want to run it a second time you have to download it again.

Using Other Programming Tools and Environments

We didn't cover all the programming environments for the NXT. There simply isn't room to cover them in depth. There are two other environments that are worth mentioning, though.

ROBOLAB is a graphical IDE that was originally developed for the RCX. It's very popular in the educational market; hundreds of thousands of students use ROBOLAB every year. In fact, it is the language of choice for RCX in the education market. ROBOLAB has been significantly enhanced and extended to support the NXT. It has also been upgraded for the RCX version.

We do not cover ROBOLAB in this chapter because, if you don't already have RCX bricks and ROBOLAB, the NXT-G IDE is the best choice for graphical programming. If you do have RCX bricks, you're already familiar with ROBOLAB and the transition to ROBOLAB on the NXT is straightforward.

Numerous evolving solutions treat the NXT as a dumb device with all the intelligence provided by programs running in the PC. The PC communicates with the NXT over the wireless Bluetooth link. We do not consider these in this book, for two reasons:

- In general, you need to have preexisting skills in writing PC-based programs. This is not the audience for this book.
- More fundamentally, the latency—30 milliseconds for a round-trip message—of the Bluetooth link is too slow for acceptable real-time performance of a robot. It takes 10 messages to read the NXT's motor encoders (three messages) and sensors (four) and update the motors (three). The cumulative time of more than 250 milliseconds is generally not fast enough.

Code Samples

Up to this point, the few programming examples you met were written in a sort of pseudocode that was very close to plain natural language. The use of pseudocode allows the programmer to "play computer" and understand what the program does. But to complete the projects in the book, some of which are a bit complex, you need a real environment with which to run and test the code. We chose to write all the examples using RobotC because it combines power with compactness, it's easy to install and learn, and it's available directly from LEGO.

This chapter concludes with two sample programs. One is for a clock, and the other is for a simple line-following robot. These examples describe some of the most important features of RobotC, but we strongly recommend that you read the available online documentation (see Appendix A). Even if you don't choose RobotC, we're sure you can easily translate our examples into your favorite programming language.

Code Sample: A Simple Clock

The logic for implementing a clock is very simple:

1. Adjust the starting time using the left and right keys to change the hour and minute values.
2. Every one second increment the "seconds" counter.
3. Every 60 seconds reset the "seconds" counter to 0 and increment the "minute" counter.

4. Every 60 minutes reset the "minute" counter to 0 and increment the "hour" counter. If the new value is 12 or greater, reset the "hour" counter to 0.

5. When the time has changed, display it on the LCD.

We could write this as a single long sequence of instructions. But it is better to break it up into several smaller "modular" units. Each unit of code is a subroutine or function. The highest-level "logic" for our program instructions is show here:

```
//
// Declare variables to hold the time and initialize them to zero
//
int second  = 0;
int minute  = 0;
int hour    = 0;

task main()
{
  setTimeBeforeStartingClock();
  runClock();
  return;
}
```

This program code is short and easy to understand. It shows a few elements of the C programming language syntax:

- Comments begin with the characters "//" and continue to the end of a line. Comments contain informative information only and are ignored by the compiler. Liberal use of comments is highly recommended for ease of understanding your program.
- Each RobotC program has at least one *task*. If you're a Windows programmer, a task is the same as a thread. A task represents a block of code that should be run. There may be cases where you want to run three tasks simultaneously and you can do that by defining more than one task in your program. The NXT interpreter will then share the computer processing among the multiple tasks. Think of this as cooking breakfast: You've got the toast in the toaster, you're frying the eggs in one pan, and you're frying the bacon in another pan—three tasks running at once!
- The *int* keyword is used to declare a variable. So *int second = 0;* declares an integer variable named *second* and gives it an initial value of *0*.
- There are two calls to user-defined functions (*setTimeBeforeStartingClock* and *runClock*) in the preceding code. A function is a block of code that can be invoked

by simply calling the function name. You use functions to split your program into smaller chunks of code.

You can also use functions as a convenient way to test your program in partial steps. You might develop the *runClock* function first to test the clock operation and just have an empty "stub" for *setTimeBeforeStartingClock*. This way, you can test the clock operation function and simply let the time always start at 00:00:00.

The *displayTime* function that follows nicely formats the time variables into a text string and then displays it on the NXT's LCD screen. This feature is used in several spots in the program so it's been defined as a function to make efficient and clean use of code. When you want to display the time you make a call to this display function. *displayTime* uses advanced formatting techniques to convert the internal format of variables into text strings—for example, the format code *%02d* will convert a variable into two characters (02) as an integer number (%d) without the fractional part.

```
void displayTime()
{
  string sTime;
  string sSecond;
  //
  // Build time string in format "hh:mm:ss"
  //
  StringFormat(sTime, "%2d:%02d", hour, minute);
  StringFormat(sSecond, ":%02d", second);
  sTime = sTime + sSecond: // concatenate the two strings;
  //
  // Now display it on the LCD
  //
  nxtDisplayBigStringAt(2, 16, sTime);
  return;
}
```

The *runClock* function is small and easy to understand. It simply updates the time every second. The minute and hour variables are updated appropriately as well, as shown here:

```
void runClock()
{
  while (true)
  {
    //
    // Loop forever incrementing second, minute, hour
    //
```

```
        displayTime();
        wait1Msec(1000); // Waits one second
        second = second + 1;
        if (second >= 60)
        {
          second = 0;
          minute = minute + 1;
          if (minute >= 60)
          {
            minute = 0;
            hour = hour + 1;
            if (hour >= 12)
              hour = 0;
          }
        }
      }
    }
```

The *setTimeBeforeStartingClock* function is used to set the start time for the clock. You use the left and right buttons to increase or decrease the *minute* value. Pressing the Enter button advances the program flow to modify the *hour* value.

```
//
// Use right / left / enter keys to adjust minute and hour
//
void setTimeBeforeStartingClock()
{
  nNxtButtonTask = -2;   // Tell NXT OS that your program will use the buttons

  // Display some help info on the LCD display

  nxtDisplayCenteredTextLine(0, "Left/Right to");// Text line 0 on LCD
  nxtDisplayCenteredTextLine(1, "Set Minute");     // Text line 1
  nxtDisplayCenteredTextLine(3, "Enter sets hour");// Text line 2
  displayTime();
  while (true)    // Repeat a block of code many times
  {
    //
    // Loop until the 'enter' key is pushed.
    //
    if (nNxtButtonPressed == kLeftButton)
```

```
    {
      if (minute < 59)
        ++minute; // "++" will increment the minute variable by '1'
      else
        minute = 0;
      displayTime();
    }
    else if (nNxtButtonPressed == kRightButton)
    {
      if (minute > 0)
        --minute; // decrease minute by one
      else
        minute = 59;
      displayTime();

    }
    wait1Msec(250);
    if (nNxtButtonPressed == kEnterButton)
      break; // exit or "break out" of the loop
  }
  //
  // Wait for the 'enter' key to be released.
  //
  while(nNxtButtonPressed == kEnterButton)
  {}

  nxtDisplayCenteredTextLine(0, "Left/Right to");
  nxtDisplayCenteredTextLine(1, "Set Hour");
  nxtDisplayCenteredTextLine(3, "Enter runs clock");
  while (true)
  {
    if (nNxtButtonPressed == kLeftButton)
    {
      if (hour < 11)
        ++hour;
      else
        hour = 0;
      displayTime();

    }
```

```
      else if (nNxtButtonPressed == kRightButton)
      {
        if (hour > 0)
          --hour;
        else
          hour = 11;
        displayTime();
      }
      wait1Msec(250);
      if (nNxtButtonPressed == kEnterButton)
        break;
    }
    while(nNxtButtonPressed == kEnterButton)
    {}

    eraseDisplay(); // Clear the LCD screen
}
```

There's one thing to watch out for if you copy the preceding program. RobotC will generate an error message if you try to call a function before it is defined. So you have to rearrange the order of the source code to ensure that functions are defined before they are called.

The preceding program is included in the sample programs distributed with RobotC. The filename is NXT Simple Clock.c.You can open this file in RobotC with the *Open Sample Program* command.

This relatively simple program is more complex to write in NXT-G. Figure 7.1 shows a screenshot of partial implementation in NXT-G.This program updates the time for minutes and hours only (not for seconds) and fills the complete display.

Figure 7.1 Partial Implementation of Preceding Code, in NXT-G

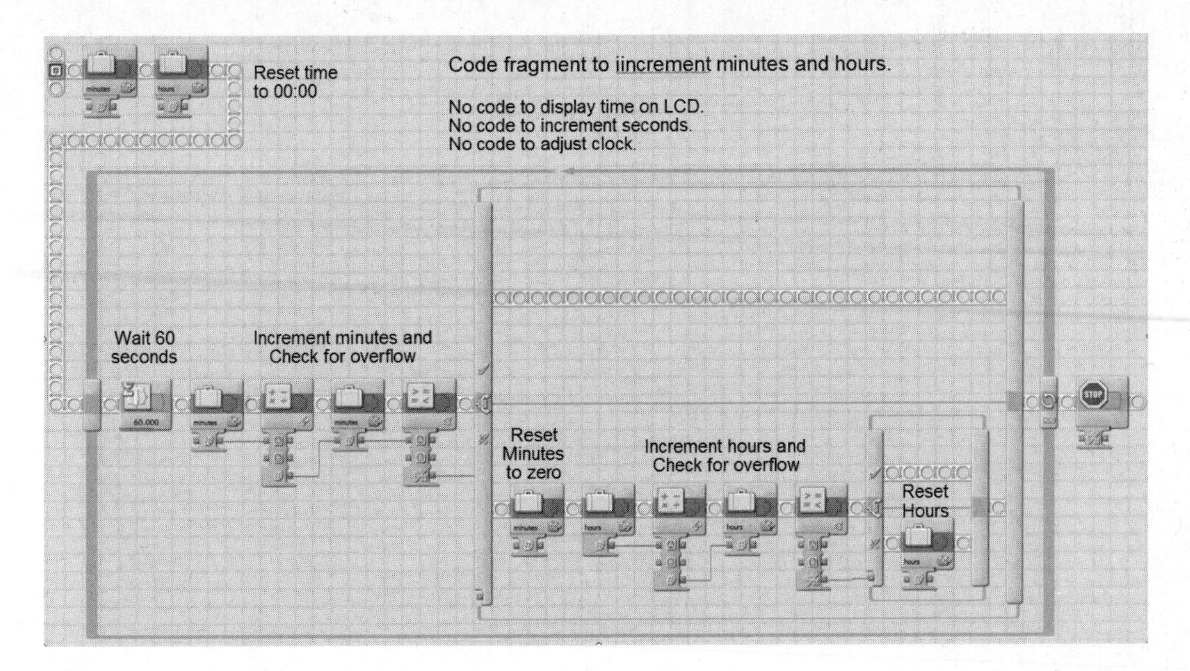

Code Sample: Following a Line

We will use another example to clarify this concept and introduce other tips. Say your robot has been designed to follow a black line, detect small obstacles with a bumper, and remove them from its path by pushing the obstacles away with some kind of arm. As we explained earlier, it's impossible to write a program without having a precise idea of how the robot is designed and what it is expected to do. For the example we are going to illustrate, we made the following assumptions about the robot and the environment:

- The line is darker than the floor.
- The robot will follow the left border of the line (e.g., it turns right to go toward the line, left to go away from line).
- Motor ports A and C control the left and right drive wheels, respectively.
- Motor port B operates the arm.
- Port S1 is attached to a touch sensor connected to the bumper. It closes (goes from 0 to 1) when the bumper is pressed.
- Port S2 is attached to a face-down light sensor that reads the line.

Here is the initial code you should write:

```
int floor;
int line;

task main()
{
  Initialize();
  Calibrate();
  Go_Straight();

  while(true)  // repeat the follow code block forever
  {
    Check_Bumper();
    Follow_Line();
  }
}
```

The main level of your program is quite simple, because at this point you're not concerned with what *Go_Straight* or the other functions mean in terms of actions. You're only concerned with the logic that connects the different situations. You are deciding the rules that affect the general behavior of the robot and you don't want to get into the details of *how* it can actually go straight. This result is achieved by encapsulating the instructions that make your robot go straight into a *function*, a small unit which "knows" what the robot requires in order to go straight. This approach has another important advantage: Your code will be more general because it doesn't depend on the architecture of the robot. For example, for one specific robot, "go straight" will mean switching motors A and C on in the forward direction, whereas for another it might mean switching on motor B in the reverse direction. When you want to adapt the program to a different architecture, you simply change the implementation details contained in the low-level subroutines, without having to intervene on the logic flow.

Let's come back to your main task to examine it in deeper detail. The first instruction is actually placed before the beginning of the task: It declares that you are going to use two variables, named *floor* and *line* and intended to contain integer numbers. A variable is like a box with a name written on it: You can place something inside; say, a specific number—that is, you can *assign* a value to the variable. Or you can watch what's inside the box, *reading* the variable. At this stage, you are neither assigning nor reading the variables; you are simply declaring that you need two of them. In other words, you are asking RobotC to prepare two boxes with the names just mentioned.

When the user starts the program on the NXT, the main task begins. After it has completed initialization and calibration procedures, the program starts the robot in straight motion, and then enters an endless loop where the program continuously manages its two tasks: removing obstacles and following the line. The *while(true)* statement repeats all the instructions delimited by the open and close braces forever. In your case, it will execute the *Check_Bumper* subroutine, then *Follow_line*, then *Check_Bumper* again, in a continuous loop, that only the user can interrupt using the **Exit** button.

Everything is clear and simple, as it should be. Now let's have a look at what happens at a lower level in our subroutines.

Any program will typically include an *initialization* section, where you set the motor power, configure the sensors, reset timers and counters, and initialize variables. This is not required when you use NXT code, because it automatically configures the input ports for you. RobotC, like the other textual environments, requires that you explicitly declare what kind of sensor you connect to each port:

```
void Initialize()
{
  SensorType[S1] = sensorTouch;
  SensorType[S2] = sensorLightActive;
}
```

The word *void* is what tells RobotC that you are describing a subroutine, and it's followed by the name you choose for it. The *SensorType* statements are used to configure input port S1 for a touch sensor and input port S2 for a light sensor.

The *calibration* routine is designed to inform your robot of the actual light readings it should expect on its path. We discussed this topic briefly in Chapter 4, explaining that keeping your program independent from particular cases is a good general programming practice. In this example, it means you should not write the light sensor thresholds into the code, but rather give your robot the possibility to read them from the environment, and this is what you have declared the *floor* and *line* variables for.

```
void WaitBumperPress()
{
  PlaySound(soundBeepBeep);
  while (SensorValue[S1] == 0);  // wait for bumper press
  while (SensorValue[S1] == 1);  // wait for bumper release
}

void Calibrate()
{
  WaitBumperPress();
  floor = SensorValue[S2];
  WaitBumperPress();
  line = SensorValue[S2];
```

```
  WaitBumperPress();
}
```

This code shows that in some situations you can recycle a sensor and use it for more than a single purpose: During the calibration process, the bumper is used as a trigger to tell the robot that it's time to read a value. It also shows that subroutines can be *nested*. In other words, you can make a subroutine call another subroutine. In this particular case, *WaitBumperPress* is a small service subroutine that produces a beep and then waits until the bumper switch gets pressed and released.

When you run the program, the calibration procedure begins and informs you with a beep that it waits for the first reading. You place your robot with the light sensor on the floor, far from the line, and push the bumper. The program reads the light sensor and stores that value as a typical "floor" value in the *floor* variable. Then it beeps again while waiting to read the line. You place the robot with the sensor just over the line and push the bumper again, making it detect the "line" light value and store it in the *line* variable. The robot finally beeps again, meaning the calibration process has finished and that the next push on the bumper will put it in motion.

This sort of prerun phase is quite useful in many other situations, such as when you need to prepare the robot for operations by either reading some environment variable or resetting mechanisms that might have been left in an unknown state by previous executions.

The *Check_Bumper* procedure is in charge of testing whether the robot has hit an obstacle, and if so, how it should react:

```
void Check_Bumper()
{
  if (SensorValue[S1] == 1)
  {
    Stop();
    Remove_Obstacle();
    Go_Straight();
  }
}
```

It checks the bumper, and, if found closed, stops the robot, calls the *Remove_Obstacle* subroutine to clear the path, and then resumes motion. Testing the bumper is as simple as checking whether *S1* has become equal to 1, which means that the touch sensor connected to port 1 has been pressed. You notice that we apply here the same concepts used at the main level: encapsulating details into routines at a lower level.

The *Follow_Line* routine is what keeps your robot close to the line edge—let's say the left edge. If the light sensor reads too much of the "floor" value, it turns right toward the line. If, on the contrary, it reads too much of the "line" value, it turns left, away from the line (see Chapter 4 for a discussion of this method).

```
void Follow_Line()
{
  const int SENSITIVITY = 5;

  if (SensorValue[S2] <= (floor + SENSITIVITY))
    Turn_Right();
  else if (SensorValue[S2] >= (line - SENSITIVITY))
    Turn_Left();
  else
    Go_Straight();
}
```

The method used in this subroutine deserves some explanation. First of all, *const int SENSITIVITY = 5;* tells RobotC that the variable *SENSITIVITY* is a constant (*const*) value that cannot be modified. It is used together with the *floor* and *line* variables to decide what the robot should do. An example with actual numbers can make things clearer. Suppose the *Calibrate* routine placed the value *55* in the *floor* variable and the value *75* in the *line* variable. The program tests whether *S1* is less than or equal to *floor + SENSITIVITY*, which results in 55 + 5 = 60, to decide whether the robot has to turn right toward the line. Similarly, it tests whether sensor *S2* is greater than or equal to *floor – SENSITIVITY*, which corresponds to 75 – 5 = 70, and if this is the case, it makes the robot turn left, away from the line. While the readings remain greater than 60 and lower than 70, the robot goes straight. You can change the value of *SENSITIVITY* to make your robot more or less reactive to readings: An increase will narrow the range of values that allow the robot to go straight; thus, your robot will make more corrections in order to remain close to the edge of the line.

The code you wrote so far is rather general and could work for a broad class of robots. Now the time has come to write the part of the program that depends on the physical architecture of your robot.

The *Go_Straight* routine will be very straightforward in most cases. You know from the initial assumptions that the robot has two side wheels (or tracks) driven by two independent motors. In Chapter 9, we will explore this configuration, called *differential drive*, in greater detail. For the moment, let's stick to the fact that if both motors go forward, the robot goes forward and straight. If one of the motors stops, the robot turns toward the side of the stationary wheel. This knowledge is enough to write the following routines, which control motion:

```
void Go_Straight()
{
  motor[motorA] = 50;    // Motor forward at 50% of full power
  motor[motorC] = 50;    // Motor forward at 50% of full power
```

```
}

void Stop()
{
  motor[motorA] =  0;    // Turn motor off
  motor[motorC] =  0;    // Turn motor off
}

void Turn_Left()
{
  motor[motorA] =  0;    // Turn motor off
  motor[motorC] = 50;    // Motor forward at 50% of full power
}

void Turn_Right()
{
  motor[motorA] = 50;    // Motor forward at 50% of full power
  motor[motorC] =  0;    // Turn motor off
}
```

There's one last routine left: *Remove_Obstacle*. Let's say your robot features a very simple arm that works with a single motor and requires only a timed activation:

```
void Remove_Obstacle()
{
  motor[motorB] =  50;  // Motor forward at 50% of full power
  wait1Msec(2000);      // Wait 2000 milliseconds or 2 seconds.
  motor[motorB] = -50;  // Negative value reverse motor at 50% of full power.
  wait1Msec (2000);
  motor[motorB] =   0;  // Turn motor off
}
```

The statement *wait1Msec(2000)* makes the program wait for 2,000 thousands of a second, or two seconds. This parameter depends on the time your mechanism needs to remove the obstacle, and it is once again related to the physical structure of the robot.

Your program is now finished and ready to be tested. We hope this example made you realize the benefits of modular and well-structured code. This program is included with the RobotC distribution as the sample program NXT FollowAndAvoid.c.

Designing & Planning ...

Benefits of Designing Modular Code

If you follow the principles illustrated in this chapter when writing modular and well-structured code, your program will result in greater readability, reusability, and testability:

Readability The program is organized into small sections that are easy to comprehend with just a quick glance. This means your program will be easier to maintain, and more easily understood by the friends with whom you share it.

Reusability Separating the logic of the program from the instruction related to the physical structure of the robot, you make your code more flexible and reusable for different architectures. The general principle is that the upper levels of the code reflect *what* the robot does, whereas the lower ones reflect *how* the robot does it.

Testability A nice side effect of well-structured code is that it speeds up your testing procedures, segmenting possible problems into small portions of code. Remove (or comment out) the call to *Follow_Line* from inside the repeat block in the main task: Your robot should simply go straight until it hits an obstacle, then activate the arm and remove the obstacle. Conversely, you can remove the call to *Check_Bumper* to turn your robot into a simple line follower!

Running Independent Tasks

All the tools you can choose from to program your NXT support some form of *multitasking*, that is, they support two or more independent tasks that run at the same time. This is not particularly evident when you use NXT code, but it's a well-documented feature in all the alternative environments.

Multitasking can be helpful in many situations and it's often a tempting approach, but you should use it with a lot of care because it will not always make your life easier. Let's go back for a moment to our previous example: Would multitasking have been a good choice? Didn't your robot have two different tasks to manage: line following and obstacle detection? Well, it did, but they were mutually exclusive—after all, your robot was not following the line *while* it removed the obstacle. In cases such as this, and in many others, your robot is asked to perform different activities *one at a time* more often than it is asked to perform different activities *at the same time*. Using multitasking, you would have made your code more complex, because of the additional instructions needed to synchronize the tasks. When the *Remove_Obstacle* task stops the robot, it should communicate the *Follow_Line* task to suspend line following, and communicate again when it can be resumed.

In designing a multitasking application, you are required to move from a sequential, step-by-step flow to an *event-driven* scheme, which usually requires additional work to keep the processes coordinated. Whereas sequential programming is like following a recipe to cook something, you can compare multitasking to preparing two or more recipes at the same time. This is quite a common practice in any kitchen, but requires some experience to manage the allocation of resources (stoves, oven, mixer, blender...), respond to the events (something's ready to be taken out of the oven), and coordinate the operations so that the tasks don't conflict with each other. You have to think in terms of *priorities*: Which dish should you put in the oven first? Programming independent tasks implies the same concerns: You must handle the situations where two tasks want to control the same motor or play two different sounds. The NXT is well equipped to manage resource allocation and to support event-driven programs, and RobotC gives you full access to these features. However, most of the effort is still on your shoulders: No tool makes up for the disadvantages inherent in a bad design.

In our experience with LEGO robotics, there are few actual situations where multitasking is absolutely necessary, or even useful. Our suggestion is that you approach it only when your robot performs some really independent activities, such as playing background music while navigating a room, or responding to messages while looking for a light source.

Summary

In this chapter, you took some first steps on your path to programming LEGO robots. We started describing the NXT, the LEGO programmable unit that's the core of your robots, to unveil some of its secrets. You discovered how its architecture can be easily understood in terms of layers: your program, its translation into bytecode, the interpreter in the firmware, and the processor which executes the operations.

To create your program on a PC, you can choose from many available tools; we briefly described NXT-G, the LEGO graphics programming environment developed by National Instruments for LEGO; and RobotC, a C programming language for the NXT available from LEGO's Education division. We also reviewed a few other environments: ROBOLAB, NBC/NXC, pbLUA, and NXT leJOS.

The second part of the chapter did for programming what the previous chapter did for building: It established some guidelines. Oddly enough, the two arenas share a lot, because layered architecture and modularity principles apply just as much to the body of the robot as they do to its brain, with the notable difference that sometimes you have good reason not to follow those principles in the hardware. In other words, there is no excuse for badly organized software! We used two short but complete programs written in RobotC to put these principles into practice, showing how they can improve the readability, reusability, and testability of your code.

Chapter 8

Playing Sounds and Music

Solutions in this chapter:

- Communicating through Tones
- Playing Music
- Converting Sound and Music Files

Introduction

The NXT features an internal speaker and the hardware necessary to drive it, thus making your robot capable of producing sounds. Those familiar with the sound system on the RCX robot will be pleasantly surprised with the NXT's superior sound system and wider range of capabilities. Perhaps most significantly, it is possible to make your NXT speak, thus providing a very simple and direct way for it to communicate with you. Do not underestimate the sound features the NXT provides. Aside from the convenience of easily getting information from your robot, which will help in testing and debugging your programs, sounds and music are a fun way to give your robots a more defined personality. The NXT also features a sound sensor, allowing a certain amount of two-way aural communication between you and your robot.

The topic of playing sounds and music with the NXT is more closely related to programming than to building techniques. However, when you are dealing with robotics, the two matters are seldom separable. For some of the robots described in the second part of this book, sounds are an important component in their interface with the external world; for others, sounds are an interesting addition that enriches their behavior.

If you are not familiar with musical terminology or audio file formats, you might find topics in this chapter a bit complex. But the prize is worth the effort, because the techniques explained here open exciting new opportunities in your robot world. You will discover how to use simple tones, how to write melodies, how to control your robot with sound, and even how to convert digital audio files into sound effects that you can incorporate into your program!

Communicating through Tones

The NXT features an internal speaker. There is little evidence of it on the outside: The NXT has four very small slits on the sides from which the sound emanates. You can alter the volume of the speaker via the **Settings** interface. The sound system of the NXT is designed to be accessed from your program; you have full control over the frequency (pitch) and the duration of the notes.

The following examples, written in RobotC, include four basic instructions on how to produce sounds, called *ClearSound*, *PlaySound*, *PlaySoundFile*, *PlayImmediateTone*, and *PlayTone*. Using the *PlaySound* statement, the NXT can output one of nine predefined sound patterns, such as a blip, a double beep, or a short sequence of tones:

```
PlaySound(soundBlip);
PlaySound(soundBeepBeep);
PlaySound(soundFastUpwardTones);
PlaySound(soundDownwardTones);
```

The *PlayTone* command plays a single note of a given pitch (in hertz) and duration (in hundredths of a second). The following statement plays a tone of 262 hertz for half a second:

```
PlayTone(262,50);
```

The NXT is capable of reproducing any frequency from 31 hertz to more than 18,000 hertz; however, you will usually limit yourself to the frequencies which correspond to the musical notes (see the table in Appendix C). All the programming languages built over the LEGO firmware offer this same feature, and most of the others include some kind of more or less sophisticated control over sound.

Sounds are the most immediate way your NXT has to inform you about a specific situation. There is, of course, the *display*, but it's not always in sight, especially when your robot is running across the room! There's also the ability to log data to a file on the NXT, but to use the file you would have to access it from your PC. Sounds, on the other hand, can be emitted by the robot without interrupting any other activities, and you can hear them even if the robot is out of sight or far away.

Through simple sound patterns, you can make your robot inform you that an operation has ended, something has gone wrong, its batteries are low, and much more. It can acknowledge the push of a button, or tell you it's waiting for specific input from you. A sound emission can even be used to communicate with nonhuman robot fans, as in the case of Katherine Anderson's SnackBot (see Appendix A), which fills a bowl with dog food before emitting two high-pitched tones to let the family pet know that his dinner is ready.

Playing Music

Sometimes a sound pattern can give your creatures a specific character. Could you imagine a *silent* reproduction of the famous R2-D2 droid from the *Star Wars* saga?

Music can enrich the personality of your robot even more than tone sequences. A wrestling robot probably appears more resolute if, while facing its opponents, it plays Wagner's "Ride of the Valkyries" rather than a Chopin piano sonata or nothing at all. And any sort of dancing robot, without musical accompaniment, becomes just a robot that is swinging its arms and moving around.

One of the easiest ways to play music on your NXT is to use a piano tool, which allows you to simply click on a graphical keyboard to choose the notes you want the NXT to play. To access the piano tool in NXT-G, create a sound block and select **Tone**; the piano tool will appear in the configuration panel. For a more sophisticated tool, including a transposition function, try the piano tool in the Bricx Command Center application created by Mark Overmars and maintained by John Hansen (see Appendix A). This will create an NXT Melody file (with an .rmd extension) which you can use in programs you write for the NXT. Note that you also can use NXT Melody files in NXT-G sound blocks, but you should save them in the appropriate directory with an .rso extension in order to get the filename to appear in the list of possible sounds that the sound block can play.

A piano tool is a good way to create a short melody. However, if you want to code a longer piece, you may find using the RobotC *PlayTone* statement to be more efficient. Every note in the song requires two attributes: *pitch* and *duration*—the first expressed by a frequency and the second by a time. You must introduce delays between the notes to let the CPU wait out the note's duration before playing the next note.

```
PlayTone(440,50);
wait1Msec(50);
PlayTone(220,100);

wait1Msec(100);
```

In this example, the NXT plays an A (440 hertz) for half a second, waits for the note to finish, and then plays another A (220 hertz) one octave below the previous note for one second.

The NXT is limited to playing a single note at a time; thus, we say it's a *monophonic* device. It is not capable of playing chords, which require two or more notes played at the same time, but you can adjust note timing to get various effects. In our previous example, the duration of the first note filled the entire interval before the second note, thus producing a *legato* effect. You can just as easily get a *staccato* effect—shortening the duration of the note inside the interval produced by the *wait1Msec* statement—by introducing a pause with no sound between the two notes:

```
PlayTone(440,10);
wait1Msec(50);
PlayTone(220,100);wait1Msec(100);
```

Coding a melody by hand is a long and tedious task. What happens if when you're finished, you discover that the execution is faster or slower than what you intended? Unfortunately, you'd have to go back and change all the time intervals. A better approach takes advantage of a feature that all textual programming environments offer: the definition of *constants*. Using constants, you can make all the intervals relative to a specific duration that controls the execution speed:

```
#define BEAT 50
PlayTone(440, BEAT);
wait1Msec(BEAT);
PlayTone(220, 2*BEAT);wait1Msec(2*BEAT);
```

The preceding code behaves exactly like our first example, but you'll see that by having defined a constant, the code is clearer and easier to maintain, simply by changing the value of *BEAT* to change the overall speed. We can extend the usage of constants to include note frequencies as well, making our code more readable:

```
#define BEAT 50
#define A3 220
#define A4 440
PlayTone(A3, BEAT);
wait1Msec(BEAT);
PlayTone(A4, 2*BEAT);
wait1Msec(2*BEAT);
```

You can also patiently define a table of constants for all the notes, so you can reuse them in many different programs:

```
#define C1  33
#define Cs1 35
#define D1  37
#define Ds1 39
//...
#define C4  262
#define Cs4 277
//...
#define B8 7902
```

The preceding table represents, for example, the D# note as *Ds* (D sharp) because most languages don't allow the use of special symbols such as # in the names of constants and variables. Don't worry about the length of this table, because constants get resolved by the compiler and don't change the length of your actual code or the space it takes up in memory.

Creating a soundtrack for your robot is a typical example of where multitasking proves to be really helpful. You will typically enclose your song in a separate *task*, starting and stopping it from the main task, as required by the situation.

Converting Sound and Music Files

If the preceding instructions for creating music for your NXT seem too involved, or if you're not familiar enough with music concepts to use them, don't despair: You can still play music on your NXT and even make it speak by using tools that convert different types of sound files into code that your NXT can understand.

MIDI and MIDIBatch

The Musical Instruments Digital Interface (MIDI) is a complex standard that includes communication protocols between instruments and computers, hardware connections, and storage formats. A MIDI file is a song stored in a file according to the format defined by this standard.

MIDI files have achieved incredible success among professionals, amateurs, and instrument manufacturers, and they are by far the most preferred way for musicians to exchange songs. For this reason, you can easily find virtually any song you're looking for already stored in a MIDI file.

But what is a MIDI file? It is simply a sequence of notes to play, their duration, their intensity, and, of course, a code that denotes the instrument to be used. Thus, a MIDI file is not an audio file. It does not contain digital music such as CDs, WAV files, MP3 files, or other common audio formats. Rather, it contains instructions for a player (either a human being or a machine) to reproduce the song, almost a score, to be performed by actual musicians. And as with a real score, the result rests heavily on who actually performs it. For MIDI files, this means that the output depends on the device which renders the music: With a professional MIDI expander, you can get impressive results, whereas execution of the notes by a low-end PC audio card will probably be very poor. What makes MIDI files so interesting to musicians is that they are easy to read and edit (with special programs) in terms of standard musical notation.

So, the key question is whether there's a way you can render MIDI files with the NXT. Though you cannot *import* them directly to the NXT, there's a very nice utility that can convert any MIDI file into the proper code: MIDIBatch, a free conversion utility that is part of the Bricx Command Center project. This utility runs on both Windows and Mac OS X machines and produces NXT Melody files, which have an .rmd extension.

Before we provide details regarding how to use MIDIBatch and what it can do for you, there's another characteristic of MIDI files you must be aware of. The notes inside a MIDI file are grouped into *channels*, and each channel is assigned to the *instrument* meant to reproduce those notes. For example, channel 1 could be assigned to an Acoustic Piano, channel 2 to a Bass Guitar, channel 3 to a Nylon String Guitar, and so on. Channel 10 is always assigned to Drums, and channel 4 is usually, but not always, assigned to the melody line—that is, the notes sung by the vocalist or played by the leading instrument. As explained earlier, the NXT has monophonic sound capabilities and cannot reproduce more than one note at a time, so you have to carefully choose the notes it plays.

Before you start converting a MIDI file into code, we suggest you explore using specific software to see which channel could better render the idea of the song. Many commercial products are capable of manipulating MIDI files in almost every possible way, but you don't actually need all the power and complexity they provide. The Internet is crammed with freeware and shareware programs perfectly suitable for the task of identifying the best single channel to be converted into instructions for the NXT. You open your MIDI file with the editor, mute all the channels except one in turn, and decide which one to use. If you feel at ease with the MIDI editor, you can cut away some notes from the selected channel, because you probably don't need the whole song, only a chunk of it (the part that contains the refrain or main theme). If you do this through editing, you will save the modified MIDI file, of course.

NOTE

You can save a lot of work if you find a MIDI file meant to be used as a ring tone. These typically have sound reproduction limits very similar to those of the NXT.

Now you're ready to use MIDIBatch. MIDIBatch can convert many files at a time into the NXT Melody format. To do this, open a **MIDIBatch window** and type the name of the file folder where the MIDI files are stored in the **Input Directory box**. Choose a directory for the output, whether to convert all channels or just the first one, and click **Convert**. Converting all of the channels is usually not a good idea: The result will be almost unrecognizable. To hear your NXT Melody file before you download it to your NXT, you can use another Bricx Command Center utility called RMDPlayer. If you're satisfied with the result, download the .rmd file to your NXT for use in your programs. Using the RobotC *PlaySoundFile* statement is one way to do this:

```
PlaySoundFile("pachelbel_canon.rmd");
```

WAV2RSO

Another handy utility that is part of the Bricx Command Center is WAV2RSO, which is an application that converts WAV files into the NXT sound file format and vice versa. Unlike MIDI files, WAV files contain digitized audio ready to be executed. If you are familiar with graphics file formats, you can think of MIDI files as *vector* graphics, whereas WAV files resemble *raster* graphics.

There are some limitations to be aware of when converting WAV files into RSO files for use on your NXT. The first is that the resulting RSO file will likely be huge by NXT standards. For example, a WAV file created using text-to-speech software with just the word *Hello*, when converted to an RSO file, may take up 20 KB. Just a couple of seconds of a recorded song could require hundreds of kilobytes of space. Converting WAV files to RSO files does decrease their size by approximately half, but space is still a major consideration. In fact, WAV2RSO does not allow WAV files larger than 64 KB to be converted, likely because the memory capacity of the NXT is only about 130.7 KB in total. You want to save a little room for your program!

Another limitation is the volume at which an RSO file is played. You may find that playing an RSO file while the NXT's motors run is pointless, because the whir of the motors drowns out the sound coming from the small speaker. This is less of a problem with .rmd files, so if you want your robot to move and sing at the same time stick with NXT Melody files rather than converted WAV files.

Despite these limitations, the ability to play sound effects and to hear your robot "talk" makes WAV2RSO a great tool for enhancing your robot's character. Look for WAV files that emulate sci-fi sound effects such as laser guns, jump sparks, and buzzing. Seek out free text-to-speech conversion software on the Web and create WAV files of phrases you want your robot to say. Then, use WAV2RSO (which you can use in a manner very similar to MIDIBatch) to convert these files into RSO files. Upload the RSO files to your NXT, use them in your programs, and watch your robot come alive.

The Sound Sensor

Aside from making sounds, the NXT is also capable of detecting sounds and reacting to them. When the sound sensor is attached, the NXT can "hear" sounds happening around it and determine the decibel level of the sounds. The NXT cannot distinguish different tones, but it can tell the difference between a soft noise and a loud one. So, for example, you can't program it to start when you say "Start" and stop when you say "Stop," but you can program it to adjust its speed (for example) according to the level of sound that it is detecting.

When the sound sensor is attached to the NXT, the robot constantly polls the sensor for data. The sound data can be normalized to a percent value, so the value of the sound sensor can range from 0 to 100. RobotC provides access to this value via the *SensorValues* array. It also provides a configuration wizard for creating the necessary constants to help you access this array. For example, if the sound sensor is plugged into sensor port 2, the wizard will place this code at the beginning of the RobotC program:

```
const tSensors soundSensor = (tSensors) S2;
```

Now the *soundSensor* constant provides an index into the *SensorValue* array, and this value can be assigned to the motors:

```
while (true) {
     motor[motorB] = SensorValue[soundSensor];
     motor[motorC] = SensorValue[soundSensor];
}
```

You can use the sound sensor to make one robot appear to control another. For example, program Robot 1 to play a sound file with a voice saying "Follow me" and have Robot 2 move forward when it detects a certain level of sound. Keep in mind that the noise of Robot 1's motors may also be enough to set Robot 2 in motion, depending on how close together the robots are and what level of sound you've programmed Robot 2 to respond to. In the following example, Robot 2 waits until the sound sensor value reaches 15 percent before following Robot 1:

```
while (true) {
     wait10Msec(100);
     currentValue  = SensorValue[sound];
```

```
if (currValue >=15) {
        motor[motorB] = 100;
      motor[motorC] = 100;
}
```

Through some skilled manipulation of NXT-G, Sivan Toledo has created a clap counter for the NXT. Using sample sound data gathered by the NXT, Toledo was able to determine a typical pattern of values that his NXT would detect when he clapped. Using this, he enabled the robot to count the number of claps it detected and show that number using the NXT graphical display. The NXT-G code for the clap counter is available on Toledo's Web site (see Appendix A).

Summary

The purpose of this short journey into the sound system of the NXT was to show that, despite its limitations, it's still an invaluable resource. It can support you in debugging, return information in the form of sounds of different patterns or frequencies, and complete the personality of your robots.

RobotC offers two commands to control the sound system: *PlaySound* to perform predefined sound patterns, and *PlayTone* to play any note of a desired pitch for the desired duration. Whereas *PlaySound* is suitable for most user interface needs, *PlayTone* offers finer control and lets you create melodies.

Thanks to the work of independent developers, you can convert some of the most common digital audio formats straight into either RSO files or NXT Melody files. Considering the hardware limitations of the NXT, MIDI files translate very well and are the ideal candidates to provide your robots with a musical soundtrack. The conversion of WAV files is somewhat problematic because of the large size of a typical WAV file. Nevertheless, this sound format can equip your robot with amazing sound effects, including speech.

The sound sensor makes the NXT capable of responding to noises, though it can distinguish sounds based only on their decibel levels and not on the nature of the sound. Still, the possibilities for utilizing the sound sensor are broad and the field is young. Sivan Toledo's clap counter hints at future work that could include teaching the NXT to detect certain sound patterns and respond in different ways.

Chapter 9

Becoming Mobile

Solutions in this chapter:

- **Building the Simple Differential Drive**
- **Building a Skid-Steer Drive**
- **Building a Steering Drive**
- **Building a Synchro Drive**
- **Other Configurations**

Introduction

Most robots are designed with some kind of mobility in mind. Motion makes your creatures animated and "alive," and offers a limitless number of interesting, fun, and challenging projects with which to test your creativity and skills. Most mobile robots belong to one of two categories: *wheeled* robots or *legged* robots. Though legs provide an effective way to move on rough terrains, wheels are generally much more efficient on smooth surfaces.

In this chapter, we will survey some common wheeled mobility configurations, discussing some of their pros and cons. Please bear in mind that the chassis shown in the following examples are designed to highlight the details of gearings and connections, and for this reason, some of them need reinforcement to be used in actual robots.

Building the Simple Differential Drive

If you have built some of the robots in the NXT-G Robo Center, or put together the test platform outlined in Chapter 6, you're already familiar with the *differential drive* architecture. It has so many advantages, particularly in its simplicity, that it's by far the most often used configuration for LEGO mobile robots.

A differential drive is made of two parallel drive wheels on either side of the robot, powered separately, with one or more casters (pivoting wheels) which help support the weight but that have no active role (see Figure 9.1). Note that it is called a differential drive because the robot motion vector results from two independent components (it's of no relation to the differential *gear*, which isn't used in this configuration).

Figure 9.1 A Simple Differential Drive

When both of the drive wheels turn in the same direction at the same speed, the robot goes straight. If the wheels rotate at the same speed but in opposite directions, the robot turns in place, pivoting around the midpoint of the line that connects the drive wheels. Table 9.1 shows the behavior of a differential drive robot according to the direction of its wheels (assuming that when it's in motion they run at the same speed).

Table 9.1 Behavior of a Differential Drive Robot According to the Direction of Its Wheels

Left Wheel	Right Wheel	Robot
Stationary	Stationary	Rests stationary
Stationary	Forward	Turns counterclockwise, pivoting around the left wheel
Stationary	Backward	Turns clockwise, pivoting around the left wheel
Forward	Stationary	Turns clockwise, pivoting around the right wheel
Forward	Forward	Goes forward
Forward	Backward	Spins clockwise in place
Backward	Stationary	Turns counterclockwise, pivoting around the right wheel
Backward	Forward	Spins counterclockwise in place
Backward	Backward	Goes backward

At different combinations of speed and direction, the robot makes turns of any possible radius. This maneuverability, the capability to turn in place in particular, makes the differential drive the ideal candidate for a broad class of projects. Add to this the fact that it is very easy to implement, and you can understand why a significant percentage of all mobile LEGO robots belong to this category.

If *tracking* the robot position is one of your goals, again the differential drive is a good candidate, requiring very simple math.

There's only one real drawback to this architecture: It's not easy to get your robot to move in a perfectly straight line. Because no two motors have exactly the same efficiency, you will always have one wheel turning a bit faster than the other, thus making your robot turn slightly left or right. In some projects, this isn't a problem, particularly those programmed for continuous route correction, such as following a line or finding a path through a maze. But when you want your robot to simply go straight in an open space, this problem can be really frustrating.

Keeping a Straight Path

One of the most significant changes moving from the MINDSTORMS Robotics Invention System (RIS) to the NXT is the inclusion of rotation sensors (encoders) in the servo motors (Chapter 4 goes into detail on this). In a nutshell, the encoders allow motor rotations to be monitored to the nearest degree. This addition provides the means for your robot to drive straight. Combined with the NXT-G *Move* block, you can now build robots that go almost perfectly straight and auto-correct themselves along the way.

Although this new feature of the NXT system has provided new ways to keep your robot tracking straight, it is still important to understand the concepts behind differential drive mechanisms. In this chapter, we will discuss alternative approaches including both the sensor-based approach as well as using gears.

Using Servo Motor Encoders to Go Straight

A more sophisticated approach that has several positive side effects requires you to introduce a feedback mechanism into your system, thus controlling each wheel with sensors and adjusting their speed according to the readings. This is what most of the "real-life" differential drives do. As noted previously, the new NXT servo motors have a high-resolution rotation sensor built right in, making this process very easy to do.

In fact, this is so easy to do that you can have a robot up and running in less than 15 minutes. Start by building a simple robot such as the TriBot from the NXT-G Robo Center. The only requirement is that you have two drive motors, each attached to a drive wheel. In the NXT-G, create a simple program using the *Move* block. This block is different from the standard *Motor* block in that it allows for two motors to be paired together so that the software can monitor both of their rotation sensors simultaneously to allow for automatic correction to keep them driving straight. Figure 9.2 shows a simple NXT-G program that will enable your robot to drive straight. The differential process is handled entirely within the *Move* block by "watching" the rotation sensors on each motor. If one of them falls behind, the NXT will increase the power on the lagging motor to allow it to catch up with the other. In addition to this, the software may also reduce the speed of the other motor if the lagging motor does not catch up in time.

Figure 9.2 NXT-G Drive Straight Sample Program

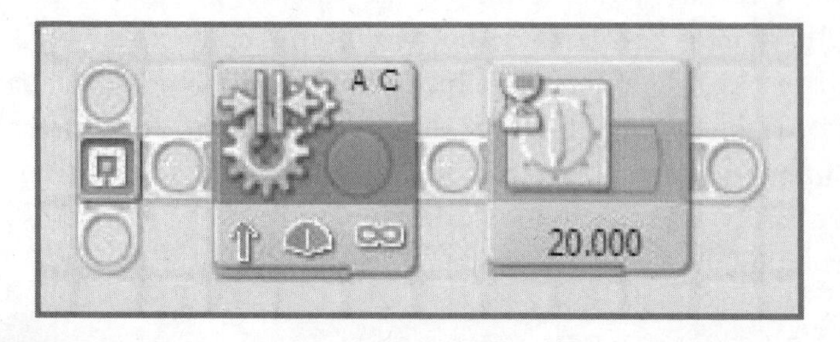

To test this feature, create the robot and set the motor speed such that you can follow along beside the robot. As it is driving forward, use your finger to slow down one of the wheels. You should see it speed up that wheel after you let it go. If you hold the wheel long enough, the NXT should stop the other drive motor.

TIP

Although we don't recommend that you stall a motor for a prolonged period of time, you don't have to worry about doing it for a brief period of time as motors are protected by a thermistor, which will automatically cut power if the motor heats up excessively.

Using Gears to Go Straight

As noted earlier, it is important to understand the concept behind a differential drive mechanism. To do this, we demonstrate using the LEGO differential gear drive mechanism in Figure 9.3.

Figure 9.3 A LEGO Differential Drive Unit

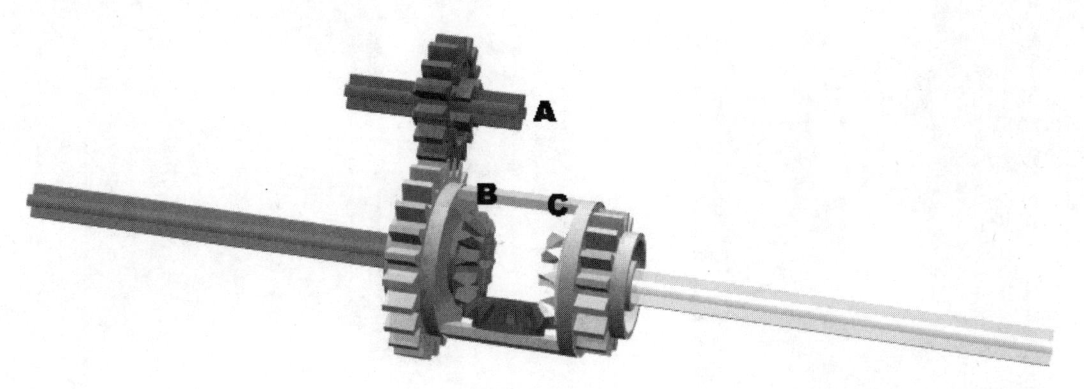

It functions in much the same way as a vehicle differential operates, the basic principle being that the motor drives the differential housing via a main drive gear, "A". The differential is geared such that both wheels turn at the same direction and speed, "B" and "C". This enables smooth transfer of power when a vehicle is turning. On the flip side, if one wheel is on a slick surface (e.g., ice), most of the power will divert to it as there is less friction. You have likely heard the term *limited slip* differential in vehicle discussions—this feature uses mechanical or hydraulic clutches to divert power away from the wheel that is slipping to the one that is not. This way, the differential functionality is still available, but it is monitored so that slip is limited. Most vehicles today use some form of this mechanism. A LEGO-only sample of this is shown later in this chapter, and Chapter 2 also has some additional information on this.

If you connect the drive wheels with a differential so that one wheel enters the differential with a direction that's inverted with respect to the other, the body of the differential itself should stay still when the wheels rotate at the same speed. Figure 9.4 shows a standard differential drive platform. Both NXT motors drive each wheel directly, and are connected to the differential unit via a series of gears such that their directions are inverted. When both motors are driving the same speed, the differential unit will remain stationary. If the motors are not in sync, the differential will turn for the duration that the motors are not in sync.

Why would you do this, you ask? Well, it provides a means to monitor when motors are driving at different speeds. When they are driving in sync, the differential will remain stationary. As their speeds vary, the differential will slowly turn.

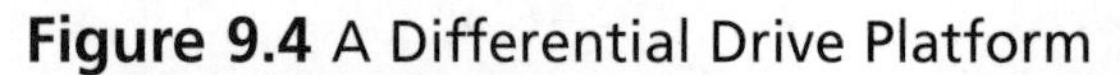

Figure 9.4 A Differential Drive Platform

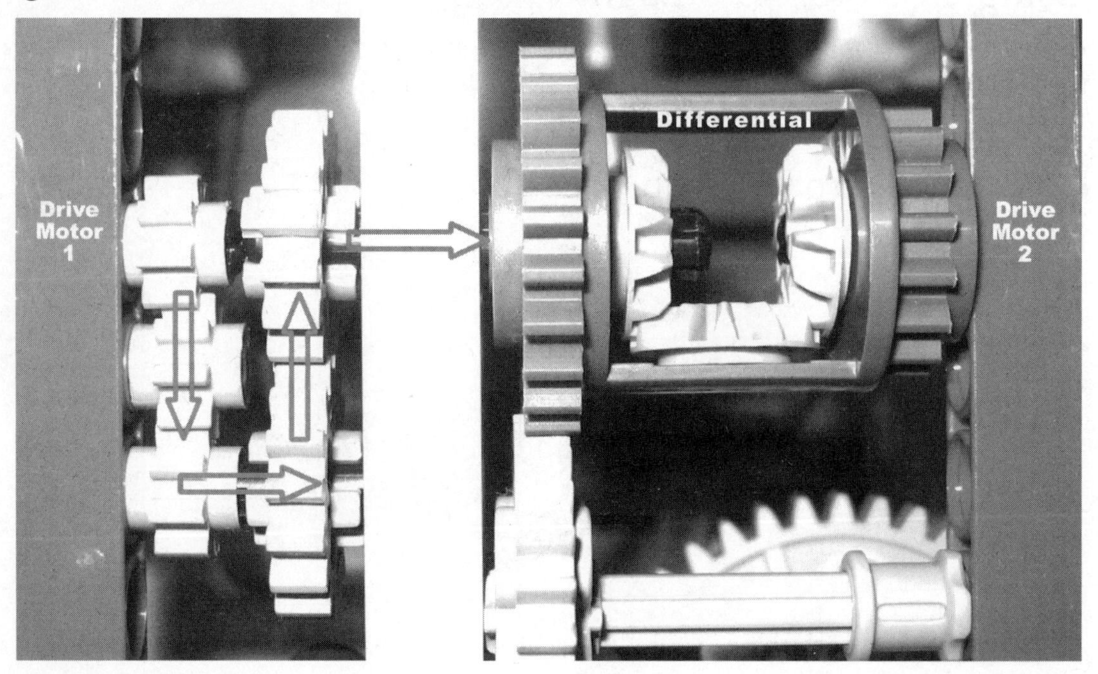

Trying to watch the differential to see any change while your robot is moving is quite difficult. Figure 9.5 shows the top of a demo robot built with the aforementioned differential drive and provides an analog method of measuring differences among motor speeds. The large 40T gear is linked to the differential unit (see Figure 9.4). On top of it is a wedge that is pointed directly at a stationary wedge. This is a crude measurement device and marks the start point for when the robot is stopped. As the robot drives along, the 40T gear will turn if one motor is slowed relative to the other. You can test this by creating the robot, and instead of using a *Move* block in the NXT-G, try using two *Motor* blocks (1 for each motor). Do a trial and force one wheel to slow down. You should notice the wedge move and not recover back to center. After this test, create another program in the NXT-G using the example from Figure 9.2. When trying to slow one motor, you may notice a temporary shift in the wedge, but through the NXT software, the *Move* block will correct this and the wedge should come back to center somewhat.

Figure 9.5 Demo Robot: Differential Drive Platform

Of course, you might ask why you should bother creating a crude analog measurement device such as this when you can use the NXT display to simply show the rotation values for each motor. Good idea! Figure 9.6 is a sample NXT-G program that will drive the aforementioned robot forward for a period of time and display the encoder values for both motors on the screen.

Figure 9.6 NXT-G: Watching the Motor Encoders

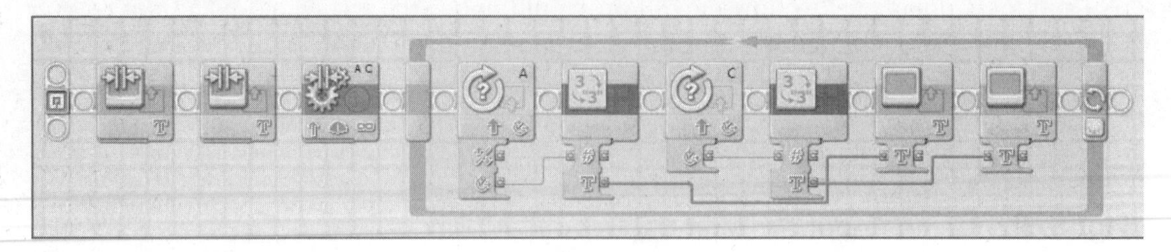

When running the robot, try slowing down one wheel while it drives, and you will see the encoder values change, but be quick, as the *Move* block will correct this. If you stop one wheel completely, you will notice the other one stop shortly afterward.

If you find yourself using some of the older RIS-based motors that don't have built-in rotation sensors, you are still in luck. A more radical solution is to lock the wheels together when you need to go straight. This system is very effective, making your robot go perfectly straight, but it requires a third motor to activate the locking system as well as some additional gearing, which makes the solution less than compact. Figure 9.7 shows an example using legacy MINDSTORMS parts of a locking mechanism that requires special parts: a dark gray *16t gear with clutch*, a *transmission driving ring*, and a *transmission changeover catch*, which combine in a sort of clutch mechanism (see Figure 9.8). That special gear has a circular hole instead of the standard cross-shaped hole; thus, it rotates freely on the axle. The driving ring should then be mounted on an axle joiner. When you push the driving ring into the gear (with the help of the changeover catch), the gear becomes solid with the axle.

Figure 9.7 A Lockable Differential Drive

Figure 9.8 The 16t Gear with Clutch, the Transmission Driving Ring, and the Transmission Changeover Catch

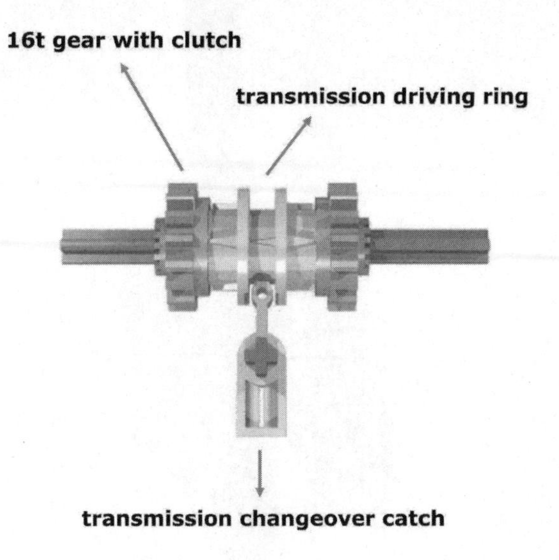

Using Casters to Go Straight

Casters are another key factor in getting your differential drive moving and turning smoothly. Most often, though, they are not given enough consideration. Figure 9.9 shows a typical coupled caster wheel assembly. Unfortunately, its design is limiting and it will skid or jam. It uses two wheels coupled on the same axle and doesn't allow the wheels to turn independently. Keep the assembly gently but firmly pressed on a table, and try to rotate it in a tight turn—it doesn't turn very well, does it? In fact, unless you let one of the wheels skid, it doesn't turn at all.

The casters shown in Figure 9.10 get much better results. The one on the left uses a single wheel, thus avoiding the problem entirely. The one on the right, which is more solid, uses two free wheels that allow the caster to turn in place without friction or slippage problems. The difference is in the wheel hubs. In the assembly on the left, the axle turns with the wheel, whereas the one on the right has the wheels spinning on the axle. Other casters are shown in Figure 9.11.

Figure 9.9 A Coupled Caster

Figure 9.10 Casters Designed to Avoid Skidding

Figure 9.11 Other Casters

The choice of using one or more casters depends on the task for which the robot is designed. A single caster is enough for most applications, but two casters at the front and rear of the robot are a better option when stability is important.

In some cases, as with a simple robot of limited weight that has a smooth surface on which to navigate, you can substitute the caster with *inverted round tiles* or other parts that provide limited friction when coming into contact with the floor (see Figure 9.12).

Figure 9.12 Inverted Round Tiles Can Replace Casters

Another less widely used method for casters is a plastic ball. If your robot is running on a smooth surface, a large ball acts quite well as a caster. Figure 9.13 shows a demonstration of this connected to two NXT motors. The caster cage unit is mounted such that it cups the ball and allows for some freedom of movement for the ball. The entire cup structure is also able to pivot up and down as well so that it handles bumps better. The advantage of this approach is that you have no caster wheels to get caught up, and the ball will turn if it can; otherwise, it will simply drag along the surface. As long as the surface is smooth, it works like a charm. This approach will not work well if you are running your robot across a surface that has a rubbery texture. Although the ball turns freely within the cup structure, it is still subject to friction within it.

Figure 9.13 Ball Caster Base

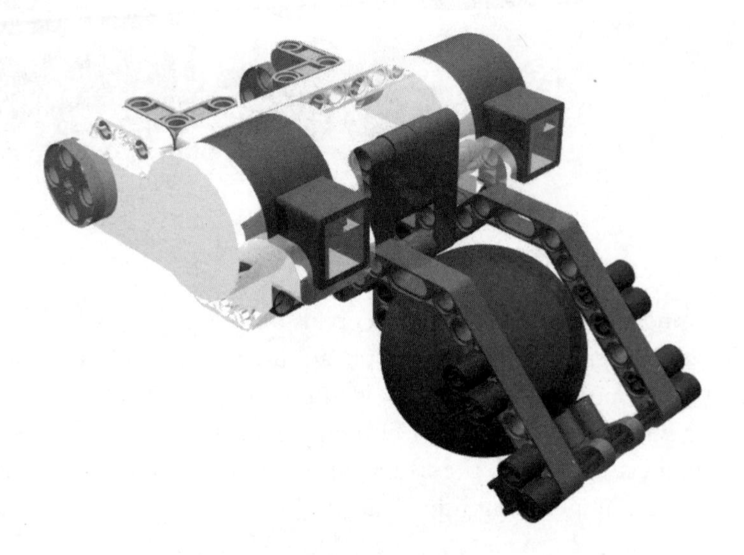

Building a Skid-Steer Drive

A *skid-steer drive* is a variation of the differential drive. It's normally used with tracked vehicles, but sometimes with four- or six-wheel platforms as well. For tracked vehicles, this drive is the only possible driving scheme. Good examples of skid-steer drives in real life are excavators, tanks, and a few high-end lawnmowers.

Figures 9.14 and 9.15 show pictures of Bryan Bonahoom's NX Tracker and Dave Astolfo's UNV tracked robots. Both use a tracked skid-steer drive with UNV employing an additional set of tracks at the front that can pivot to provide additional navigation functionality. Each track is powered by its independent motor.

The advantage of a skid-steer drive is its capability to turn on the spot, which allows vehicles of this type to operate in tight areas. The downside to this is that when turning there is a lot of friction on surfaces such as carpets. If you are running rubber tracks, you also get the added advantage of climbing capability, but if you are running ones such as those shown here, expect some slippage—unless, of course, you find some rubber grommets that just happen to fit the TECHNIC-size pin holes in each link.

Figure 9.14 Bryan Bonahoom's NX Tracker: A Tracked Skid-Steer Drive

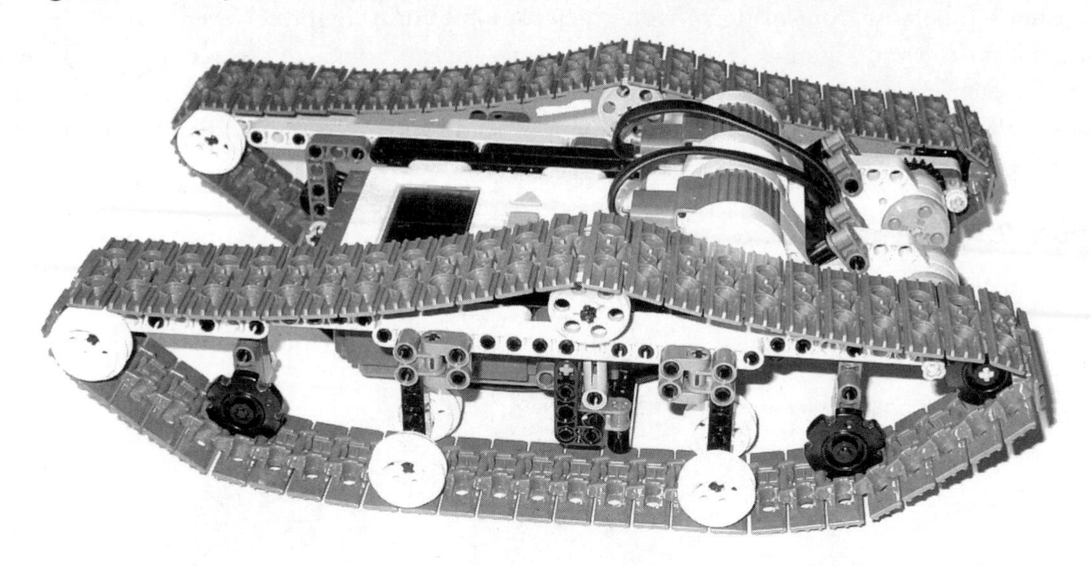

Figure 9.15 Dave Astolfo's UNV: A Variable-Tracked Skid-Steer Drive

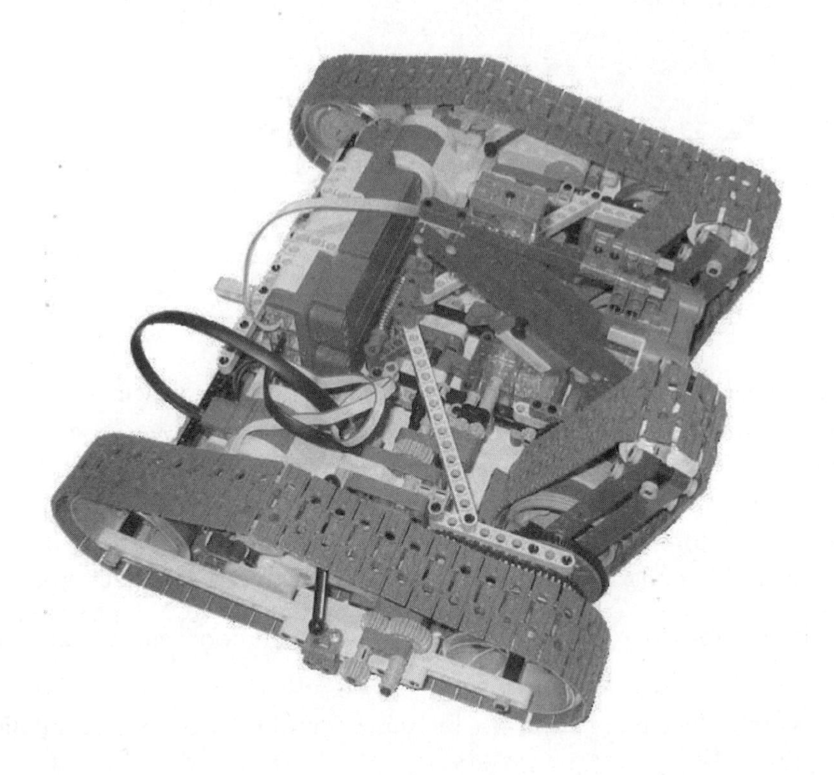

A wheeled skid-steer drive requires a trickier setup. You must transmit the power to all the wheels; otherwise, your platform won't turn smoothly, or it might not even turn at all. The model shown in Figure 9.16 uses a number of 24t gears connected by a common chain drive for each side receiving power from two motors, as in the tracked version. This platform serves well as it is strong, fast, compact, and light. You can add an NXT, and there is plenty of room for sensors, or even a grabber arm.

Figure 9.16 A Wheeled Skid-Steer Drive

Tracked robots are easy to build and fun to see in action, thus placing them among the favorites of many builders. Just as with differential drives, when the tracks go in the same direction, the robot goes forward; differences in their speeds or directions make the robot turn; in-place steering is possible too. Skid-steer drives also share with differential drives the same difficulties in getting them to move in a straight line.

Here is where the similarities end, and some peculiarities of skid-steer emerge:

- Tracks have a better grip than wheels do on rough floors and terrains, but this is not true on smooth surfaces.
- Tracks introduce more friction which uses up some of the power supplied by the motors.

- The unavoidable skidding intrinsic in the nature of these vehicles makes them absolutely unsuitable for applications where you need to determine the position by utilizing the motion of the robot.

Building a Steering Drive

A *steering drive* is the standard configuration used in cars and most other vehicles that features two front steering wheels and two fixed rear wheels. Thankfully, it's suitable for robots too. You can drive either the rear or the front wheels, or all four of them, but the first is by far the easiest solution to implement with LEGO parts, so this is what we'll cover here. Though less versatile than differential drives, and impossible to steer in place or in very tight turns, this configuration has many advantages: It's very easy to drive straight, and it's very stable on rough terrain. Figure 9.17 shows a simple steering platform created by Laurens Valk (Appendix A).

Figure 9.17 Laurens Valk's "NXT-Only" Steering Drive

The mechanics are straightforward. Two motors drive the rear wheels (one for each) and one motor drives the steering mechanism. Unlike a skid-steer drive, this setup provides separate control of the steering and drive motors, allowing for more precision in steering, but lacking the capability to turn in place. The steering unit (see Figure 9.18) is geared to offer fine control over the steering and uses a *rack and pinion* structure where a series of gears combine with a special plate with teeth, a sort of "unrolled gear" (the *rack*) which is connected to the wheels to provide steering. This is also a nice demonstration of studless building techniques.

Figure 9.18 "NXT-Only" Steering Drive: Close-Up

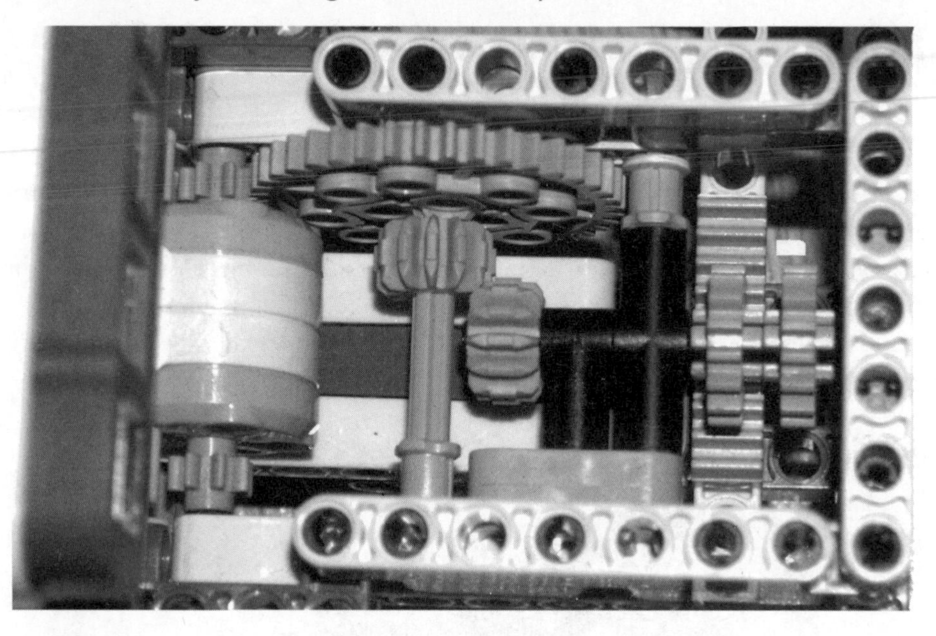

Designing & Planning...

Using Ackerman Steering for Smooth Turns

True-life steering vehicles implement a more sophisticated scheme called *Ackerman steering* (from the name of the person who first studied it). In our simple design, the steering wheels turn at the same angle, but this is not entirely correct—during turns, the inner wheel goes along a tighter bend than the outer one. During large radius turns, the difference is small and its effect negligible. In tight turns, however, the effect becomes quite noticeable, causing one of the steering wheels to skid. Ackerman's steering system is designed to compensate for the different turning angle of the inside wheel, thus eliminating any skidding. The theory says that the vehicle turns smoothly when the "lines" extended from every wheel axle meet and revolve around one common point (see Figure 9.19).

Building an Ackerman scheme with LEGO is definitely possible with a little thought and planning.

Figure 9.19 Ackerman Steering Scheme: The Inner Wheel Turns More than the Outer One

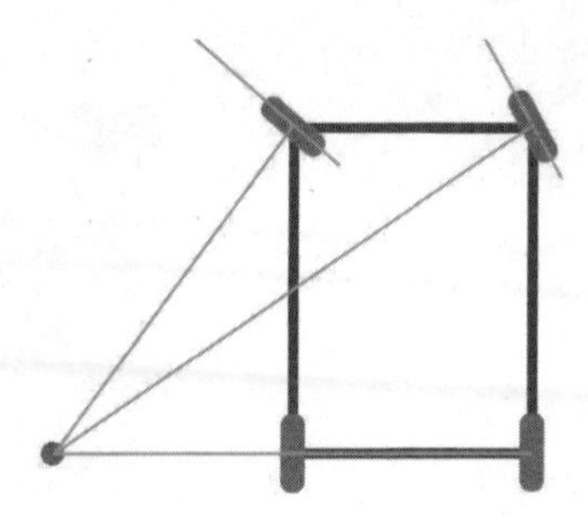

When you build the steering assembly, you can move the wheel behind its pivoting axle for self-centering steering (an advisable property in many situations). In version "a" in Figure 9.20, you see a wheel mounted just below the pivoting axle, which does not affect the steering. If you mount the wheel behind its steering column, friction causes the dynamic forward motion of the car to push the wheels toward the rear, resulting in a self-centering action. Look at the design of a shopping cart, and you will see that the actual wheel contact area is behind the pivoting axis. The more you move the wheel behind the pivoting axis, as in versions "b" and "c," the more self-centering you get. Don't ever mount the wheel in front of the pivoting axle, as in version "d." This will make your steering unstable. In fact, the wheel will tend to go toward the rear, causing your car to turn spontaneously.

Figure 9.20 Moving the Wheel from the Pivoting Axle

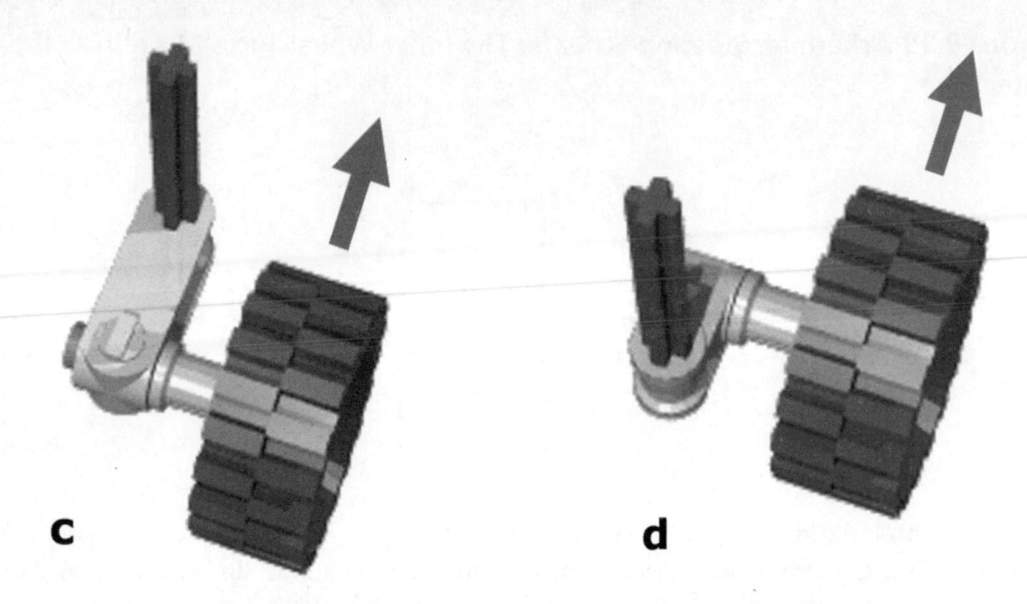

We encourage you to experiment with these concepts, building a simple chassis and exploring the properties of the various assemblies shown in Figure 9.20.

The steering drive is a suitable configuration for rough terrains, because it's very stable on its four wheels. You can improve the grip of the wheels on the ground by using some kind of suspension. It's very important that none of the drive wheels permanently loses contact with the ground; otherwise, the differential would find the path of least resistance and transfer all the power to that wheel, resulting in the wheel spinning and your robot becoming immobilized.

A *limited slip differential* can help reduce this problem (see Figure 9.21) by connecting the wheel axles to a common supplementary axle through pulleys and belts. The belts tend to keep the driven axles rotating at the same speed, but during turns they slip a bit on their pulleys, allowing the wheel to adjust their speeds. Should a wheel lose contact with the ground, the belts will still be able to transfer a good portion of power to the other wheel. Alternatively, you could try to integrate two TECHNIC clutch gears (see Figure 9.21, top right). Simply replace the two wheel hubs with the clutch gears, move the axle closer, and substitute 2x24T gears in place of the pulleys attached to the differential. The TECHNIC clutch gears have a measured amount of slip built into them which will cause the outer (white) section to turn independently from the inner (dark gray) axle unit when enough force is applied.

Figure 9.21 A Limited Slip Differential

Building a Synchro Drive

A *synchro drive* uses three or more wheels, all of them driven and steering. They all turn together in sync, always remaining parallel; thus, the robot changes its direction of motion without changing its orientation.

Synchro drives are quite challenging to build with LEGO parts. Years ago, these types of robots were few, but, if you navigate the Internet now, you can find many well-designed LEGO synchro drives out there.

To make a full 360-degree synchro drive and avoid any limitations in its turning capability, the key point is to transfer motion along the pivoting axle of each wheel. The simplest approach requires a special part called the *turntable*, a large, round, rotating platform usually employed in LEGO models to support revolving cranes or excavators (see Figure 9.22).

Figure 9.22 The LEGO Turntable

You can attach two pulley wheels and drive it with an axle that passes through the center of the turntable. In Figure 9.23, you can see an example of this technique. Notice the orientation of the turntable—the black side is connected to the wheel assembly. This is necessary because the wheel must be connected to the part of the turntable that gets rotated by the external gear. This way, you can attach the top (gray) side to the stationary portion of the robot with the black section being driven by the common steering motor.

Figure 9.23 A Wheel Assembly for a Synchro Drive

To build a complete synchro drive, you need at least three of these turntables. Then you have to connect them so that one motor can drive all the axles at the same time, while another can turn all the wheels in sync.

In Figure 9.24, you see the bottom view of a four-wheeled synchro drive created by Preston Hervey. These robots tend to have a large number of gears and pieces in small areas, so it is difficult to point out the mechanics here. However, the principle behind driving them is quite simple. One motor is connected to a common geartrain that drives an axle down the center of the turntable to turn the small drive wheels which give it motion, while the other motor is connected to a common geartrain that drives the outer turntables (black sections seen here) to allow it to turn.

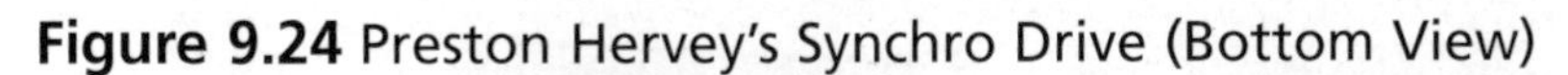

Figure 9.24 Preston Hervey's Synchro Drive (Bottom View)

Figure 9.25 shows a side view of the same platform: You can see here that each turntable top portion (gray) is connected to the stationary part of the robot. This allows the bottom (black) section to be turned. The common drive motor geartrain can be seen with the gears and chain drive (top middle), then each of these transfers into a worm drive, then to a 24T gear, then down an axle into the center of the turntable. This mechanism is what sets it in motion.

Regarding a synchro drive robot, there is no real "front," "back," or "sides." Because it can turn on the spot, the front simply becomes the side that is moving forward. However, for programming, you can arbitrarily pick a side that you consider the "front" to provide something to start with.

Figure 9.25 Preston Hervey's Synchro Drive (Side View)

Synchro drives are quite amazing to see in action, and yours will be no exception. But if you expect it to navigate the room detecting obstacles, your challenge isn't quite over yet: You still have to manage bumpers. Because there is no concept of "front" and "rear," you have to place bumpers all around it. Or you could simply use a single omnidirectional sensor such as the one shown in Figure 9.26; the touch sensor is normally closed, but it opens whenever the upper axle departs from its default position (kept by the rubber bands). Note that the figure shown here was built as a sample for demonstration purposes. It is not complete and would require additional modifications to integrate fully into your robot. Surround your robot with a ring of tubes or axles, connect the ring to the omnidirectional sensor, and that's it!

Figure 9.26 An Omnidirectional Touch Sensor

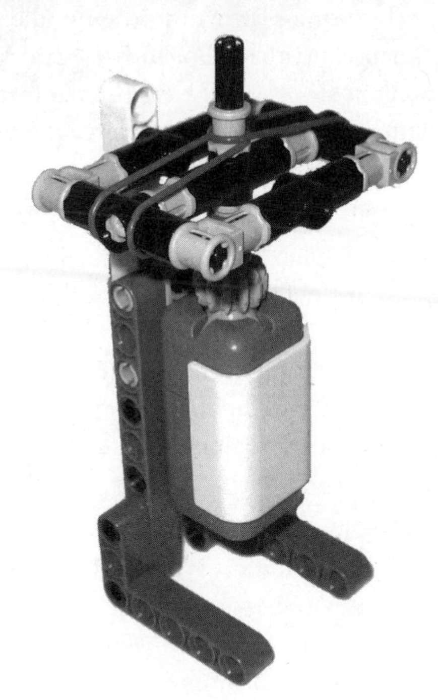

Other Configurations

Our roundup doesn't cover all the possible mobile configurations. There are other, more sophisticated or specialized types:

- **Multi-Degree-of-Freedom (MDOF) vehicles** MDOF vehicles have three or more wheels, or groups of wheels, both independently turned and driven. Imagine a synchro drive where each wheel can change its speed and direction with no connection to the others: Such a robot would be able to behave like a differential drive, a steering drive, or a synchro drive just by controlling its configuration from the software. Though interesting to study and very versatile in their use, they are also extremely difficult to build and control. In fact, not all of their possible configurations result in a coordinated motion!
- **Articulated drive** This is very similar to the steering drive, but instead of steering the wheels, it steers a whole section of the vehicle. The front wheels always remain parallel to the front part of the chassis, and the same applies to the rear wheels in regard to the rear portion of the chassis. Nevertheless, the two sections connect through an articulation point that lets them pivot in the middle. This configuration is common in wheeled excavators and other construction equipment.

- **Pivot drive** Keith Kotay defines a pivot drive as a configuration made of a chassis with nonpivoting wheels with a platform in the middle that can be lowered or raised. When the platform is up, the robot moves perfectly straight on its wheels. When it requires turning, it stops and lowers the platform until the wheels don't touch the ground anymore. At this point, it rotates the platform to change its heading, and then raises the platform again and resumes a straight motion.
- **Tri-Star wheel drive** The Tri-Star configuration has been designed for high-mobility, all-terrain vehicles. Each "wheel" is actually an equilateral triangle with wheels in each vertex; the vehicle features three of them for a total of 12 wheels. The wheels turn, and the triangles can also turn like larger wheels. During normal motion, two wheels of each triangle touch the ground, but when a wheel sticks against an obstacle, a complex gearing system transfers motion to the triangular structure, which turns and places its upper wheel past the obstacle. As complicated to build as it is interesting!
- **Killough platform** Developed by Francois Pin and Stephen Killough, the official name of this mechanical configuration is Omnidirectional Holonomic Platform (OHP). Holonomy is the capability of a system to move toward any given direction while simultaneously rotating. Although conventional wheeled vehicles aren't holonomic at all, this platform allows for unprecedented mobility. Seen from the top, a Killough drive shows three wheels placed at the vertices of an equilateral triangle. Each "wheel" is a sort of sphere made of actual wheels combined together and used in a quite unconventional way: on their side! For a nice demonstration of this, check out Steve Hassenplug's OMNI (Appendix A).

We hope we've made you curious about these configurations, and we invite you to find out more about them using the reference material provided in Appendix A. You can build all of them from LEGO parts, and they'll give you further challenges for when the standard configurations shown in this chapter have become old hat.

Summary

This chapter was quite dense, but we hope we were able to help you to choose a drive configuration. When building a mobile robot, different architectures are relevant to its resulting shape and, most important, to its performance.

The differential drive is simple and versatile, but it can't go straight. The steering drive, meanwhile, goes straight but cannot turn in place. The skid-steer drive can do both, but its drive is not as smooth. Robotics is like cooking: There are many recipes for the same dish, but to be successful you still must know the ingredients well and use them in the right proportions. Of course, don't forget to add the most important ingredient of all: your creativity.

Chapter 10

Getting Pumped: Pneumatics

Solutions in this chapter:

- **Recalling Some Basic Science**
- **Pumps and Cylinders**
- **Controlling the Airflow**
- **Building Air Compressors**
- **Building a Pneumatic Engine**

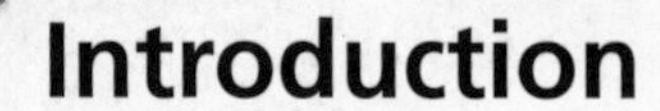

Introduction

Pneumatics is the discipline that describes gas flow and how to use the properties of gas to transmit energy or convert the same into force and motion. Most pneumatic applications use that gaseous mixture most widely available—air—and the LEGO world is no exception.

Pneumatics is a great tool for robotics, and is especially useful when your mechanisms need linear motion or an elastic behavior. It's also a very functional way to store energy for subsequent uses. We will briefly cover the basic concepts of pneumatics, and then put those theories into practice, explaining how LEGO pneumatic components work and what you can do with them, and along the way showing you how to stop and start airflow in order to produce motion in your robot. By the end of the chapter, you should be up to speed on many pneumatic components, including valves, pumps, cylinders, compressors, and pneumatic engines.

Recalling Some Basic Science

To understand pneumatics, you have to recall the properties of gases. The most important property is that gases have neither specific shape nor volume, because they expand and fill all available space within a container. This means the quantity of gas inside a tank does not solely depend on the tank's volume. The greater the quantity of gas in a given volume, the higher its *pressure*.

NOTE

The science that describes the properties of gases is called *thermodynamics*. Its *Ideal Gas Law* relates four quantities: volume, pressure, temperature, and mass (expressed in moles). In our simplified discussion, we will deliberately ignore temperature because, in our situation, it shall essentially remain constant throughout.

We all have the opportunities to experiment with pneumatics using everyday objects. The tires of a bicycle are a good example: Their inner volume is constant, but you can increase their pressure by pumping air in. The more air inside, the greater the pressure, and the more it opposes external forces—in other words, the tires become "harder."

This example leads to a second important property of compressed gases: their pushing outward on the walls of their containers illustrates their *elasticity*. Elasticity is the property of an object that allows it to return to its original shape after deformation. The greater the elasticity, the more precisely it returns to its original configuration. In the example of the bicycle tire, if you push your finger against it, you can temporarily create a dimple in the surface, but as soon as you remove your finger, the tire resumes its shape—the greater the pressure inside, the higher the resistance to deformation.

The fact that gases are so easy to compress is what makes pneumatics different from *hydraulics* (the science of liquid flow). Essentially, liquids are uncompressible.

When you compress a gas into a tank, increasing its pressure, you are essentially storing energy. Pressure can be interpreted also as a *density of energy*, that is, the quantity of energy per unit volume. This leads to a very interesting application of pneumatics: You can use tanks to accumulate energy, which can be released later when needed. You pump gas in to increase the pressure in the tank, storing energy, and you draw gas out to use that energy, converting it into motion.

A flow of air or gas in general is produced by a difference in pressure: The air flows from the container with the higher pressure into the one with the lower pressure, until the two equalize. (In this context, we've giving the term *container* the widest possible meaning. It can be a tank, a pipe, or the inner chambers of a pump or cylinder.)

Pumps and Cylinders

LEGO introduced the first pneumatic devices in the TECHNIC line during the mid-1980s, and then a few years later modified the system to make it more complete and efficient. The TECHNIC line has a long tradition of impressive pneumatic sets, including trucks, cranes, excavators, and bulldozers.

The basic components of the LEGO pneumatic systems are *pumps* and *cylinders* (see Figure 10.1). The function of a pump is to convert mechanical work into air pressure. Pumps come in two kinds: the large variety, designed to be used by hand, and its smaller cousin, suitable for operation with a motor. Cylinders, on the other hand, convert air pressure back into mechanical work, and come in two different sizes as well.

Figure 10.1 Pumps and Cylinders

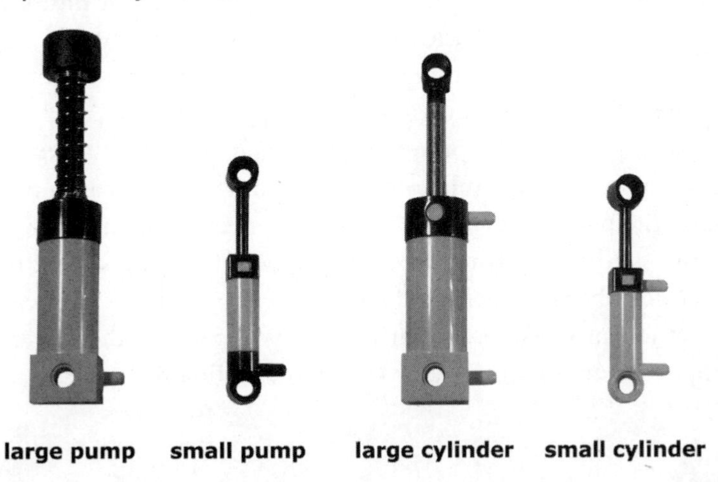

Figure 10.2 shows a cutaway of the large pump in action. When you press its piston down, you reduce the volume of the interior section, thus increasing the pressure and forcing air to exit the output port until the inner pressure equals that outside. When you release the piston, the spring pushes the piston up again; a valve closes the output port so as not to let the compressed air come back inside the pump, while another valve lets new air come in around the piston rod. The small pump follows the same working scheme exactly, with the difference being that it doesn't contain a spring and its piston needs to be pulled after having been pushed. It's designed to be operated through an electric motor.

Figure 10.2 Cutaway of the Large Pump in Action

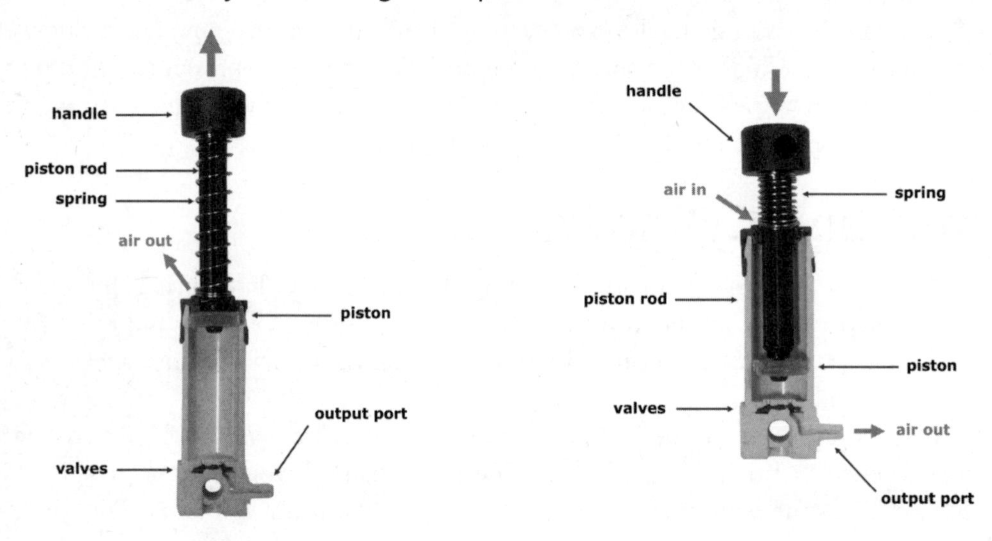

Cylinders are slightly different from pumps. A cylinder's top is airtight and doesn't let air escape from around the piston rod. The piston divides the cylinder into both a lower and an upper chamber, each one provided with a port. The basic property of a pneumatic cylinder is that its piston tends to move according to the difference in pressure between the chambers, expanding the volume of the one with higher pressure and reducing the other until the two pressures equalize, or until the piston comes to the end of its stroke. When you connect the lower port to a pump using a tube, and supply compressed air into the lower chamber, its pressure pushes the piston up. Doing this, the volume of the chamber increases, and this lowers the pressure until it's equal to that of the upper chamber. During the operation, the port of the upper chamber has been left open, so its air can freely escape, reaching equilibrium with the outside air pressure. Similarly, when you connect the upper port to the pump, and supply compressed air, the piston moves down (see Figure 10.3).

Figure 10.3 Cutaway of the Large Cylinder in Action

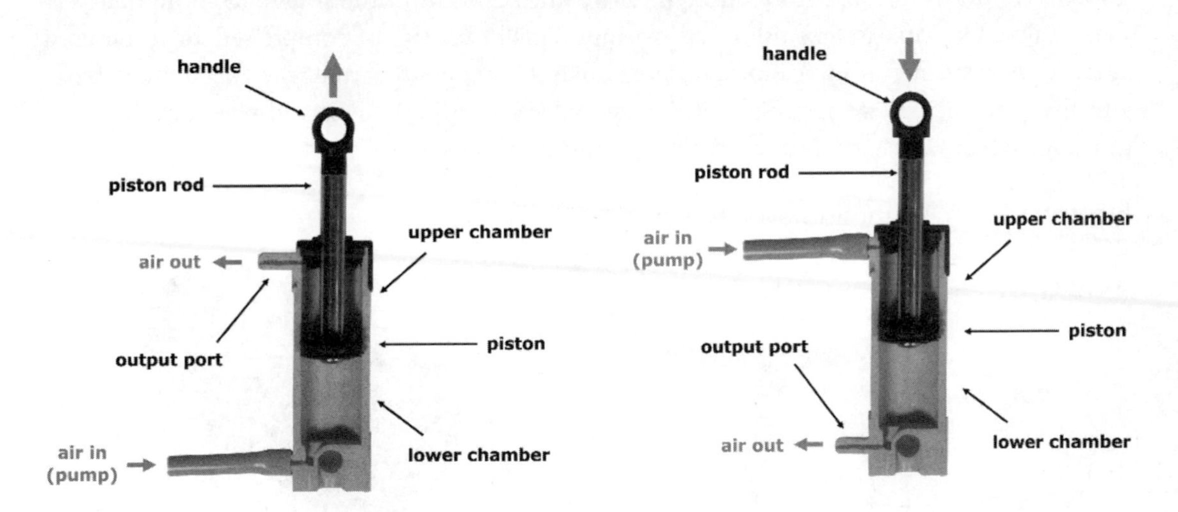

Surely you don't want to move the tube from one port or the other to operate the cylinder: It may work, but it's not very practical. The LEGO *valve* has been designed precisely for this task: It can direct the airflow coming from a pump to either one of the two ports of a cylinder, while at the same time let the pressure from the other chamber of the cylinder discharge into the atmosphere (see Figure 10.4). The valve also has a central (neutral) position, which traps the air in the system so that the cylinder can move neither up nor down.

Figure 10.4 The Basic Pneumatic Connection

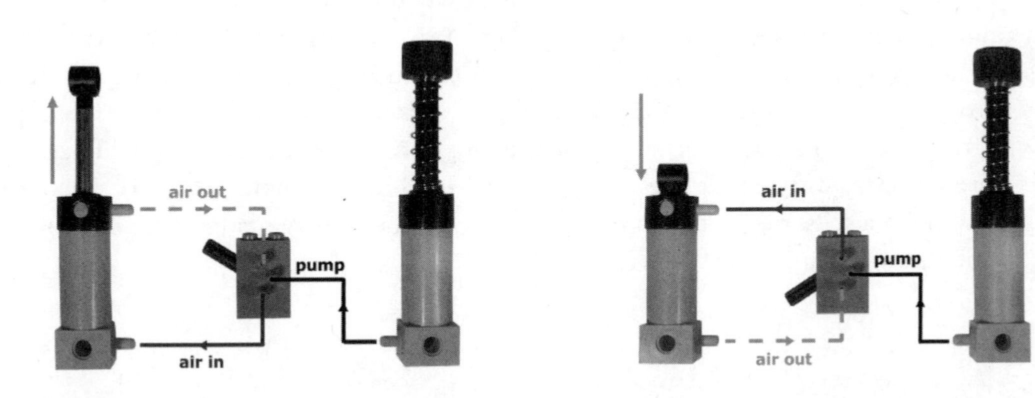

The LEGO tubing system is completed by a *T-junction* and a *tank* (see Figure 10.5). T-junctions allow you to branch tubes, typically to bring air from the source to more than a single valve. The tank is very useful for storing a small quantity of compressed air to be used later. We explained that increasing pressure is like storing energy; thus, the air tank can be effectively considered an accumulator: Charge it with compressed air and release it through the valve when necessary to convert that energy into mechanical work.

Figure 10.5 A T-Junction and a Tank

Pneumatic cylinders provide high-power linear motion, and thus are the ideal choice for a broad range of applications: articulated arms or legs, hands, pliers, cranes, and much more. In describing the basic concepts of pneumatics, we told you that compressed gases tend to make their containers react elastically to external forces. You can test this property with LEGO cylinders, too: Connect a cylinder to a pump and operate the pump until the piston of the cylinder extends in full. Now, press the rod of the cylinder. You can push it down, but as soon as you stop applying force, the rod comes back up again. This property is quite desirable in many situations.

Let's suppose you're going to build a robotic hand. If you try to use an electric motor to open and close the hand, you must somehow know when to stop it. To do this, you can use some kind of sensor as a feedback control system that tells your NXT the object has been grabbed and the motor can be stopped. However, a pneumatic cylinder, in most cases, needs no feedback. The air pressure closes the hand until it encounters enough resistance to stop it. This approach works in a wide variety of objects (if your robot is designed to hold eggs, make sure it exerts a very gentle pressure!). Figure 10.6 shows a simple pneumatic hand. You see that we used a scissorlike setup that gives our hand a rather large range in regard to the size of the things it can handle.

Figure 10.6 A Simple Pneumatic Hand

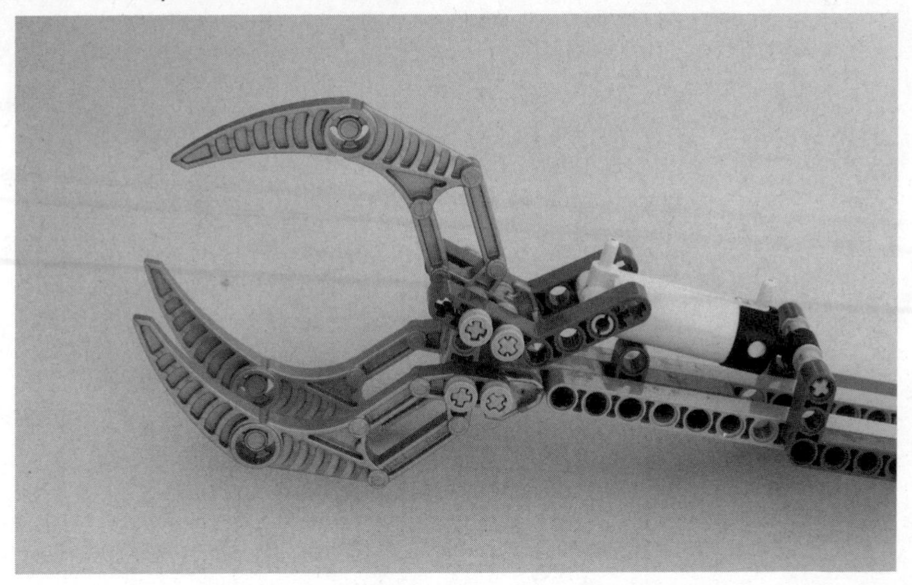

The pneumatic hand shown in Figures 10.7 and 10.8 opens even wider and closes even tighter than the one shown in Figure 10.6. To achieve this wider range of motions we fixed the cylinder so that it remains parallel to the long beams. The parts that enclose the cylinder and keep it parallel to the beams are called *cylinder brackets*. They are designed mainly for attaching two cylinders back-to-back to obtain a longer linear motion than what's possible with a single cylinder, but it also works well as a way to prevent a cylinder from rotating.

Figure 10.7 A Better Pneumatic Hand (Open)

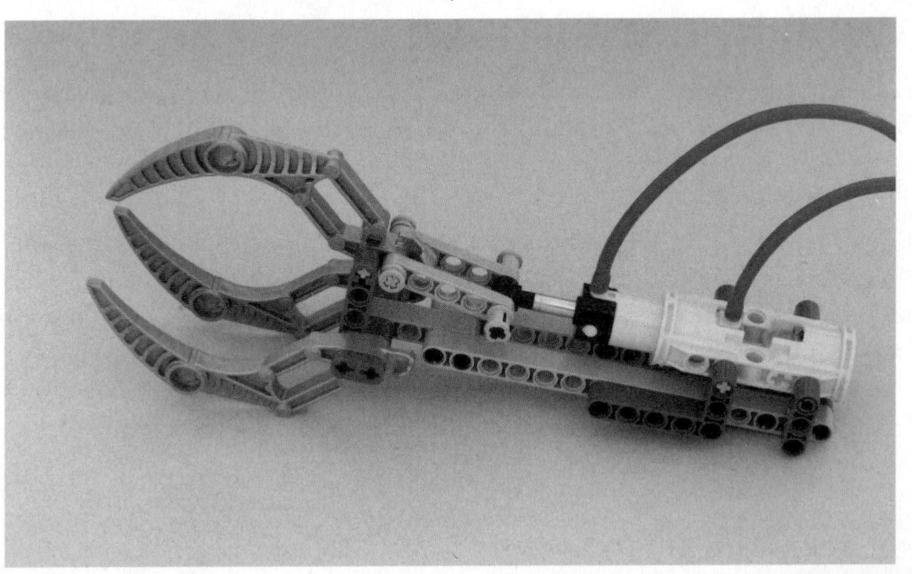

Figure 10.8 A Better Pneumatic Hand (Closed)

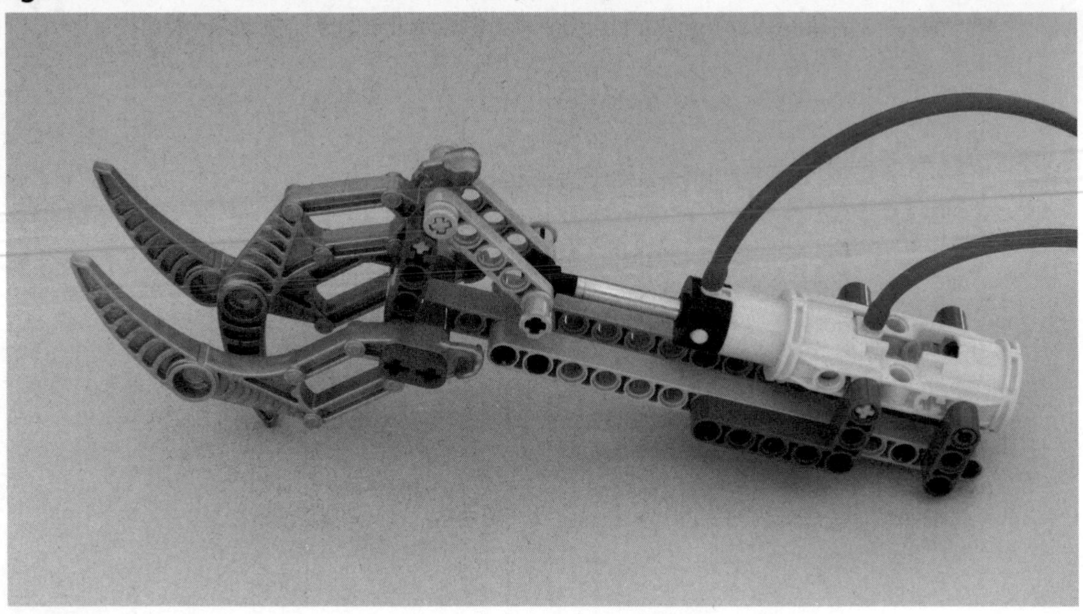

The preceding example gives you an idea of what pneumatics can be used for. Likely, you're already imagining other interesting applications. Unfortunately, the LEGO pneumatic system was not designed to be electrically controlled, so to effectively use it in your robotic projects you need an interface that allows your NXT to open and close valves. And unless you plan to run behind your robot pumping like crazy, you probably want to provide it with an automatic compressor.

Controlling the Airflow

Industrial pneumatic controllers usually operate valves using *solenoids*, which are electromagnets that can pull or push valves. LEGO has not released solenoids, but we can use a motor to actuate a valve.

Figure 10.9 shows one of many possible solutions: The motor turns a 12t gear, which turns a 40t gear. When the 40t gear turns, it moves a beam that pulls or pushes the valve. We exploit here the capability of the NXT motor to turn a specific angle. To move the valve from the center position to either extreme, we turn the motor 180 degrees. To return to the center, we turn the motor 180 degrees in the opposite direction.

Figure 10.9 An Electric Valve

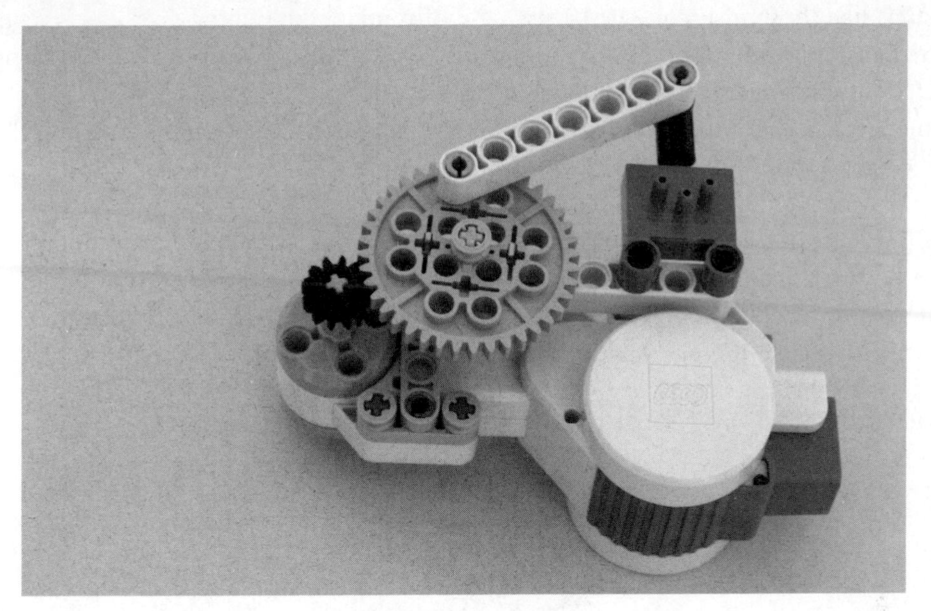

This electric valve is not very compact, but there's not much more you can do considering the size of the motor. Just the same, it works well, and you may feel satisfied with it. But could you make something better? Try applying some of the tricks you learned in previous chapters. For example, you know you can control more than one valve with a single motor. You have seen that, using a differential, it's possible to separate the two turning directions of a motor on two different axles. Now you only need to connect each axle to a valve so that the valve cycles between its positions using only one turning direction. You can do this using a liftarm as a connecting rod; like in old steam locomotives (see Figure 10.10). We used a bent liftarm as a connecting rod because the distance between its extreme holes is about 6.5 LEGO units, a fractional distance that cannot be achieved with a straight beam.

Figure 10.10 A Cycling Valve

Figure 10.11 shows a prototype of a complete double-electric valve, which combines two setups like those of Figure 10.10 with a motor and a differential gear. This mechanism requires a lot of torque to be operated, but the NXT motor has enough torque to operate it without additional gearing. The differential splits the power onto two 36t gears, each one featuring a ratchet beam that lets it rotate only in a specific direction. Thus, when the motor turns clockwise, one valve moves, and if it turns counterclockwise, the other does, each one cycling between all positions.

As in Figure 10.9, the servo motor can move the valves accurately from one position to another by turning a specific number of degrees. The fact that the NXT motors include high-resolution encoders allows the motors to easily and precisely operate mechanisms that require accuracy, like the pneumatic valves.

Figure 10.11 A Single-Motor Dual-Electric Valve

Building Air Compressors

Now that you have discovered a way to operate pneumatic cylinders from your NXT, the next step is to provide them with a good supply of compressed air. Some applications require only a small quantity of air for each motion, in which case you have the option to preload a tank by pumping it manually before you run the robot. A good example is a robot that blows out a candle: All it has to do is find the candle in the room, and then release its air supply to blow it out. You can extend the range using more tanks, but for most practical applications, you will need something more substantial: an unlimited source of compressed air.

You can achieve this goal easily by building an electric compressor, such as the one shown in Figure 10.12. The small pump is connected to an NXT motor and to an L-shaped liftarm using frictionless pins. The liftarm is mounted directly onto the motor. This compressor is probably the smallest you can build with the NXT motor. There are many possible setups, but it's very important that you design yours to take advantage of the entire stroke of the pump, because this will make it more efficient. In fact, if your compressor, for example, uses half of the stroke of the pump, it will release only half the maximum quantity of air it could potentially release. When extended, the small pump is two LEGO units longer than when it is retracted; the distance between opposing holes on the NXT motor's orange shaft is two LEGO units. Thus, the geometry is perfect—if there is a motor position in which the pump is fully retracted, when the motor rotates 180 degrees the pump is extended by exactly two units to its full extension. The only remaining issue, which the L-shaped liftarm takes care of, is to ensure that the pump is fully retracted at some motor position.

Figure 10.12 A Simple Compressor

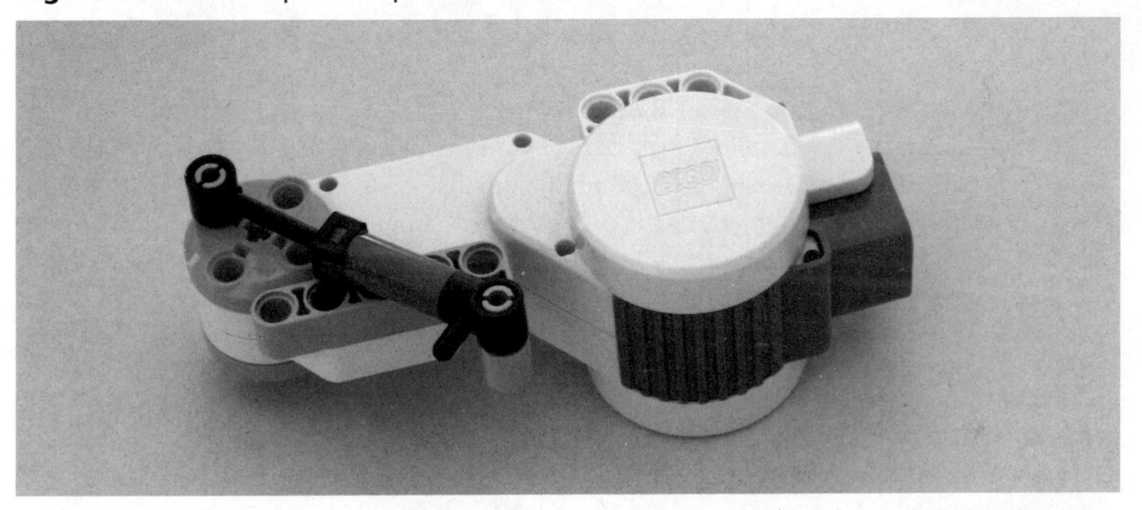

A *double-acting* compressor uses two pumps to provide more airflow. While one of the pumps is pumping, the other takes air, thus providing continuous airflow. Figure 10.13 shows one design, which was inspired by Ralph Hempel's compressor (his used an older LEGO motor). Three 24t gears are mounted on a beam, which is attached to the motor using a 2 x 4 L-shaped liftarm. The motor rotates the center gear, and two pumps are connected to the outer gears. Note that the pumps are mounted such that when one is fully extended, the other is completely retracted. This ensures that one of the pumps is supplying air almost all the time.

Figure 10.13 A Double-Acting Compressor

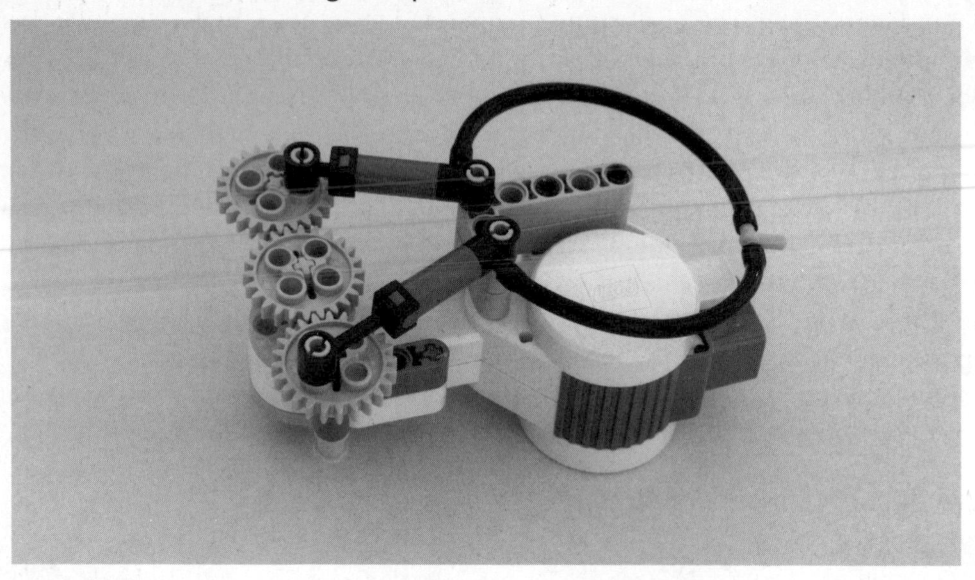

You can improve the compressor shown in Figure 10.13 in two ways. First, the NXT motor does not turn very fast. The small LEGO pumps work well at high speed, so you can increase the air supply by gearing up. Second, the distance between opposing holes on the 24t gears is smaller than two units, so the pumps do not supply all the air they can in each stroke. The variant shown in Figure 10.14 fixes both problems. A 36t gear turns two smaller 12t gears, increasing the speed of the compressor by a factor of three. Also, the two thin 2 x 1 liftarms that push and pull the pumps provide the maximal stroke length (pairs of medium pulleys will achieve exactly the same effect).

Figure 10.14 A More Powerful Double-Acting Compressor

You also can use the large pump to build compressors, but they will be less efficient and bulkier than compressors that use the small pump. The compressor shown in Figure 10.15 is inspired by a design by Christopher R. Smith. The large pump requires much more force to operate, so it must be mounted securely in a stiff structure to avoid twisting. This is usually achieved by using gears on both sides of the pump, as we have done here. If you remove the spring, the compressor will run a bit more smoothly and the stroke can be longer, but the compressor works even if you leave the spring in place.

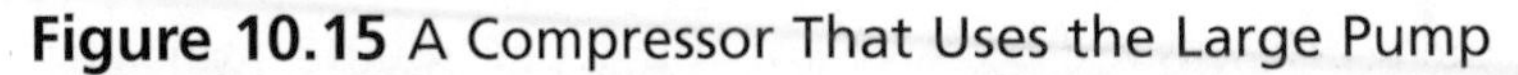

Figure 10.15 A Compressor That Uses the Large Pump

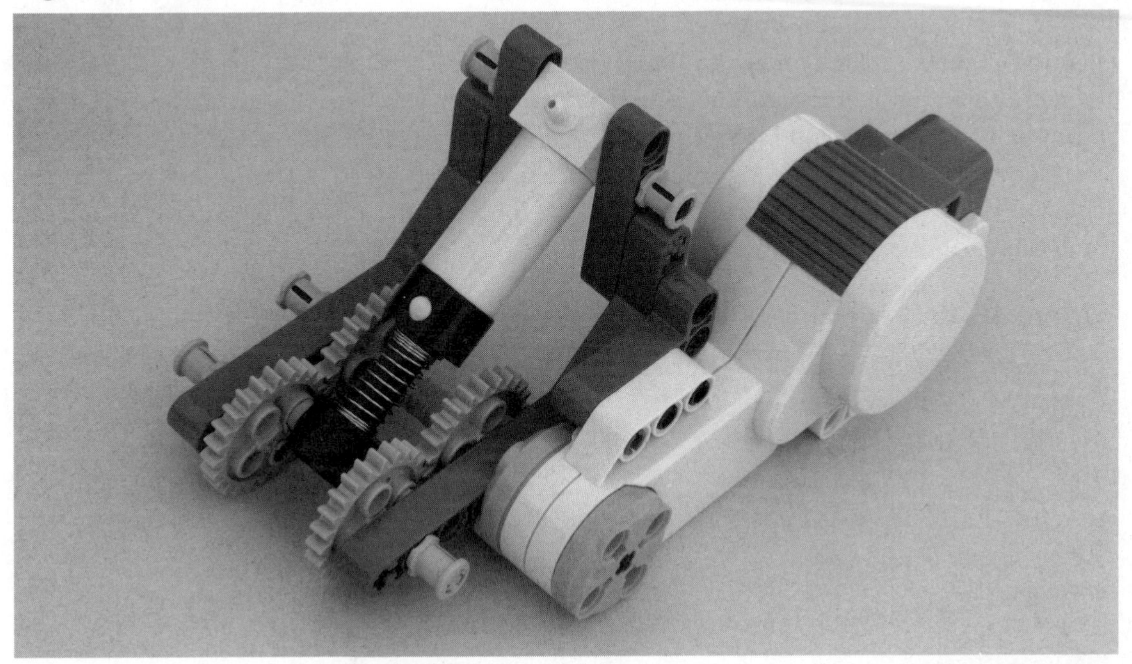

The direction of turning is completely irrelevant to compressors: You can always turn the motor that operates the compressor in the same direction. This opens the possibility of operating a compressor and controlling a valve with a single servo motor. To do so, we would replace one valve in the direction-separation mechanism shown in Figure 10.11 with a compressor, such as the one shown in Figure 10.15. Because of the ratchets, the operation of the compressor is somewhat noisy and inefficient (the motor needs not only compressed air, but also to stretch the rubber bands, which perform no useful work when they shrink back). Therefore, the large pump would work in this setup better than the small pump, because it compresses more air in each stroke.

The nice thing about compressors is that they don't need to be wired to one of the precious output ports of your NXT: A battery box is enough to run them. If you use the NXT motor with a battery box, you will need a converter cable, or you can use a battery box and some other LEGO motor. You can easily adapt the compressors shown in this chapter to use the RC Buggy motor, and there are also good compressor designs for older LEGO motors.

But you might wonder when you should stop your compressor, and how. The simplest option is not to stop it. Instead, you can place a torque-limiting component in the gearing, such as a pulley or a clutch gear, so that when the pressure reaches a given level, the gearing idles. Figure 10.6 shows a much more elegant solution. Rubber bands pull a cylinder in. When the pressure builds up in the bottom inlet of the cylinder, it pushes the cylinder out. Eventually the cylinder presses the touch sensor. When pressure drops, the rubber bands pull the cylinder back in, releasing the touch sensor. The NXT uses the input from the touch sensor to start and stop the compressor. We have essentially constructed a simple pressure sensor. By adjusting the number and strength of the rubber bands, you can set your pressure switch for the desired pressure threshold. The mechanism shown in Figure 10.16 is loosely based in a design by Ralph Hempel.

Another option is to use a genuine pressure sensor that uses a pressure-sensing electronic chip. An NXT pressure sensor is available from Mindsensors, but we have also successfully built homemade ones.

Figure 10.16 A Pressure Switch

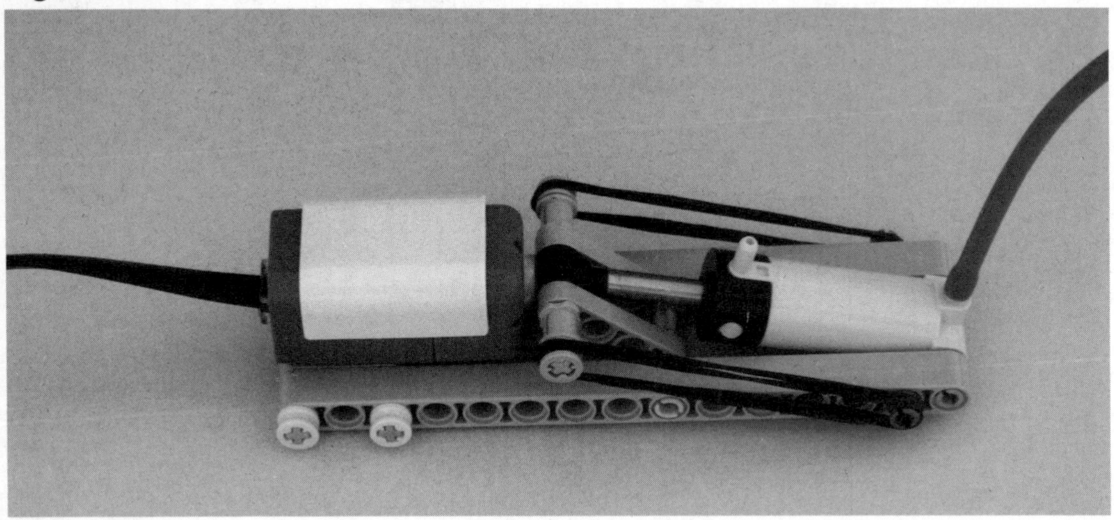

Building a Pneumatic Engine

We mentioned before that you can make cylinders control other cylinders. You accomplish this by making a cylinder operate the valve that controls a second cylinder. This is not useful in itself, but you can make a cylinder do something *and* move a valve. A very interesting case is one in which you connect two cylinders in a loop where each one controls the other, resulting in an unstable system that continuously, and automatically, changes its state (see

Figure 10.17). Provided that you have a supply of compressed air, you can take advantage of this feature to make your robot perform an action.

Figure 10.17 An Unstable Pneumatic System

Figure 10.18 shows a diagram of this pneumatic circuit. Cylinder 1 operates valve 1, which controls cylinder 2, which operates valve 2, which controls cylinder 1!

Figure 10.18 Diagram of the Cyclic Pneumatic System

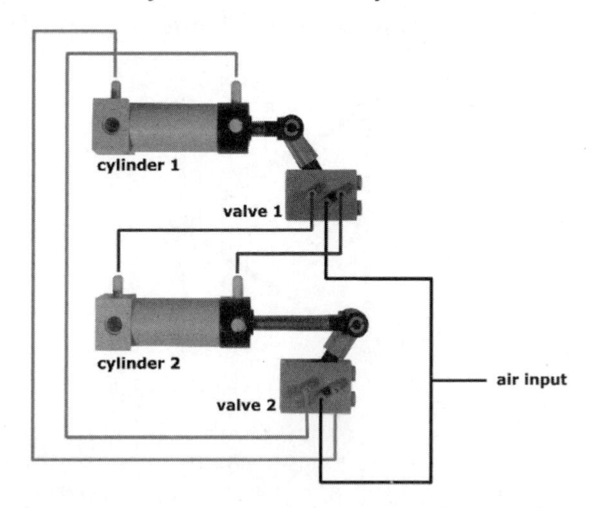

Probably the first robot based on this system to appear publicly on the Internet was Bert van Dam's pneumatic insect. Figure 10.19 shows our slightly modified replica.

Figure 10.19 Bert van Dam's Pneumatic Insect

The complicated tubing hides the same basic circuit shown in Figure 10.16—one of the control cylinders moves the three leg assemblies forward and backward, while the other moves the legs up and down. These are made of six cylinders, split into two groups of three, controlled by the same valve. Each group has a leg in a central position on one side, and one leg front and one leg rear on the other side (see Figure 10.20).

Figure 10.20 Leg Connection Scheme for the Pneumatic Insect

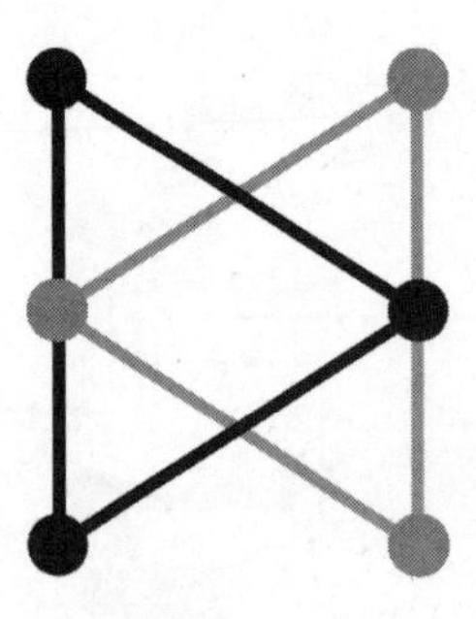

Though rather complicated to build, and more academic an example than practical, van Dam's insect is quite amazing to see in action.

When you use the same principle, it's possible to build a true pneumatic engine, where the push of the cylinders is converted into rotary motion exactly like in steam engines. Figure 10.21 shows our implementation of a LEGO pneumatic engine designed by C. S. Soh (whose Web site contains a wealth of information on LEGO pneumatics, including fascinating performance measurements of various compressors). The key points about this engine are:

- Each cylinder has a dead point in its cycle, when it is either fully extended or retracted. In this position, the cylinder is not able to perform any work, as its push/pull force cannot be converted into rotary motion. This happens because the two connection points of the cylinder (on the chassis and the wheel) and the fulcrum of the wheel align along the same line. For this reason, a pneumatic engine with a single cylinder would not work. The addition of a second cylinder solves the problem: You must mount it with a difference of 90 degrees in its phase against the first one, so when one reaches a dead point, the other is at midstroke.
- The phasing of the valves is very important: You must take care to position them precisely; otherwise, your engine won't work. Mount the 36t gears on the axles in such a way as to align one of their holes with the holes on the cams. Attach the liftarms to that hole with a gray pin. Connect the tubing exactly as shown in Figure 10.21.

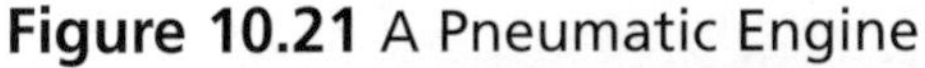

Figure 10.21 A Pneumatic Engine

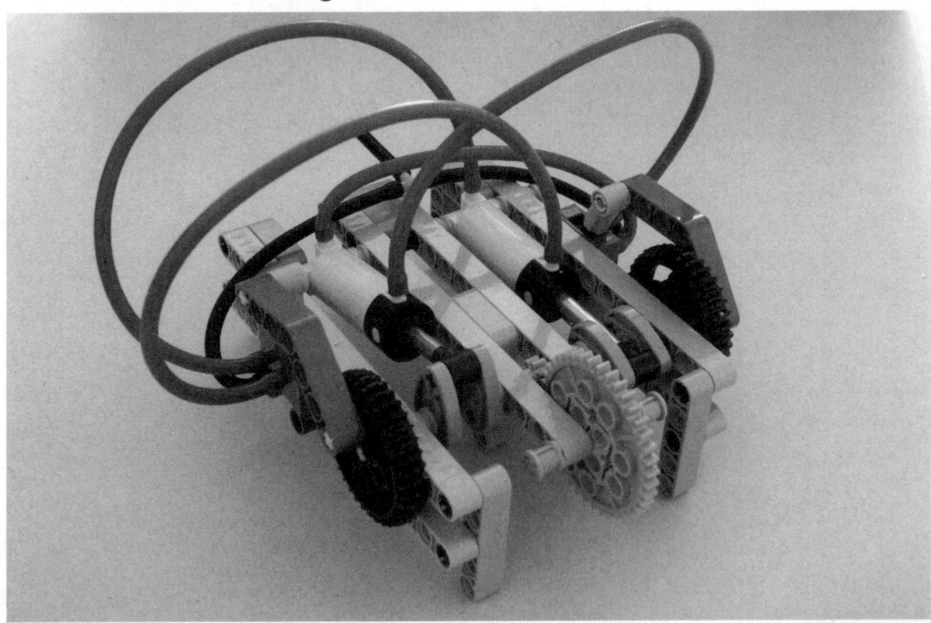

Pneumatic engines are capable of high torque, but due to their intrinsic friction they are not suitable for high-speed applications. Most of the friction comes from the cylinders themselves, which, in order to be airtight, are a bit stiff to move.

Generally speaking, a vehicle moved by this engine, and supplied by an onboard compressor, is not very efficient. But it's indeed fun to see in action and might have its special uses, too.

Summary

Beyond the fascinating sight of all those tubes, and the dramatic hissing of the air coming out of the valves, pneumatics has its practical strong points. In this chapter, you reviewed some basic concepts about the properties of gases, and learned how to exploit these when building your robots. Cylinders are definitely a better choice than electric motors for performing particular tasks and, most significantly, have the capability to grab objects and create linear motion.

Electric compressors can provide a constant airflow to supply your cylinders, and can be used to control this flow from the NXT. Unfortunately, interfacing pneumatics to the NXT is not so simple, and requires a bulky assembly that includes an electric motor and some gearing. Perhaps in the future, the LEGO Company will produce a smart and compact interface capable of controlling many valves from a single output port.

Pneumatics also offers the opportunity to implement simple automation based on cyclical operation, as we showed in the six-legged walker and with the pneumatic engine.

Chapter 11

Finding and Grabbing Objects

Solutions in this chapter:

- **Operating Hands and Grabbers**
- **Finding Objects**
- **Distinguishing Different Objects**

Introduction

It's always great fun and very satisfying to see your robot pick things up from the ground, or take an object when you offer it. In this chapter, we'll illustrate some ways to build arms, hands, clamps, pliers, and other tools to grab and handle objects. One of the basic measurements of movement we'll explore is the *degree of freedom* (*DOF*), or the number of directions in which an object (such as a robotic arm) has a range of motion. In the last part of the chapter, we'll show you methods by which your robot can find the objects, the most challenging part of the job, and distinguish among target objects (those you wish to pick) and walls or other obstacles.

We divide the process of grabbing an object into four steps:

1. Find an object.
2. Distinguish target objects from walls, other obstacles, or "dummy" objects.
3. Position the grabber and/or object in correct orientation.
4. Operate the grabber to catch the object.

The order of these steps may vary, as sometimes your robot must capture the object before it can distinguish between a "target" object and "dummy" objects (for example, by a color sensor mounted on top of the grabber). We'll start from the last step which is more technical, and then discuss the other steps which rely mainly on programming.

Operating Hands and Grabbers

In Chapter 10, we illustrated that pneumatic cylinders are generally the ideal choice for making grabbing devices, or *grabbers.* We will explain the advantage of pneumatic grabbers in more detail in the next section. Unfortunately, pneumatics is not always a possible option. You might not have LEGO pneumatic parts or you don't have room on your robot to fit a pneumatic compressor (a pressure switch and some motor-driven valve switches). We've seen that NXT-to-pneumatics interfaces are rather cumbersome. We will defer the discussion of pneumatic grabbers until after we explain how to use the NXT servo motors to drive your grabber.

The problem with using motors is not in opening or closing the hand; it's in getting the hand to apply continuous pressure on the object to prevent it from falling. This means you cannot just position the fingers around it. You must also exert a force that tightens around the object even though you are not moving the fingers anymore. If you are familiar with (or still use with your NXT) the old electric motors of the RCX, you probably know that stalling them (having them powered but their movements blocked) could cause their internal gears to wear and the motor to overheat. We have explained in Chapters 2 and 3 that the new NXT servo motors are more robust and can withstand short-term stalling. They are, however, extremely power-hungry and would waste a lot of your batteries' power doing so. Long-term stall may still heat up the servo motors, and although they have a protection mechanism which will shut them down before they are damaged (dropping your object...), overheating is still undesirable. When you know you're going to handle a soft object that has some intrinsic elasticity, you can sometimes simply brake the motor, which can, combined with friction among gears, keep the fingers against the object. Try this with TriBot and a sponge ball! If you have heavier objects, however, you will need to find ways to reduce the power forcing the fingers to open. Again, using the TriBot example, try catching the NXT plastic ball. You will not be able to do this without constantly powering the motor, forcing it to close the claws on the ball. A simple way to overcome this problem is to use a worm gear. The simple grabber in Figure 11.1 uses a worm gear that drives the fingers. The worm gear prevents them from releasing the ball when the motor is not powered (either in "brake" or in "coast" mode). Recall from Chapter 2 that the worm gear is a one-way gear: It can turn a meshing gear but cannot be turned by it.

Figure 11.1 A Simple Hand-Operated Grabber with a Worm Gear

Figure 11.2 shows a different design, where the rotary motion from the motor gets converted into linear motion through a worm gear and two translating axles. It's this motion that operates the movable fingers of the grabber. The translation of rotary motion to linear motion is performed by the two half-bushings with mesh the teeth of the worm gear; when

the worm gear rotates, the bushings get pushed or pulled, and the axles where they are mounted move accordingly.

Figure 11.2 This Hand Uses Linear Motion

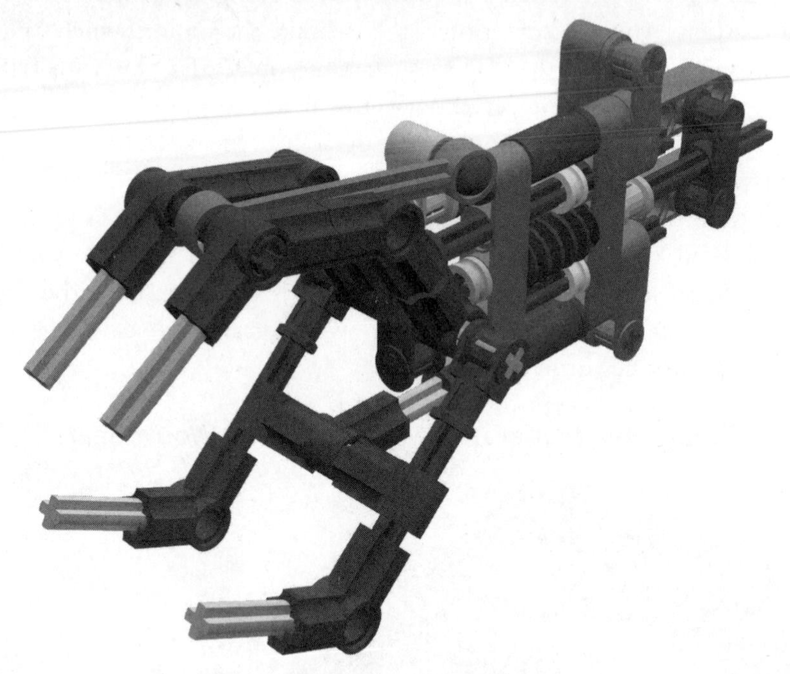

The same principle is used also in the grabber shown in Figure 11.3. Here only the two cone-toed fingers move, with a very compact worm gear pushing them to close on the third finger.

Figure 11.3 A Compact Grabber with a Worm Gear

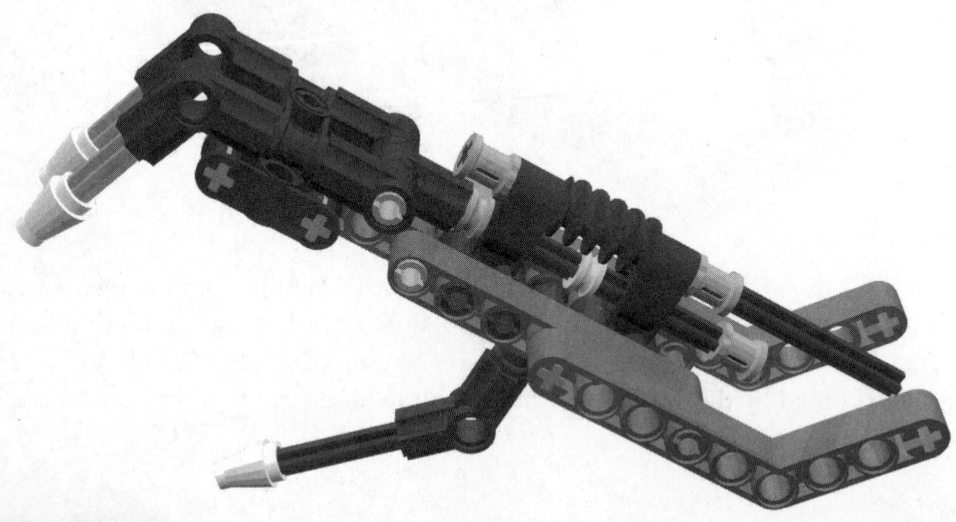

We've repeatedly said that pneumatic cylinders are your best choice in this field, but let's analyze what makes them so good to see whether we can learn something and replicate the same behavior. A pneumatic cylinder can be considered a two-state system: The cylinder is either extended or retracted (we are deliberately ignoring that you can somehow manually stop the cylinder in an intermediate position, centering the switch, and assuming that the switch is in one of its extreme positions). If something prevents the cylinder from actually reaching one of these states, it can, however, continue to push in that direction. Its natural behavior is to move until it finds resistance that balances its inner pressure. This pressure is what keeps the fingers applying a force to the object, thus making your robotic hand hold it firmly.

The point now is to replicate this behavior in a nonpneumatic device. Is it possible? Yes. Figure 11.4 shows an example of a simple *bi-stable* system, so-called because it has two default states, or two possible rest positions to which it tends to go. A rubber band forces the liftarm to stay against one of the two black pegs, either in A or in B. If you move the liftarm slightly from the peg and then release it, it goes back against the peg. If you move it a bit more and pass the midpoint between A and B, it goes to the other peg. You need to provide only enough force to make the system switch from one to the other; the rubber band will do the rest.

Figure 11.4 A Simple Bi-Stable Mechanism

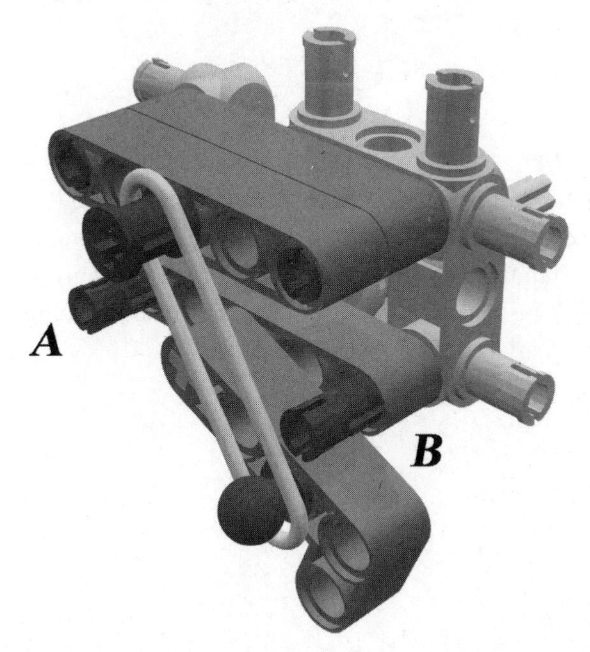

Applying this principle, you can design the pliers shown in Figure 11.5, which are suitable for grabbing very small objects such as a 1 x 2 brick (seen at the bottom of the figure).

To actually use them in a robot you must add a motor that, through brief impulses, rotates the axle connected to the liftarm, shifting the pliers into their open or closed state.

Figure 11.5 Bi-Stable Pliers

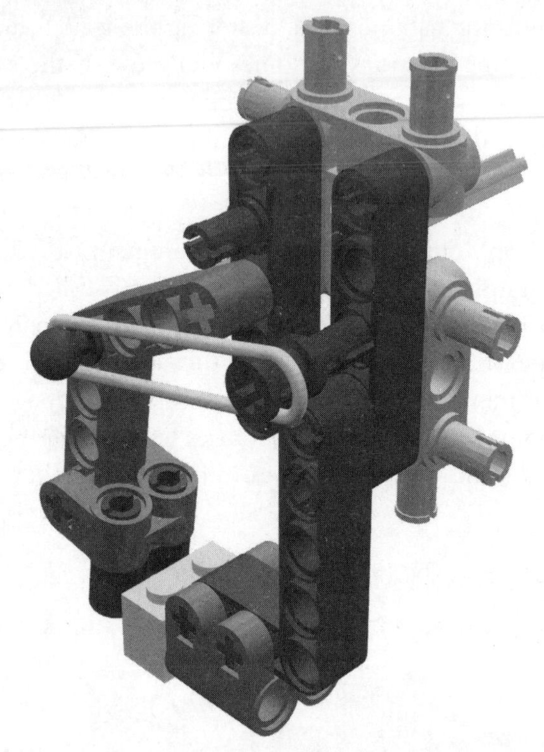

The same approach can be used for larger hands. In Figure 11.6, we demonstrate how one can make a TriBot-like grabber which uses a bi-stable mechanism. Here the rubber band between the two hands keeps them forcing on the object in the closed state, while you can still open the hand with the motor (low power with speed regulation works best for this design), and the rubber band keeps it open afterward.

Figure 11.6 A TriBot-like Bi-Stable Hand

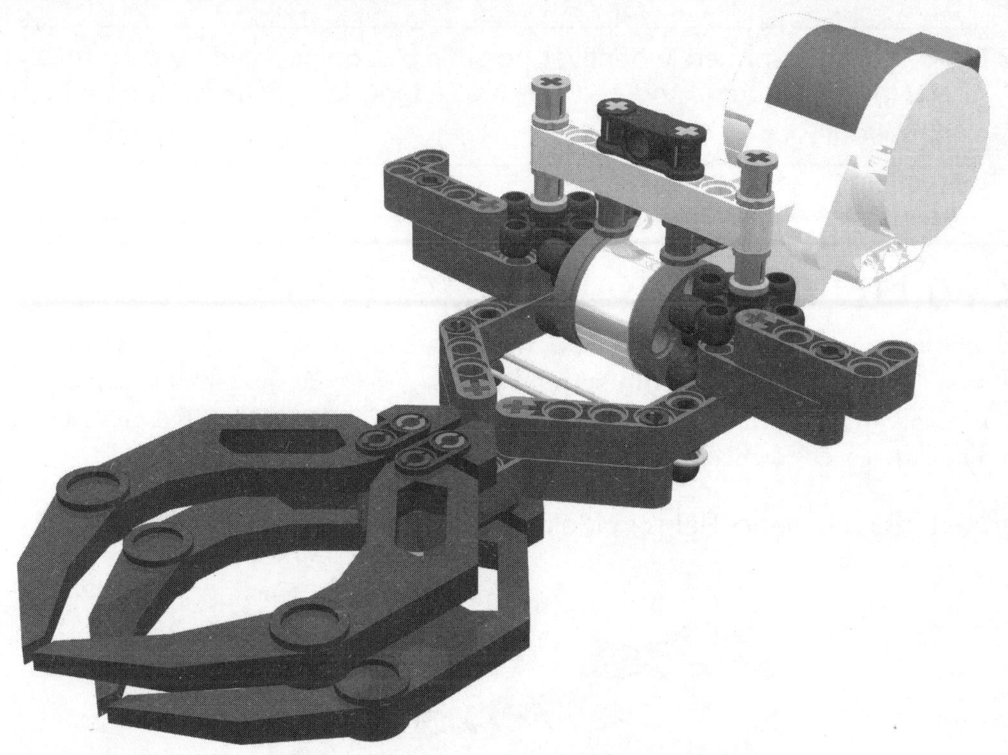

What if you need a compact, lightweight grabber? The flex system we briefly described in Chapter 9 allows you to transfer motion to distant parts, away from the motors and gears. Figure 11.7 shows a small operating hand based on this technique. A pair of opposing rubber bands introduces a degree of elasticity into the system, and helps the fingers return to their default setting once the hand comes to rest in its open position.

Figure 11.7 The Flex System Helps in Making Lightweight Hands

TIP

Try to use short fingers whenever possible. An object held by long fingers exerts more torque on the gears, just as a long lever makes lifting a heavy weight easier.

Using Pneumatics to Drive Your Grabber

In discussing the advantages of pneumatics when grabbing objects, we must also mention that tubing provides a simple way to keep bulky things far from the movable parts. Compare the simplicity of the pliers in Figure 11.8 with the complex gearings of the previous examples. The difference is dramatic.

Figure 11.8 Pneumatics Helps in Making Essential and Clean Assemblies

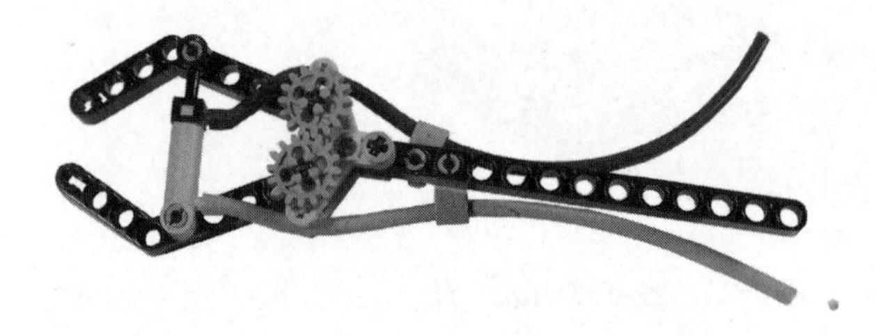

Pneumatics makes it possible to build more versatile hands than gear-driven grabbers. Figure 11.9 shows a three-joint pneumatic finger. This is a nice design, but it's a pity that it requires all three ports of your NXT to be fully controlled. How could you control more than a finger if you are already out of ports? To make the system simpler, though still useful, you can connect all the cylinders together (you won't be able to move a single segment of the finger by itself, but the finger can still adapt well to the shape of many different objects). This is the technique we used in the three-finger pneumatic hand shown in Figures 11.10 and 11.11, which is controlled by a single valve switch.

Figure 11.9 A Three-Degrees-of-Freedom Pneumatic Finger

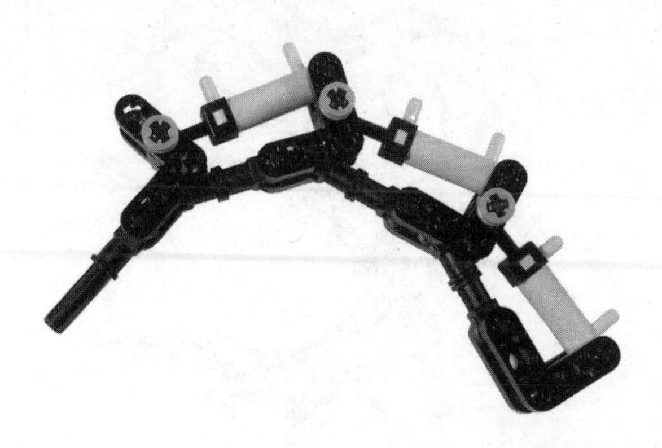

Figure 11.10 A Three-Finger Pneumatic Hand

Figure 11.11 The Three-Finger Pneumatic Hand with Complete Tubing

Bricks & Chips...

Understanding Degrees of Freedom

If you look carefully at your hands, you'll discover that they are an incredible piece of machinery, capable of handling a wide array of objects of every size and shape. Just think about the long list of verbs describing all the things hands can do: grab, handle, hold, take, squeeze, grip, point, pinch, shake, roll, press, grasp, push, pull ... and those are only a few of the terms. Where does all this versatility come from?

Observe your finger while you move it. You'll notice four individual movements: three for the joints—from the fingertip to the hand—that let you bend the finger, and a fourth that allows for slight left–right motion where the finger joins the hand. Although you usually operate more than one muscle at the same time, you can also move them individually (say, with your other hand). Multiply these four movements by five (for a hand's five fingers) and add the mobility given by the wrist, and 25 movements or so come to mind, which, in turn, lead to a huge number of combinations and configurations. This is what makes your hand able to conform to the shape of the object you want to handle. To complete the picture, consider that you can control the strength of each muscle so finely that you can pick up a delicate wine glass without damaging it, yet so firmly grip a baseball bat that you can send a ball over the right-field wall.

Continued

Every independent movement represents a degree of freedom (DOF), something that can happen without affecting and being affected by other movements in the same device. Most of the previous examples were very simple mechanisms with just one degree of freedom, being that all the possible positions of the "fingers" were determined by a single motor or pneumatic cylinder. The DOF concept helps you understand in terms of numbers why those simple hands diverged so widely from the flexibility a human hand has.

The DOF concept will help us understand the advantage of the pneumatic hand we showed in Figure 11.11. When we couple the movement of several "fingers" by a geartrain (such as the simple hand in Figure 11.1), all fingers move by a fixed amount (linear or angular) for each revolution of the motor, as determined by the gear ratios. If one of the fingers gets stuck, all other fingers get stuck too. On the other hand, when we increase the internal pressure in the cylinders of Figure 11.11, the individual cylinders extend and close the fingers. Here, if one finger gets stuck, the others will still extend until they balance the internal pressure. Thus, even though it seems that all cylinders move concurrently in the absence of a load, they are in fact mutually independent and the number of DOFs equals the number of cylinders.

The DOF concept applies not only to hands, but also to any mechanical device. The RoboArm T-56 in the Robo Center has three DOFs—one motor rotates the turntable, another motor extends the arm, and a third motor opens and closes the grabber. Figure 11.12 shows another type of robotic arm. Here one motor moves the grabber forward and backward using a long worm gear, and the other motor opens and closes the grabber (the same grabber as in Figure 11.3). This arm has two degrees of freedom.

Obviously, you cannot aim to make a robotic hand with 25 degrees of freedom using your MINDSTORMS NXT kit. Each degree of freedom will typically require a dedicated motor or pneumatic cylinder, and this puts the task out of reach. You should stay with something much simpler and consequently reduce the range of objects your mechanical hand will be able to grab. This is sometimes limiting, but in many situations you will know in advance the types and shapes of objects your robot will be expected to handle, making your task less demanding. In contests that involve collecting things, for example, your robot usually will deal with very specific objects, such as soda cans, small LEGO cubes, or marbles, and because of this you can design it to target those types of objects.

Generally speaking, locating a point in a plane requires two DOFs, whereas locating a point in space requires three. There are many examples of two-DOF and three-DOF mechanisms in everyday objects: An ink-jet printer has two DOFs, one corresponding to the head movement and the other to the paper feeding. A construction crane is an example of a machine with three DOFs: The hook can go up and down, it's attached to a carriage that moves back and forth along the boom, and finally, the boom can rotate. With the three output ports of your NXT, you can drive a robotic arm that addresses any point inside a delimited space, called the *operating envelope*, exactly like the crane of the preceding example. If you also want to pick up and drop objects, you would need another port.

Figure 11.12 A Two-DOF Robotic Arm

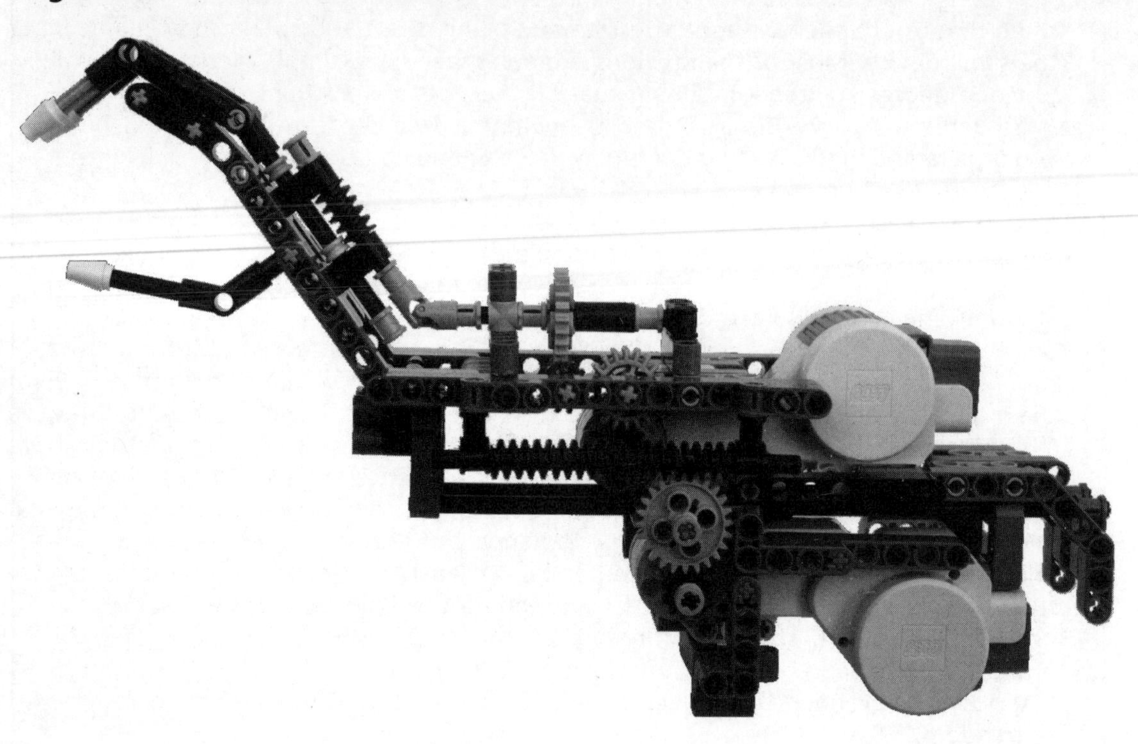

Finding Objects

Building robotic arms and hands is the easy part of the job. The hardest part is finding the objects to grab. We will skip the case where your robot *knows* the position of the objects, because this brings into play a general navigation problem we'll discuss in Chapter 13. So, for the time being, we'll stick with the fact that the robot knows nothing about the location of the objects.

As we explained when talking about bumpers in Chapter 4, navigation in real environments is quite a tough task, and distinguishing a specific object from others makes things much harder. So the second assumption we make here is that you know what kind of object you're expected to handle, as well as all the details of the environment where your robot moves (typically an artificial one prepared for the task). You might think that we are introducing too many simplifications here, but even in these conditions, the task remains quite hard. It's very important that you progress in short steps. The most common mistake of beginning builders is to start out with goals too difficult for their robots, where mechanical and programming difficulties add to navigation problems. As a general approach, we suggest you apply the "divide and conquer" strategy and solve the problems one by one.

Let's make an example: a simple variation on line following that might involve removing objects placed along the path. A very simple bumper is probably enough to detect objects. The arm will be more or less sophisticated depending on whether you have to collect them or just move them out of the way.

In wider environments, things become trickier. Imagine you have to find things in a delimited space with no walls (how could a space be delimited without having walls? By using different colors on the floor and reading them with a light sensor facing down!). You can still use a bumper, and make your robot move around at random or follow some kind of scheme. Depending on whether you are participating in a contest with specific rules, you could make this approach more efficient using a sort of funnel to convey the objects against the bumper, or some long antennas (connected to a touch sensor) to help you detect objects in a wider area.

If the objects in your environment are large enough, you can use the ultrasonic (US) sensor to find them. Figure 11.13 shows a modified TriBot that uses ultrasound to find soda cans and plastic cups. There are several important issues to remember when using ultrasound:

- It works well for relatively large, solid objects with more or less flat surfaces. This is due to the nature of ultrasonic sound waves—they do not reflect back (or are attenuated drastically) from round or spongy surfaces. An alternative can be the Mindsensors IR distance sensor, which is relatively unaffected by the object's composition and shape.
- Putting the US sensor too high can cause your robot to lose track of the object when it gets closer. Putting it too low can lead the sensor to receive echoes from ground reflections, thereby confusing the measurement.
- The ultrasound beam is about 30 degrees wide. This means that even a "point" object such as a standing marker would seems to occupy 30 degrees around your NXT. This also means that your robot may be facing an object farther away, but because there is a closer object a little to its left (or right), it will see this one instead.
- Approaching a flat surface in an oblique angle may result in a weird US readout, basically due to weak reflection back toward the sensor (most of the sound is reflected away from the sensor, just like the reflection of a flashlight from a mirror).
- Using more than one US sensor in a room (say, in a competition) may be problematic, as the sensor cannot distinguish between its own generated waves and other sensors' ultrasound waves. It is possible to synchronize US sensors (as we discuss later on), but this requires cooperation with your opponents…

Figure 11.13 The NXT CanFinder

How can you cover the whole space while searching with a US sensor? One way is to perform a 360-degree scan (typically by rotating the whole robot), looking for anything within, say, 30 inches. If you find something, go toward it one-half its distance; if not, pick a random direction and move 15 inches (half the target threshold distance). Repeating this basic step will give you good coverage of the area around you. Of course, you can improve on this simple scheme by remembering your previous "step," and so on.

You can use other sensors to help you find your target. For example, if your target emits light (which is strong enough to be detected from a distance, or alternatively you dim the lights in the room) or makes a sound, you can use the light or sound sensor to locate the target. A good practice in such cases is to have your robot move some distance, stop, rotate 360 degrees while measuring the light/sound level and keeping a record of the motor encoder value of the largest level, and then rotate to the orientation of maximal intensity and move some fixed distance again. Repeating this procedure again and again will finally lead you close to your target. If the target blinks or makes a nonconstant sound (like another NXT playing music), you can use two identical sensors separated a few inches apart and use the difference between them to decide whether the target is to the left or to the right, similar to how your own ears identify the direction of noises. Another, more specific, example is the IRSeeker sensor from HiTechnic, aimed at locating the IR-emitting Junior Cup ball. This sensor gives a 1–9 value telling the NXT in which sector (out of 240 degrees) the ball is, and another 1–9 value giving the signal strength (corresponding to the distance to the ball).

Positioning the Grabber

If you did the NXT-G Robo Center TriBot missions, you might have noticed that placing the ball not directly in front of the TriBot will sometimes cause the grabber to fail to catch it. This demonstrates another important issue in grabbing objects: aligning the grabber and the object so that the object can be caught by it (i.e., the object must be within the *operating envelope* of the grabber, as we discussed earlier).

The easiest way, of course, is to force the object to reach the grabber. You can do this, for instance, by having a "funnel"-like construction in front of your robot so that the object is pushed to a defined position where the grabber catches it. However, you can do this only for relatively small objects. Another approach is to adjust your robot orientation and position to have the grabber in the right place relative to the object.

Consider the example we gave earlier of an NXT using US to find its target. Once the robot is relatively close to the target, we can use a more accurate distance measurement during the US scan to pinpoint the object's direction. The CanFinder shown in Figure 11.13, for example, used a sophisticated procedure of this sort. It scanned clockwise over the object, and whenever the distance was smaller than the previous minimal distance found, it reset a motor encoder. If the distance was *the same* as the previous distance, it stored in memory the *current* encoder value (let's call this variable A). After the scan was complete, it rotated counterclockwise the final encoder value minus A, and then rotated counterclockwise A / 2 degrees. This placed the robot pointing directly toward the can.

Another alternative is to use a short-range sensor to check whether the object is positioned correctly. For example, you can put a light sensor near your robot's hands in such a way that you'll get a signal only when an object is at the right position.

Distinguishing Objects and Obstacles

Another important problem is distinguishing between "target" objects and "dummy" objects, and between objects and other obstacles such as chairs, walls, and so forth. The former is typically performed by some sort of sensor—a light sensor or HiTechnic color sensor, for example. The latter, however, is more difficult. As we already discussed, you typically wish to test your robots in a well-defined environment (divide and conquer, remember?). However, walls are hard to avoid, even in the most controlled conditions. How can you tell whether you found an object or hit the wall? Well, here are a few possible tricks:

- Try moving it. If it's a wall, it won't move. You can check whether you've moved it by monitoring the motor encoders.
- If you are using a US sensor, scan clockwise over the "object." Reset the encoder once you "see" the object, and keep track of the encoder value once you pass it to the other side. The difference between these two values (or better yet, the difference divided by the minimal distance to the object) can be compared to a threshold you find to distinguish your object and other (typically larger) obstacles.
- Usually the walls are taller than the soda cans or marbles you have to find, so you can prepare two bumpers at different heights and see which one closes to decide what your robot ran into.
- You can add a second US sensor above the object's expected height, or pointing upward at an angle (just remember that oblique angles are troublesome). Using two US sensors on a single NXT requires a little custom programming to ensure that they do not send sound waves together. Fortunately, the LEGO US sensors have two modes: a continuous mode (the default used by NXT-G and RobotC) and a single-shot mode in which it sends a "ping" of sound only by command. Using the single-shot mode you can alternate between two US sensors, pinging each one in turn. The following RobotC code displays two US readouts on the NXT LCD:

```
void InitializeSonar(tSensors nPort) {
  static const byte kSonarInitialize[] = {3, 0x02, 0x41, 0x01};

  SensorType[nPort] = sensorI2CCustomStd9V;
  sendI2CMsg(nPort, kSonarInitialize[0], 0);
  wait10Msec(5);
}

void PingSonar(tSensors nPort) {
  static const byte kSonarPing[] = {3, 0x02, 0x41, 0x01};
  static const byte kSonarRead[] = {2, 0x02, 0x42};
  const int nSonarReplySize = 1;

  byte replyMsg[1];

  sendI2CMsg(nPort, kSonarPing[0], 0);
```

```
  wait1Msec(10); // wait 10ms, enough for echoes to return to sensor.

  nI2CBytesReady[nPort] = 0; // Clear any pending bytes
  sendI2CMsg(nPort, kSonarRead[0], nSonarReplySize);
  while (nI2CStatus[nPort]==STAT_COMM_PENDING)
    wait1Msec(2); // Wait till I2C communication ends

  readI2CReply(nPort, replyMsg[0], nSonarReplySize);
  SensorValue[nPort] = replyMsg[0];
}

const tSensors kUS1 = S4;
const tSensors kUS2 = S3;

task main() {
 InitializeSonar(kUS1);
 InitializeSonar(kUS2);

 while (true)  {
    PingSonar(kUS1);
   PingSonar(kUS2);
    nxtDisplayTextLine(2, "US1: %d", SensorValue[kUS1]);
   nxtDisplayTextLine(7, "US2: %d", SensorValue[kUS2]);
   wait1Msec(50);
 }
 return;
}
```

A different case is when you want to manually trigger your robot to grab or release objects. This is very easy to implement with a touch sensor, a push button that you press when you want your robot to open or close its hand. US detection makes your robot even more impressive to see in action: You can, for instance, build a robot that navigates the room, and that, when you offer it an object, stops to grab it. This technique is a bit tricky to use if your robot is expected to navigate a room with walls and other obstacles, because it won't be able to tell what triggered its distance detection. One way to overcome this is to continuously monitor the distance and interpret a sudden radical change in its movement as a request to grab or release objects.

Summary

Designing a good robotic hand or arm is more of an art than a technique. There are indeed technical issues when it comes to gearing and pneumatics that you must know and consider to successfully position the grabbers or hands, apply the right amount of pressure, troubleshoot the elasticity of the object to be grabbed, and not allow your robot to drop the ball (or object, rather). Even then, there's still a lot of space for good intuitions and heavy prototyping. You can choose pneumatic or nonpneumatic approaches, design for different degrees of freedom in your gripping arm, use a flex system with tubing for lightweight designs, and create solutions that reserve ports for additional functions.

To make an easy start, target your first projects toward a specific type of object, and then progress to more versatile grabbers only when you feel experienced and confident enough to meet the challenge.

We also explained that finding the object is the hardest part of the job, but there are cases where you can use a random search pattern, or where the object sits on the robot's path, as in the line-following example. We discussed the pros and cons of using the ultrasonic sensor for finding objects, as well as other techniques such as sound and light. Finally, we discussed how to distinguish real objects and other obstacles your robot may encounter.

Chapter 12

Doing the Math

Solutions in this chapter:

- **Multiplying and Dividing**
- **Averaging Data**
- **Using Interpolation**
- **Understanding Hysteresis**
- **Higher Math**

Introduction

You may be surprised to find a chapter about mathematics in a book aimed at explaining building techniques. However, just as we can't put programming aside totally, so too we cannot neglect an introduction to some basic mathematical techniques. As we've explained, robotics involves many different disciplines, and it's almost impossible to design a robot without considering its programming issues together with the mechanical aspects. For this reason, some of the projects we are going to describe in this book include sample code, and we want to provide here the basic foundations for the math you will find in that code. Don't worry' the math we'll discuss in this chapter doesn't require anything more sophisticated than the four basic operations of adding, subtracting, multiplying, and dividing. The first section, about multiplying and dividing, explains in brief how computers deal with integer numbers, focusing on the NXT in particular. This topic is very important, because if you are not familiar with the logic behind computer math, you are bound to run into some unwanted results, which will make your robot behave in unexpected ways.

The three subsequent sections deal with *averages*, *interpolation*, and *hysteresis*, and a fourth explains strategies to deal with more complex calculations. Though averages, interpolation, and hysteresis are not essential, you should consider learning these basic mathematical techniques, because they can make your robot more effective while at the same time keep its programming code simpler. Averages cover those cases where you want a single number to represent a sequence of values. School grades are a good example of this: They are often averaged to express the results of students with a single value (as in a grade point average). Robotics can benefit from averages on many occasions, especially those situations where you don't want to put too much importance on a single reading from a sensor, but rather observe the tendency shown by a group of spaced readings.

Interpolation deals with the estimating, in numerical terms, the value of an unknown quantity that lies between two known values. Everyday life is full of practical examples—when the minute hand on your watch is between the three and four marks, you interpolate that data and deduce that it means, let's say, 18 minutes. When a car's gas gauge reads half a tank, and you know that with the full tank the car can cover about 400 miles, you make the assessment that the car can currently travel approximately 200 miles before needing refueling. Similarly in robotics, you will benefit from interpolation when you want to estimate the time you have to operate a motor in order to set a mechanism in a specific position, or when you want to interpret readings from a sensor that fall between values corresponding to known situations.

The last tool we are going to explore is *hysteresis*. Hysteresis defines the property of a variable for which its transition from state A to state B follows different rules than its transition from state B to state A. Hysteresis is also a programmed behavior in many automatic control devices because it can improve the efficiency of the system, and it's this facet that interests us. Think of hysteresis as being similar to the word *tolerance*, describing, in other words, the amount of fluctuation you allow your system before undertaking a corrective action. The hysteresis section of the chapter will explain how and why you might add hysteresis to the behavior of your robots.

If your robot requires more complex calculations, two simple strategies can simplify the task. The first strategy is to use a type of arithmetic that can represent fractional numbers, not only integers. The second is to rely on existing building blocks from which you can construct more complex calculations, such as NXT-G blocks and RobotC functions that compute trigonometric functions, square roots, and so on.

Multiplying and Dividing

If you are not an experienced programmer, first of all we want to warn you that in the world of computers, mathematics may be a bit different from what you've been taught in school, and some expressions may not result in what you expect. The math you need to know to program the small NXT is no exception.

Computers are generally very good at dealing with *integer* numbers, that is, whole numbers (1, 2, 3...) with the addition of zero and negative whole numbers. In Chapter 7, we introduced *variables*, and explained that variables are like boxes aimed at containing numbers. An *integer variable* is a variable that can contain an integer number. What we didn't say in Chapter 7 is that variables put limits on the size of the numbers you can store in them, the same way that real boxes can contain only objects that fit inside. You must know and respect these limits; otherwise, your calculations will lead to unexpected results. If you try to pour more water in a glass than what it can contain, the exceeding water will overflow. The same happens to variables if you try to assign them a number that is greater than their capacity—the variable will retain only a part of it.

NXT-G manipulates integer numbers in the range –2,147,483,648 through 2,147,483,647 (a little more than 2 billion). RobotC gives you more options. In the program fragment that follows, the variables *a*, *b*, and *c* can store any integer between –2,147,483,648 and 2,147,483,647 (just like NXT-G numbers). The variables *x*, *y*, and *z* can store only integers between –32,768 and 32,767. The type of the variable determines the range of numbers that it can store: *Long* variables can store values between –2,147,483,648 and 2,147,483,647, but *short* variables can store only values between –32,768 and 32,767.

```
long  a,b,c;
short x,y,z;
```

Why would you declare variables that can represent only a small range of values? Because they use less memory, so there is room for more of them. Usually, there is no need to worry too much about which number type is best: *Long* is often a good choice because it provides a large range of numbers and because the NXT has plenty of memory for storing numbers. If your program uses very large arrays that never need to store values greater than

32,767 or less than –32,768, you should declare them as *short*. RobotC even supports *floating-point* numbers, which can store fractional values, such as 3.14; we discuss these later in this chapter.

Whatever number type you choose, you must keep the results of your calculations inside the range of the type. This rule applies also to any intermediate result, and entails that you learn to be in control of your mathematics. If your numbers are outside this range, your calculations will return incorrect results and your robot will not perform as expected; in technical terms, this means you must know the *domain* of the numbers you are going to use. Multiplication and division, for different reasons, are the most likely to give you trouble.

Let's explain this statement with an example. You build a robot that mounts wheels with a circumference of 231mm. Attached to one wheel is a sensor geared to count 105 ticks per each turn of the wheel. Knowing that the sensor reads a count of 385, you want to compute the covered distance. Recall from Chapter 4 that the distance results from the circumference of the wheel multiplied by the number of counts and divided by the counts per turn:

231 x 385 / 105 = 847

This simple expression has obviously only one proper result: 847. But if you try to compute it in RobotC using *short* variables, you will find you *cannot* get that result. If you perform the multiplication first, that is, if the expression were written as follows:

(231 x 385) / 105

you get 222! If you try to change the order of the operations this way:

231 x (385 / 105)

you get 693, which is closer but still wrong! What happened? In the first case, the result of performing the multiplication first (88,935) was outside the upper limit of the allowed range, which is only 32,767. The NXT couldn't handle it properly and this led to an unexpected result. In the second case, in performing the division operation first, you faced a different problem: The NXT handles only integers, which cannot represent fractions or decimal numbers; the result of 385 / 105 should have been 3 2/3, or 3.66666..., but the processor truncated it to 3 and this explains the result you got.

Unfortunately, there is no general solution to this problem. A dedicated branch of mathematics, called *numerical analysis*, studies how to limit the side effects of mathematical operations on computers and quantify the expected errors and their propagation along calculations. Numerical analysis teaches that the same error can be expressed in two ways: *absolute errors* and *relative errors*. An absolute error is simply the difference between the result you get and the true value. For example, 4355 / 4 should result in 1,088.75; the NXT truncates it to 1,088, and the absolute error is 1,088.75 – 1,088 = 0.75. The division of 7 by 4 leads to the same absolute error: The right result is 1.75, it gets truncated to 1, and the absolute error is again 0.75. To express an error in a relative way, you divide the absolute error by the number to which it refers. Usually, relative errors get converted into *percentage errors* by multiplying them by 100. The percentage errors of our previous examples are quite different one from the other: 0.07 percent for the first one (0.75 / 1,088.75 x 100) and an impressive 42.85 percent error for the latter (0.75 / 1.75 x 100)! Here are some useful tips to remember from this complex study:

- You have seen that integer division will result in a certain loss of precision when decimals get truncated. Generally speaking, you should perform divisions as the *last step* of an expression. Thus, the form (A x B) / C is better than A x (B / C), and (A + B) / C is better than its equivalent, A / C + B / C.
- Although integer divisions lead to small but predictable errors, operations that go off-range (called *overflows* and *underflows*) result in gross mistakes (as you discovered in the example where we multiplied 231 by 385). You must avoid them at all costs. We said that the form (A x B) / C is better than A x (B / C), but *only* if you're sure A x B doesn't overflow the established range! If you use NXT-G or if you declare your variables in RobotC as *long* (like we declared *a*, *b*, and *c* earlier), overflows and underflows are not very likely to occur. If you declare your RobotC variables as *short* to save memory, beware of over/underflows.
- When dividing, the larger the dividend over the divisor, the smaller the relative error. This is another reason that (A x B) / C is better than A x (B / C): The first multiplication makes the dividend bigger.
- Fixed-point and floating-point numbers, which we cover later in the chapter, can help you avoid loss of accuracy at the cost of slower computational speed.

Averaging Data

In some situations, you may prefer that your robot base its decisions not on a single sensor reading, but on a *group* of them, to achieve more stable behavior. For example, if your robot has to navigate a pad composed of colored areas rather than just black and white, you would want it to change its route only when it finds a different color, ignoring transition areas between two adjacent colors (or even dirt particles that could be "read" by accident).

Another case is when you want to measure ambient light, ignoring strong lights and shadows. *Averaging* provides a simple way to solve these problems.

Simple Averages

You're probably already familiar with the simple average, the result of adding two or more values and dividing the sum by the number of addends. Let's say you read three consecutive light values of 65, 68, and 59. Their simple average would be:

(65 + 68 + 59) / 3 = 64

which is expressed in the following formula:

$$A = (V_1 + V_2 + ... + V_n) / n$$

The main property of the average, what actually makes it useful to our purpose, is that it smoothes single peak values. The larger the amount of data averaged, the smaller the effect of a single value on its result. Look at the following sequence:

60, 61, 59, 58, 60, 62, 61, 75

The first seven values fall in the range of 58 to 62, and the eighth one stands out with a 75. The simple average of this series is 62; thus, you see that this last reading doesn't have a strong influence (see Figure 12.1).

Figure 12.1 How Averaging Smoothes Peaks and Valleys in the Data

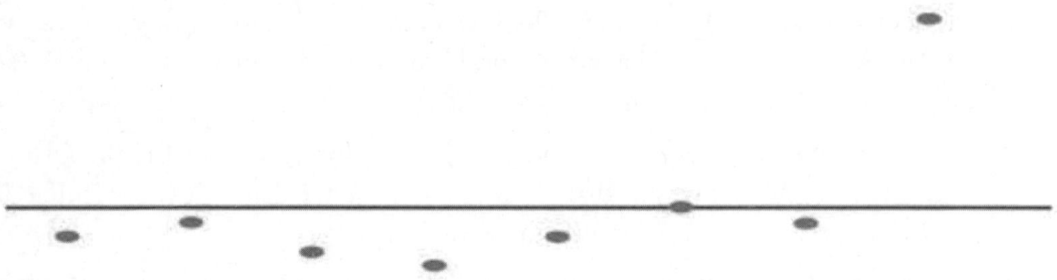

In your practical applications, you won't average all the readings of a sensor, usually just the last *n* ones. It is like saying you want to benefit from the smoothing property of an average, but you want to consider only more recent data because older readings are no longer relevant.

Every time you read a new value, you discard the oldest one with a technique called the *moving average*. It's also known as the boxcar average. Computing a moving average in a program requires you to keep all the last *n* values stored in variables, and then properly initialize them before the actual execution begins. Think of a sequence of sensor values in a long line. Your "boxcar" is another piece of paper with a rectangular cutout in it, and you can see exactly *n* consecutive values at any one time. As you move the boxcar along the line of sensor values, you average the readings you see in the cutout. It is clear that as you move the boxcar by one value from left to right along the line, the leftmost value drops off and the rightmost value can be added to the total for the average.

Going back to the series from our previous example, we'll now show you how to build a moving average for three values. You need the first three numbers to start: 60, 61, and 59. Their average is (60 + 61 + 59) / 3 = 60. When you receive a new value from your sensor, you discard the oldest one (60) and add the newest (58). The average now becomes (61 + 59 + 58) / 3 = 59.333... Figure 12.2 shows what happens to the moving average for three values applied to all the values of the example.

Figure 12.2 A Moving Average for Three Values

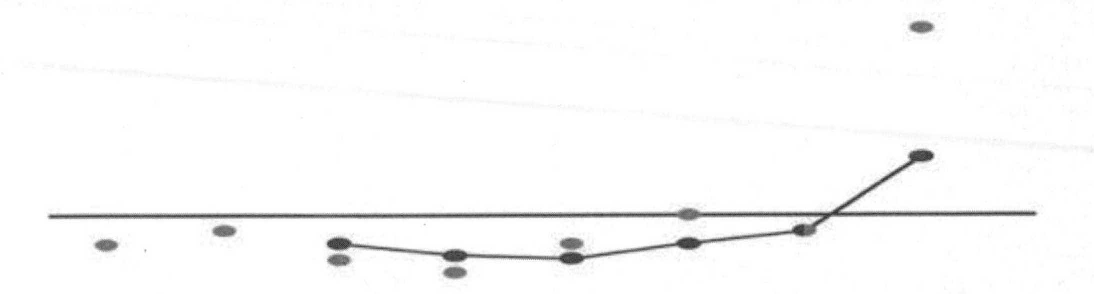

When raw data shows a trend, moving averages acknowledge this trend with a "lag." If the data increases, the average will increase as well, but at a slower pace. The higher the number of values used to compile the average, the longer the lag. Suppose you want to use a moving average for three values in a program. Your RobotC code could be as follows:

```
long ave, v1, v2, v3;

v2 = SensorValue[S1];
v3 = SensorValue[S1];

while (true)
{
  v1 = v2;
  v2 = v3;
  v3 = SensorValue[S1];
  ave = (v1+v2+v3) / 3;
  // other instructions...
}
```

Note the mechanism in this code that drops the oldest value (*v1*), replacing it with the subsequent one (*v2*), and that shifts all the values until the last one is replaced with a fresh reading from the sensor (in *v3*). The average can be computed through a series of additions and a division.

When the number of readings being averaged is large, you can make your code more efficient using *arrays*, adopting a trick to improve the computation time and keep the number of operations to a minimum. If you followed the description of the boxcar cutout as it moved along the line, you would realize that the total of the values being averaged did not have to be calculated every time. We just need to subtract the leftmost value, and add the rightmost value to get the new total!

A circular pointer, for example, can be used to address a single element of the array to substitute, without shifting all the others down. The number of additions, meanwhile, can be drastically decreased, keeping the total in a variable, subtracting the value that "exits," and adding the entering one. The following RobotC code provides an example of how you can implement this technique:

```
const short SIZE = 3;
long v[SIZE];
long i,sum,ave;

// initialize the array
sum = 0;
for (i=0;i<SIZE-1;i++)
{
  v[i] = SensorValue[S1];
  sum += v[i];
}

// first element to assign is the last of the array
i=SIZE-1;
v[i]=0;

// compute moving average
while (true)
{
  sum -= v[i]; //
  v[i] = SensorValue[S1];
  sum += v[i];
  ave = sum / SIZE;
  i = (i+1) % SIZE;
  // other instructions...
}
```

The constant *SIZE* defines the number of values you want to use in your moving average. In this example, it is set to 3, but you can change it to a different number. The statements that start with *long* declare the variables; *v[SIZE]* means that the variable *v* is an *array*, a container with multiple "boxes" rather than a single "box." Each *element* of the array works exactly like a simple variable, and can be addressed specifying its position in the array. Array elements are numbered starting from 0; thus, in an array with three elements they are numbered 0, 1, and 2. For example, the second element of the array *v* is *v[1]*.

This program starts initializing the array with readings from the sensor. It uses the *for* control statement to loop *SIZE-1* times, at the same time incrementing the *i* variable from 0 to *SIZE-1*. Inside the loop, you assign readings from the sensor to the first *SIZE-1* elements of the array. At the same time, you add those values to the *sum* variable. Supposing that the first readings are 72 and 74; after initialization, *v[0]* contains 72, *v[1]* contains 74, and *sum* contains 146. The initialization process ends assigning to the variable *i* the number of the first array element to replace, which corresponds to *SIZE-1*, which is 2 in our example.

Let's see what happens inside the loop that computes the moving average. Before reading a new value from the sensor, we remove the oldest value from *sum*. The first time *i* is 2 and *v[i]*, that is *v[2]*, is 0; thus, *sum* remains unchanged. *v[i]* receives a new reading from the sensor and this is added to *sum*, too. Supposing it is 75, *sum* now contains 146 + 75 = 221. Now you can compute the average *ave*, which results in 221 / 3 = 73.666..., and which is truncated to 73. The following instruction prepares the pointer *i* to the address of the next element that will be replaced. The symbol % in RobotC corresponds to the *modulo* operator, which returns the remainder of the division. This is what we call a *circular pointer*, because the expression keeps the value of *i* in the range from 0 to *SIZE-1*. It is equivalent to the code:

```
i = i+1;
if (i==SIZE) i=0;
```

which resets *i* to 0 when it reaches the upper bound. The resulting effect is that *i* cycles among the values 0, 1, and 2.

During the second loop *i* is 0, so *sum* gets decreased to *v[0]*, that is 72, and counts 221 - 72 = 149. *v[0]* is now assigned a new reading—for example, 73—and *sum* becomes equal to 149 + 73 = 222. The average results 222 / 3 = 74, and *i* is incremented to 1. Then the cycle starts again, and it's time for *v[1]* to be replaced with a new value.

This program is definitely much more complicated than the previous one, but it has the advantage of being more general: It can compute moving averages for any number of values by just changing the *SIZE* constant.

Weighted Averages

We explained that simple averages have the property of smoothing peaks and valleys in your data. Sometimes, though, you want to smooth data to reduce the effect of single readings, yet at the same time put more importance on recent values than on older ones. In fact, the more recent the data, the more representative the possible trend in the readings.

Let's suppose your robot is navigating a black-and-white pad, and that it's crossing the border between the two areas. The last three readings of its light sensor are 60, 62, and 67, which result in a simple average of 63. Can you tell the difference between that situation and one in which the readings are 66, 64, and 59 using just the simple average? You can't, because both series have the same average. However, there's an evident diversity between the two cases—in the first, the readings are increasing, and in the second, they are decreasing but the simple average cannot separate them. In this case, you need a *weighted* average, that is, an average where the single values get multiplied by a factor that represents their importance.

The general formula is:

$$A = (V_1 \times W_1 + V_2 \times W_2 + \ldots + V_n \times W_n) \; / \; (W_1 + W_2 + \ldots + W_n)$$

Suppose you want to give a weight of 1 to the oldest of three readings, 2 to the middle, and 4 to the latest one. Let's apply the formula to the series of our example:

$$(60 \times 1 + 62 \times 2 + 67 \times 4) / (1 + 2 + 4) = 64.57$$

$$(66 \times 1 + 64 \times 2 + 59 \times 4) / (1 + 2 + 4) = 61.43$$

You notice that the results are very different in the two cases: The weighted average reflects the trend of the data. For this reason, weighted averages seem ideal in many cases: They allow you to balance multiple readings, at the same time taking more recent ones into greater consideration. Unfortunately, they are memory-intensive and time-consuming when computed by a program, especially when you want to use a large number of values.

Now, there is a particular class of weighted averages that can be of help, providing a simple and efficient way of storing historical readings and calculating new values. They rely on a method called *exponential smoothing* (don't let the name frighten you!).

The trick is simple: You take the new reading and the previous average value, and combine these into a new average value using two weights that together represent 100 percent. For example, you can take 40 percent of the new reading and 60 percent of the previous average, or instead take only 10 percent of the new reading and 90 percent of the previous average. The less weight you put on the new value, the more stable and slow to change the average will be.

The general equation for exponential smoothing is expressed as follows:

$$A_n = (V_n \times W_1 + A_{n-1} \times W_2) \; / \; (W_1 + W_2)$$

You can choose W1 and W2 to add to 100 so that you can easily read them as a percentage. For example:

$$A_n = (V_n \times 20 + A_{n-1} \times 80) \; / \; 100$$

Let's apply this formula to the series of the previous example. The first number in the first series was 60. There is no previous value for the average, so we simply take this number:

$$A_1 = 60$$

When the next reading (62) arrives, you compute a new value for the average using the whole formula:

$A_2 = (62 \times 20 + 60 \times 80) / 100 = 60.4$

Then you apply the rule again for the third value:

$A_3 = (67 \times 20 + 60.4 \times 80) / 100 = 61.72$

The result tells you that the average is *slowly* acknowledging the trend in the data. This happens because the last reading counts only for 20 percent, whereas 80 percent comes from the previous value. If you want to make your average more reactive to recent values, you must increase the weight of the last factor. Let's see what happens by changing 20 percent to 60 percent:

$A_1 = 60$

$A_2 = (62 \times 60 + 60 \times 40) / 100 = 61.2$

$A_3 = (67 \times 60 + 61.2 \times 40) / 100 = 64.68$

You notice that the formula is still smoothing the values, but it gives much more importance to recent values. One of the advantages of exponential smoothing is that it is very easy to implement. The following is an example of RobotC code:

```
long ave;

// initialize the average
ave = SensorValue[S1];

// compute average
while (true)
{
  ave = (SensorValue[S1] * 20 + ave * 80) / 100;
  // other instructions...
}
```

Simple, isn't it? You could be tempted to *reduce* the mathematical expression, but be careful; remember what we said about multiplying and dividing integer numbers. These are okay:

```
  ave = (SensorValue[S1] * 2 + ave * 8) / 10;
  ave = (SensorValue[S1] + ave * 4) / 5;
```

But this, though mathematically equivalent, leads to a worse approximation:

```
  ave = SENSOR_1 / 5 + ave * 4 / 5;
```

Designing & Planning ...

Exponential Smoothing

Those of you with a gift for math might be interested in understanding where exponential smoothing got its name. Let's try to analyze the equation:

$$A_n = (V_n \times W_1 + A_{n-1} \times W_2) \ / \ (W_1 + W_2)$$

We can rewrite the weights W_1 and W_2 as fractions: $w_1 = W_1 / (W_1 + W_2)$ and $w_2 = W_2 / (W_1 + W_2)$, where w_1 and w_2 result in the range of 0 to 1. As $w_1 + w_2 = 1$, we can substitute w_2 with $(1 - w_1)$. Our equation then becomes:

$$A_n = V_n \times w_1 + A_{n-1} \times (1 - w_1)$$

Expanding the term A_{n-1} we get:

$$A_{n-1} = V_{n-1} \times w_1 + A_{n-2} \times (1 - w_1)$$

and substituting in the previous:

$$A_n = V_n \times w_1 + V_{n-1} \times w_1 \times (1 - w_1) + A_{n-2} \times (1 - w_1)^2$$

Continuing this expansion, we get the general form:

$$A_n = V_n \times w_1 + V_{n-1} \times w_1 \ (1 - w_1) + V_{n-2} \times w_1 \times (1 - w_1)^2 + ... + V_{n-m} \times w_1 \times (1 - w_1)^m + ... + V_1 \cdot w_1 \cdot (1 - w_1)^n$$

This average is thus equivalent to an average of all the values, where the older they are the more they get smoothed by the exponential term $(1 - w_1)^m$. The term $(1 - w_1)$ being less than zero means the higher the exponent, the smaller the result.

Using Interpolation

You've built a custom temperature sensor that returns a raw value of 200 at 0° C and 450 at 50° C. What temperature corresponds to a raw value of 315? Your robotic crane takes 10 seconds to lift a load of 100g, and 13 seconds for 200g. How long will it take to lift 180g? To answer these and similar questions, you would turn to *interpolation*, a class of mathematical tools designed to estimate values from known data.

Interpolation has a simple geometric interpretation: If you plot your known data as points on a graph, you can draw a line or a curve that connects them. You then use the points on the line to guess the values that fall inside your data. There are many kinds of interpolation, that is, you can use many different equations corresponding to any possible mathematical curve to interpolate your data. The simplest and most commonly used one is *linear interpolation*, for which you connect two points with a straight line, and this is what we are going to explain here (see Figure 12.3).

Figure 12.3 Linear Interpolation

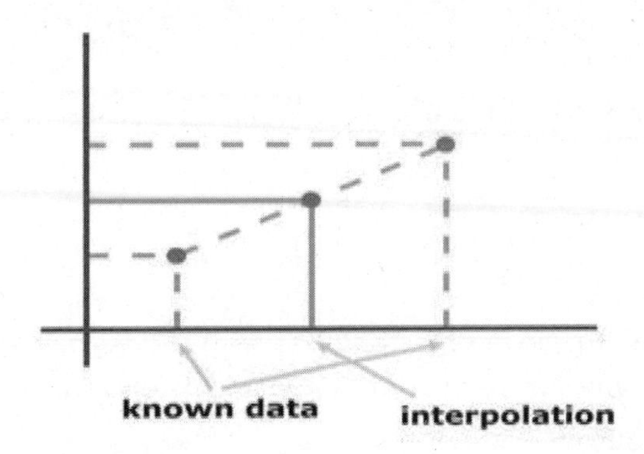

Please be aware that many physical systems don't follow a linear behavior, so linear interpolation will not be the best choice for all situations. However, linear interpolation is usually fine even for nonlinear systems, if you can break down the ranges into almost linear sections.

In following standard terminology, we will call the parameter we change the *independent variable*, and the one that results from the value of the first, the *dependent variable*. With a very traditional choice, we will use the letter X for the first and Y for the second. The general equation for linear interpolation is:

$$(Y - Y_a) / (Y_b - Y_a) = (X - X_a) / (X_b - X_a)$$

where Y_a is the value of Y we measured for $X = X_a$ and Y_b the one for X_b. With some simple work, we can isolate the Y and transform the previous equation into:

$$Y = (X - X_a) \cdot (Y_b - Y_a) / (X_b - X_a) + Y_a$$

This is very simple to use, and allows you to answer your question about the custom temperature sensor. The raw value is your independent variable X, the one you know. The terms of the problem are:

$X_a = 200$ $Y_a = 0$

$X_b = 450$ $Y_b = 50$

$X = 315$ $Y = ?$

We apply the formula and get:

$$Y = (315 - 200) \times (50 - 0) / (450 - 200) + 0 = 23$$

To make our formula a bit more practical to use, we can transform it again. We define:

$$m = (Y_b - Y_a) / (X_b - X_a)$$

If you are familiar with college math, you will recognize in *m* the slope of the straight line that connects two points. Now our equation becomes:

$$Y = m \times X - m \times X_a + Y_a$$

As X_a and Y_a are known constants, we compute a new term *k* as:

$$b = Y_a - m \times X_a$$

So our final equation becomes:

$$Y = m \times X + b$$

This is the standard equation of a straight line in the Cartesian plane. Looking back to our previous example, you can now compute *m* and *b* for your temperature sensor:

$$m = (50 - 0) / (450 - 200) = 0.2$$

$$b = 0 - 0.2 \times 200 = -40$$

$$Y = 0.2 \times X - 40$$

You can confirm your previous result:

$$Y = 0.2 \times 315 - 40 = 23$$

Implementing this equation inside a program for the NXT will require that you convert the decimal value 0.2 into a multiplication and a division, as follows:

```
temp = (raw * 2) / 10 - 40;
```

Interpolation is also a good tool when you want to relocate the output from a system in a different range of values. This is what the NXT firmware does when converting raw values from the light and sound sensors into a percentage. You can do the same in your application. Suppose you want to change the way raw values from the light sensor get converted into a percentage. The raw values that the analog sensors return are always in the range 0 to 1,023, but extreme raw values are rare. Let's say you want to fix an arbitrary range of 900 –> 0 percent and 500 –> 100 percent, and this is what you get from the interpolation formula:

$$m = (0 - 100) / (900 - 500) = -0.25$$

$$b = 100 + 0.25 \times 500 = 225$$

$$Y = -0.25 \times X + 225$$

Multiplying by 0.25 is like dividing by 4, so we can write this expression in code as:

```
perc = - raw / 4 + 225;
```

Understanding Hysteresis

Hysteresis is actually more a physical than a mathematical concept. We say that a system has some hysteresis when it follows a path to go from state A to state B, and a different path when going *back* from state B to state A. Graphing the state of the system on a chart shows two different curves that connect the points A and B, one for going out and one for coming back (see Figure 12.4).

Figure 12.4 Hysteresis in Physical Systems

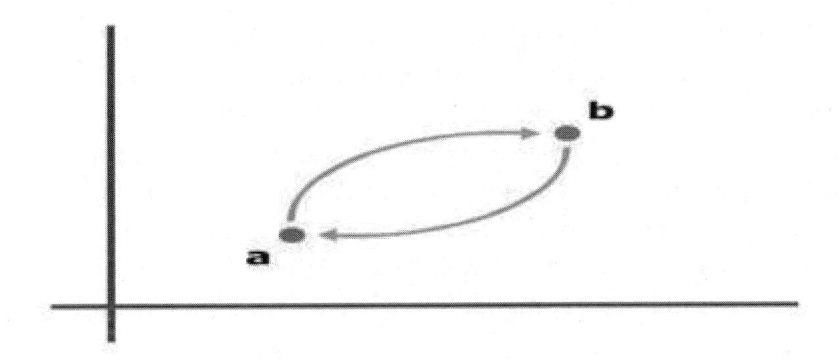

Hysteresis is a common property of many natural phenomena, magnetism above all, but our interest here is in introducing some hysteresis in our robotics programs. Why should you do it? First of all, let us say this is quite a common practice. In fact, many automation devices based on some kind of feedback have been equipped with artificial hysteresis.

A very handy example comes from the thermostat that controls the heating in your house. Imagine that your heating system relies on a thermostat designed to maintain an exact temperature, and that during a cold winter you program your desired home temperature to 21° C (70° F). As soon as the ambient temperature goes below 21° C, the heater starts. In a few minutes, the temperature reaches 21° C and the heater stops, then a few minutes later it starts again, and so on, all day long. The heater would turn off and on constantly as the temperature varies around that exact point. This approach is not the best one, because every start phase requires some time to bring the system to its maximum efficiency, just about when it gets stopped again. In introducing some hysteresis, the system can run more smoothly: We can let the temperature go down to 20.5° C, and then heat up the house until it reaches 21.5° C. When the temperature in the house is 21° C, the heater can be either on or off, depending on whether it's going from on to off or vice versa.

Hysteresis can reduce the number of corrective actions a system has to take, thus improving stability and efficiency at the price of some tolerance. Autopilots for boats and planes are another good example. Could you think of a task for your robots that could benefit from hysteresis? Line following is a good example.

In Chapter 4, in talking about light sensors, we explained that the best way to follow a strip on the floor is to actually follow one of its edges, the borderline between white and black. In that area, your sensor will read a "gray" value, some intermediate number between the white and black readings. Having chosen that value for *gray*, a robot with no hysteresis may correct left when the reading is greater than *gray* and right when the reading is less than *gray*. To introduce some hysteresis, you can tell your robot to turn left when reading *gray+h* and right when reading *gray-h*, where *h* is a constant properly chosen to optimize the performance of your robot. There isn't a general rule that is valid for any system; you must find the optimal value for *h* by experimenting: Start with a value of about 1/6 or 1/8 of the total white-black difference. This way, the interval *gray-h* to *gray+h* will cover 1/3 or 1/4 of the total range. Then start increasing or decreasing its value, observing what happens to your robot, until you are happy with its behavior. You will discover that by reducing *h* your robot will narrow the range of its oscillations, but will perform more frequent corrections. Increasing *h*, on the other hand, will make your robot perform fewer corrections but with oscillations of larger amplitude (see Figure 12.5).

Figure 12.5 How Hysteresis Affects Line Following

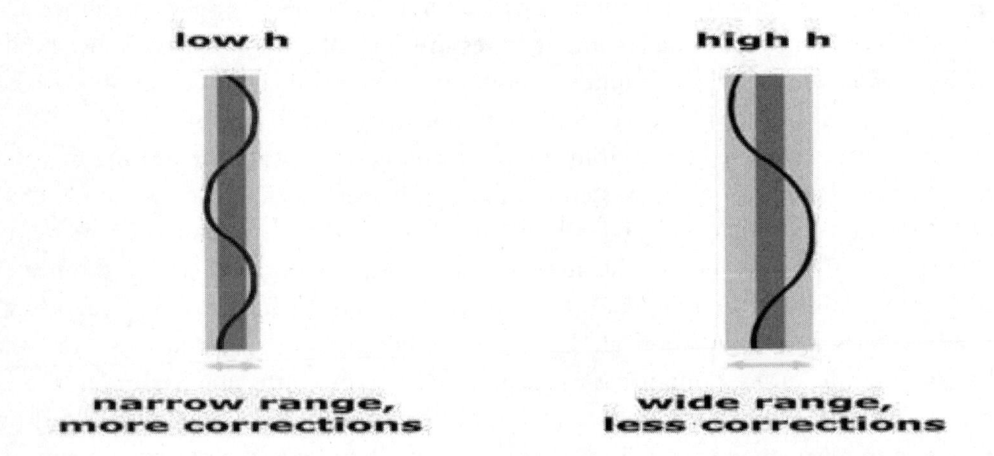

We suggest a simple experiment that will help you put these concepts into practice by building a real sensor setup that you can manipulate by hand to get a feel for how the robot would behave. Write a simple program that plays tones to ask you to turn left or right. For example, it can beep high when you have to turn left and low to turn right. The RobotC code that follows shows a possible implementation:

```
const long GRAY = 50;
const long H    = 3 ;

task main()
{
  SetSensorType(S1,sensorLightActive);

  while (true)
  {
    if (SensorValue[S1]>GRAY+H)
      PlayTone(440,20);
    else if (SensorValue[S1]<GRAY-H)
      PlayTone(1760,20);
    wait10Msec(2);
  }
}
```

Equip your NXT with a light sensor attached to input port 1 and you are ready to go. You should tune the value of the *GRAY* constant to what your sensor actually reads when placed exactly over the borderline between the white and the black. When the program runs, you can move the sensor toward or away from the line until you hear the beep that asks you to correct the direction (keep the sensor always at the same distance from the pad). Experiment with different values for *H* to see how the accepted range of readings gets wider or narrower.

If you keep a pencil in your hand together with the light sensor, you can even perform this experiment blindfolded! Try to follow the line by just listening to the "instructions" coming from your NXT and compare the lines drawn by your pencil for different values of *H*.

Higher Math

Robots perform mathematical computations for two reasons: to understand the world around them and to plan their actions. Consider a robot that seeks a certain object—say, a blue ball. The robot uses its sensors to obtain information about the world around it and about its own position. It uses this information to try to estimate where the blue ball is and where it is. Once the robot decides where the ball is and where it is, it needs to plan how to reach the ball. If the environment contains obstacles (such as walls or other robots), finding a path may be a challenging task that requires interesting computations. In some robots, the estimation and planning are simple and require no math or just a bit of math. But a robot may require more sophisticated calculations than we have shown in this chapter.

Integers are not always suitable for complex mathematical calculations, because integer division can generate absolute errors of up to ±1/2, which may be too high. For example, dividing 301 by 2 gives the integer 150, whereas the exact answer is 150 1/2. Computers use several other number types to reduce these errors. The most important noninteger number types are *fixed-point* numbers and *floating-point* numbers. We explain these number types with decimal examples, but keep in mind that computers always represent numbers in binary format (base 2). Suppose that each word in the computer's memory can store a number with a sign and four decimal digits, and suppose that a certain word stores the digits 3759. In an integer, the decimal point is to the right of the rightmost digit, so if we view these digits as an integer, the number that the word represents is 3,759.0. We can also decide that the decimal point is at some arbitrary fixed position, such as between the second and third digits. The computer stores exactly the same information, but now the word in memory represents 37.59. This is called a fixed-point representation. In fixed point, we cannot represent very large numbers, but we have more accuracy. For example, fixed-point numbers are ideal for averaging sensor readings in percentage mode. On the one hand, the readings (and the averages) never exceed 100, so we never need very large numbers, and on the other hand, fixed-point numbers allow us to represent fractional averages more accurately than integers. To view the digits stored in a memory word as a floating-point number, we split the digits into two groups, a fraction and an exponent. Let's assume that we use the three leftmost digits as a fraction and the last digit as an exponent. In this case, our memory word would represent the number 0.375×10^9, or 375,000,000. We now have less accuracy, because our numbers have only three significant digits, but we have a much larger range of numbers: These four decimal digits allow us to represent numbers that approach 1 billion.

The processor in every desktop or laptop computer can perform arithmetic on both integers and floating-point numbers very quickly. Therefore, complex mathematical calculations that can produce noninteger results are usually performed on such computers using floating-point arithmetic. The processor of the NXT can perform only integer arithmetic. It can be programmed to perform floating-point arithmetic, but then the arithmetic operations are carried out in software, so they are fairly slow. RobotC supports floating-point calculations in this way. Neither the NXT nor your PC supports fixed-point arithmetic in the processor itself, but fixed-point arithmetic that is implemented in software is only slightly slower than integer arithmetic. Therefore, fixed-point arithmetic is a good choice for performing more complex calculations on the NXT.

Addition and subtraction on fixed-point numbers are performed as though they were integers. For example, 37.59 + 0.63 = 38.22; we can obtain this result by adding 3,759 + 63, which gives 3,822. In other words, for addition and subtraction, it does not matter where the point is, as long as it is in the same fixed positions in all the numbers. Multiplication and division are a bit more complex. The product 37.59 x 0.63 should give 23.6817, but the integer product 3,759 X 63 gives 236,817. To get the correct answer, we need to multiply the input numbers as though they were integers, and then shift the integer product to the right by two digits (because our fixed-point numbers have two digits to the right of the point). This truncates the digits 17 and gives 2368, which we interpret as 23.68, which is the correct fixed-point answer. The answer is not exact, but then there is really no way to represent the 23.6817 with only four decimal digits! Fixed-point multiplication done this way can easily overflow, so you need to either temporarily store the integer product in a number type with more digits (say, *long* rather than *short*), or to resort to a more complex multiplication algorithm. Division is somewhat similar to multiplication: To get the correct fixed-point answer, you need to shift the dividend to the left before you divide them by integers. For example, the integer division 375,900 / 63 gives 5,966, which is the correct fixed-point representation of 37.59 / 0.63 = 59.66. Here too you need a temporary variable with more digits.

Fixed-point arithmetic allows you to strike the right balance between accuracy and range, and it is just a bit slower than integer arithmetic. Floating-point arithmetic is even more convenient than fixed-point, because the computer chooses the right range for you after every arithmetic operation, but it is not more accurate and it is slow on the NXT (and is not available at all in NXT-G). Use integers if the accuracy is good enough for you, use fixed-point if you need more accuracy, and use floating-point if speed is not important, you need the convenience, and your programming environment supports floating-point.

Another point to keep in mind if your robot must perform sophisticated calculations is that software components are available that can help you implement these calculations. For example, there are MyBlock and third-party native blocks that compute square roots and trigonometric functions, such as sine and cosine. RobotC also includes functions that compute these functions. As time goes by, more and more such building blocks will become available, both for NXT-G and for RobotC. Expressing calculations in terms of preexisting building blocks simplifies programming considerably; this is how virtually any complex piece of software is built.

Summary

Math is the kind of subject that people either love or hate. If you fall in the latter group, we can't blame you for having skipped most of the content of this chapter. Don't worry; there was nothing you can't live without. Just make an effort to understand the part about multiplication and division, because if you ignore the possible side effects, you could end up with some bad surprises in your calculations.

Consider the other topics—averages, interpolation, and hysteresis—to be like tools in your toolbox. Averages are a useful instrument to soften the differences between single readings and to ignore temporary peaks. They allow you to group a set of readings and consider it as a single value. When you are dealing with a flow of data coming from a sensor, the moving average is the right tool to process the last *n* readings. The larger *n* is, the more the smoothing effect on the data.

Weighted averages have an advantage over simple averages in that they can show the trend in the data: You can assign the weights to put more importance on more recent data. Exponential smoothing is a special case of weighted averages, the results of which are particularly convenient on the implementation side, because they allow you to write compact and efficient code.

The interpolation technique proves useful when you want to estimate the value of a quantity that falls between two known limits. We described linear interpolation, which corresponds to tracing a straight line across two points in a graph. You then can use that line to calculate any value in the interval.

Hysteresis, a concept borrowed from physics, will help you in reducing the number of corrections your robots have to make to keep within a required behavior. By adding some hysteresis to your algorithms, your robots will be less reactive to changes. Hysteresis can also increase the efficiency of your system.

Some robots may require more sophisticated calculations. This may involve using fixed-point or floating-point numbers and more complex mathematical calculations. If your robot needs complex calculations, try to find building blocks that perform parts of the computation. These building blocks can include MyBlock and third-party blocks that others have made available, or library functions in textual programming languages.

It's not necessary that you remember all the equations, just what they're useful for! You can always refer back to the text when you find a problem that might benefit from a mathematical tool that you read about.

Chapter 13

Knowing Where You Are

Solutions in this chapter:

- Choosing Internal or External Guidance
- Looking for Landmarks: Absolute Positioning
- Map Matching Using Ultrasonic Sensor
- Combining Compass Sensor to Increase Precision
- Measuring Movement: Relative Positioning
- Measuring Movement: Acceleration Sensor

Introduction

After our first few months of experimenting with robotics using the Mindstorms NXT kit, we began to wonder if there was a simple way to make our robot know where it was and where it was going—in other words, we wanted to create some kind of navigation system able to establish its position and direction. We started reading books and searching the Internet, and discovered that this is still one of most demanding tasks in robotics and that there really isn't any single or simple solution.

In this chapter, we will introduce the concept of navigation, which can get very complex. We will start describing various positioning methods. Then we will provide some examples for these methods, showing solutions and tricks that suit the possibilities of the Mindstorms NXT system. In this discussion, we will introduce navigation on pads equipped with grids or gradients, use of laser beams to locate your robot in a room and explain how to equip your robot with sensors for the various measurements, and will provide the math to convert those measurements to determine your position.

Choosing Internal or External Guidance

There is no single method for determining the position and orientation of a robot, but you can combine several different techniques to get useful and reliable results. All these techniques can be classified into two general categories: *absolute* and *relative* positioning methods. This classification refers to whether the robot looks to the surrounding environment for tracking progress, or just to its own course of movement.

Absolute positioning refers to the robot using some external reference point to figure out its own position. These can be landmarks or obstacles in the environment, either natural landmarks or obstacles recognized through sensory inputs such as the touch, ultrasonic, magnetic compass sensors, or more often, artificial landmarks easily identified by your robot (such as colored tape on the floor). Another common approach includes using radio (or light) beacons as landmarks, like the systems used by planes and ships to find the route under any weather condition. Absolute positioning requires a lot of effort: you need a prepared environment, or some special equipment, or both.

Relative positioning, on the other hand, doesn't require the robot to know anything about the environment. It deduces its position from its previous (known) position and the movements it made since the last known position. This usually is achieved through the use of encoders that precisely monitor the turns of the wheels, but there are also inertial systems that measure changes in speed and direction. This method also is called *dead reckoning* (short for *deduced reckoning*).

Relative positioning is quite simple to implement, and applies to our NXT robots, too. Unfortunately, it has an intrinsic, unavoidable problem that makes it impossible to use by itself: it accumulates errors. Even if you put all possible care into calibrating your system, there will always be some very small difference due to slippage, load, or tire deformation that will introduce errors into your measurements. These errors accumulate very quickly, thus relegating the utility of relative positioning to very short movements. Imagine you have to measure the length of a table using a very short ruler: you have to put it down several times, every time starting from the point where you ended the previous measurement. Every placement of the ruler introduces a small error, and the final result is usually very different from the real length of the table.

The solution employed by ships and planes, which use beacons like Loran or Global Positioning Systems (GPS) systems or reference earth's magnetic field using Compass, is to combine methods from the two groups: to use dead reckoning to continuously monitor movements and, from time to time, some kind of absolute positioning to zero the accumulated error and restart computations from a known location. This is essentially what human beings do: when you walk down a street while talking to a friend, you don't look around continuously to find reference points and evaluate your position; instead, you walk a few steps looking at your friend, then back to the street for an instant to get your bearings and make sure you haven't veered off course, then you look back to your friend again.

You're even able to safely move a few steps in a room with your eyes shut, because you can deduce your position from your last known one. But if you walk for more than a few steps without seeing or touching any familiar object, you will soon lose your orientation.

In the rest of the chapter, we will explore some methods for implementing absolute and relative positioning in NXT robots. It's up to you to decide whether or not to use any one of them or a combination in your applications. Either way, you will discover that this undertaking is quite a challenge!

Looking for Landmarks: Absolute Positioning

The most convenient way to place artificial landmarks is to put them flat on the floor, since they won't obstruct the mobility of your robot and it can read them with a light sensor without any strong interference from ambient light. You can stick some self-adhesive tape directly on the floor of your room, or use a sheet of cardboard or other material over which you make your robot navigate.

Line following, which we have talked about in this book, is probably the simplest example of navigation based on using an artificial landmark. In the case of line following, your robot knows nothing about where it is, because its knowledge is based solely on whether it is to the right or left of the line. But lines are indeed an effective system to steer a robot from one place to another. Feel free to experiment with line following; for example, create some interruptions in a straight line and see if you are able to program your robot to find the line again after the break. It isn't easy. When the line ends, a simple line follower would turn around and go back to the other side of the line. You have to make your software more sophisticated to detect the sudden change and, instead of applying a standard route correction, start a new searching algorithm that drives the robot toward a piece of line further on. Your robot will have to go forward for a specific distance (or time) corresponding to the approximate length of the break, then turn left and right a bit to find the line again and resume standard navigation.

When you're done and satisfied with the result, you can make the task even more challenging: place a second line parallel to the first, with the same interruptions, and see if you can program the robot to turn 90 degrees, intercept the second line, and follow that one. If you succeed in the task, you're ready to navigate a grid of short segments, either following along the lines or crossing over them like a bar code.

You can improve your robot navigation capabilities, and reduce the complexity in the software, using more elaborate markers. The NXT light sensor is not very good at distinguishing different colors, but is able to distinguish between differences in the intensity of the reflected light. You can play with black and gray tapes on a white pad, and use their color as a source of information for the robot. Remember that a reading at the border between black and white can return the same value of another on plain gray. Move and turn your robot a bit to decode the situation properly, or employ more than a single light sensor if you have them. Alternatively, you could use the color sensor (sold separately by LEGO) with your own color-coded markers.

Instead of placing marks on the pad, you can also print on it with a special black and white gradient. For example, you can print a series of dots with an intensity proportional to their distance from a given point A. The closer to A, the darker the point; A is on plain black (see Figure 13.1). On such a pad, your robot will be able to return to A from any point, by simply following the route until it reads the minimum intensity.

Figure 13.1 A Gradient Pad with a Single Attractor

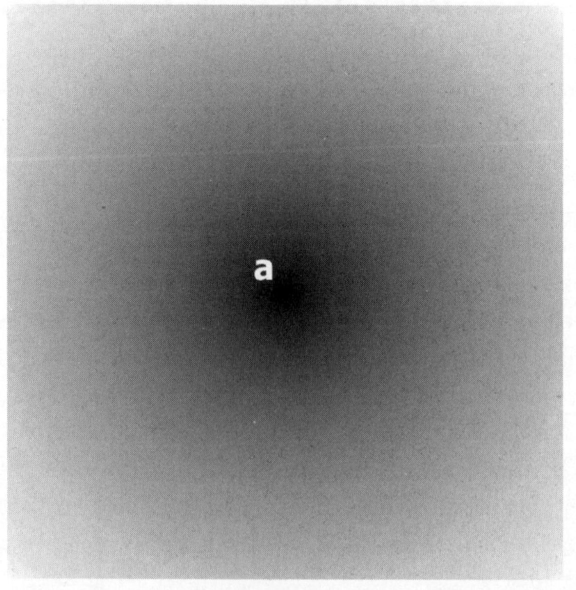

The same approach can be used with two points A and B, one being white and the other black. Searching for the whitest route, the robot arrives at A, whereas following the darkest it goes to B (see Figure 13.2).

RoboCup Junior uses a gradient pad for the soccer field. While searching for and chasing the ball, the robot could be anywhere on the field. When it gets the ball, it can refer to the pad below and with a light sensor, read the gray values and deduce how far it is from the goal. The same technique can then be used to detect which side is the opponent's goal. Say the opponent's goal is on the dark edge of the mat; the robot can spin around itself while measuring the gray values below, and deduce which side is the opponent's goal.

Figure 13.2 A Gradient Pad with Two Attractors

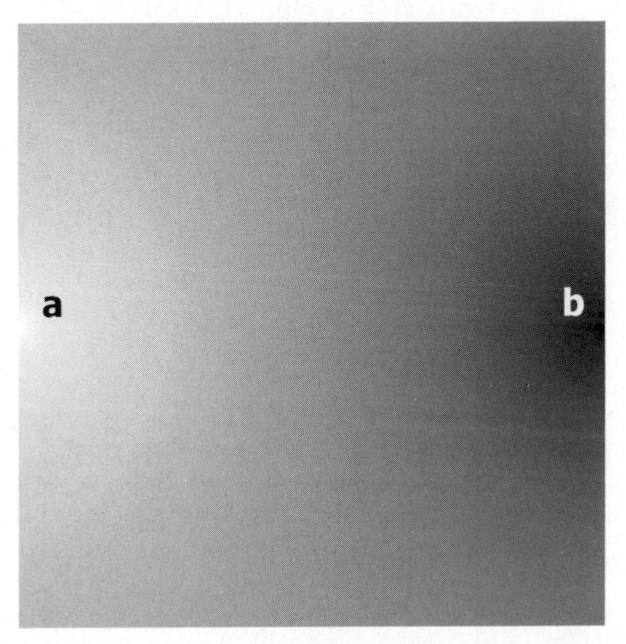

Designing & Planning ...

Making a Gradient Pad

You can make your own gradient pad in a Microsoft Excel spreadsheet, by writing a Visual Basic macro to color each cell background with appropriate intensity. Depending on your size, you can print this gradient pad on multiple papers and join them together. The following macro would make a gradient pad with two attractors (see Figure 13.2) where Point A (*whitePointX, whitePointY*) and Point B (*blackPointX, blackPointY*) are, respectively, white and black.

Continued

```
Sub gradient_pad()
  padWidth = 50
  padHeight = 50
  wPointX = padHeight / 2
  wPointY = 3
  bPointX = padHeight / 2
  bPointY = padWidth - 3

   For y = 1 To padWidth
      For x = 1 To padHeight
        whiteDist = Sqr((x - wPointX) * (x - wPointX) + (y - wPointY) * (y -
wPointY))
        blackDist = Sqr((x - bPointX) * (x - bPointX) + (y - bPointY) * (y -
bPointY))
        cValue = blackDist / (whiteDist + blackDist) * 255
        Cells(x, y).Interior.Color = RGB(cValue, cValue, cValue)
      Next x
    Next y
End Sub
```

Try variation of the formula get a different gradient pad. For example the following variation creates a gradient pad with one edge of the pad being black and the other edge white. Change the padWidth and padHeight to suit your environment. Try other variations to create color pads. You could possibly use other tools and follow a similar concept.

```
Sub gradient_pad()
  padWidth = 50
  padHeight = 50

   For y = 1 To padWidth
      For x = 1 To padHeight
        cValue = y * (255 / padWidth)
        Cells(x, y).Interior.Color = RGB(cValue, cValue, cValue)
      Next x
    Next y

End Sub
```

The MINDSTORMS NXT kit comes with a pad that has gradient printed on the edge, for black and white as well as color use. This is a good place to start and experiment with.

There are other possibilities. People have suggested using bar codes on the floor: when the robot finds one, it aligns and reads it, decoding its position from the value. Others tried complex grids made out of stripes of different colors. Unfortunately, there isn't a simple solution valid for all cases, and you will very likely be forced to use some dead reckoning after a landmark to improve the search.

Following the Beam

In the real world, most positioning systems rely on beacons of some kind, typically radio beacons. By using at least three beacons, you can determine your position on a two-dimensional plane, and with four or more beacons you can compute your position in a three-dimensional space. Generally speaking, there are two kinds of information a beacon can supply to a vehicle: its distance and its heading (direction of travel). Distances are computed using the amount of time that a radio pulse takes to go from the source to the receiver: the longer the delay, the larger the distance. This is the technique used in the Loran and the GPS systems. Figure 13.3 shows why two stations are not enough to determine position: because there are always two locations A and B that have the same distance from the two sources.

Figure 13.3 Ambiguous Positioning with Two Stations

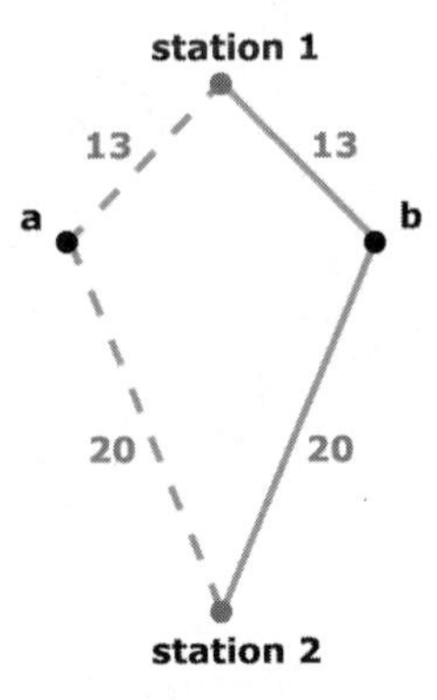

If you add a third station, the system can solve the ambiguity, provided that this third station does not lie along the line that connects the previous two stations (see Figure 13.4).

Figure 13.4 Positioning with Three Stations Using Distances

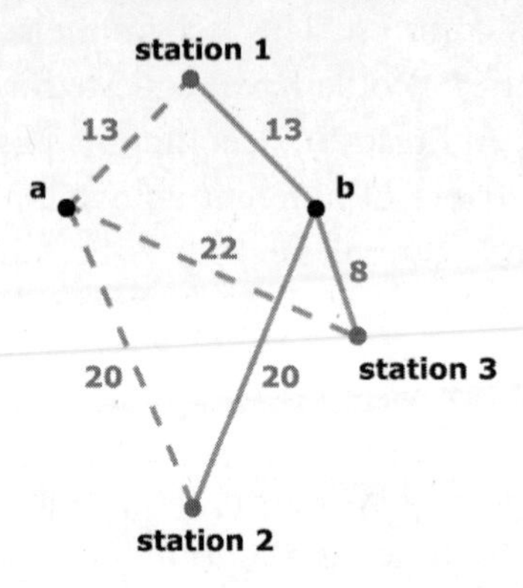

The stations of the VHF Omnidirectional Range system (VOR) cannot tell you the distance from the source of the beacon, but they do tell you their heading; that is, the direction of the route you should go to reach each station. Provided that you also know you're heading north, two VOR stations are enough to locate your vehicle in most cases. Three of them are required to cover the case where the vehicle is positioned along the line that connects the stations, and as for the Loran and GPS systems, it's essential that the third station itself does not lay along that line (see Figure 13.5).

Figure 13.5 Positioning with Three Stations Using Headings

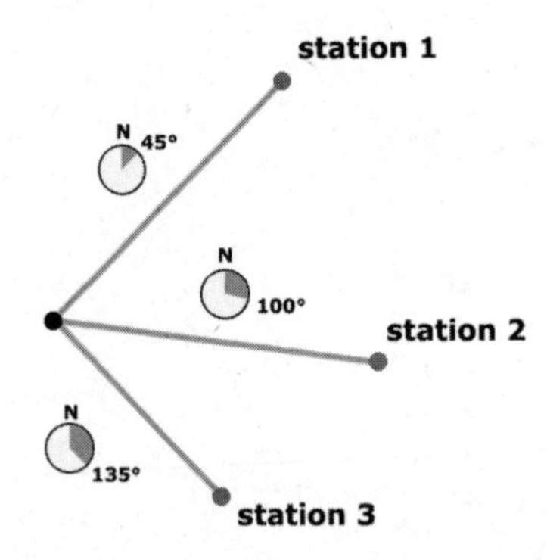

Using three stations, you can do without a compass; that is, you don't need to know you're heading north. The method requires that you know only the angles between the stations as you see them from your position (see Figure 13.6).

Figure 13.6 Positioning with Three Stations Using Angles

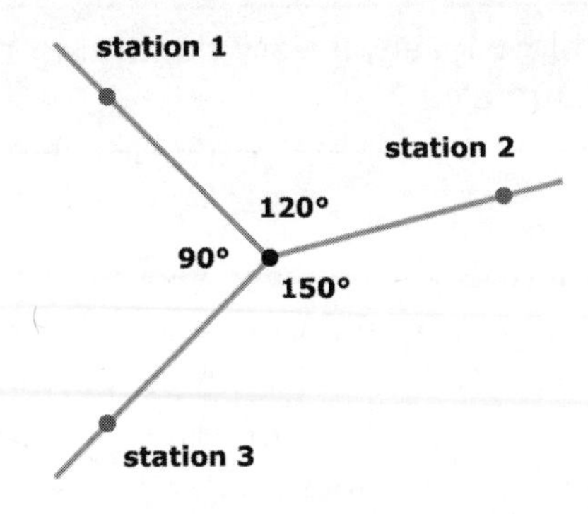

To understand how the method works, you can perform a simple experiment: take a sheet of paper and mark three points on it that correspond to the three stations. Now take a sheet of transparent material and put it over the previous sheet. Spot a point anywhere on it that represents your vehicle, and draw three lines from it to the stations, extending the lines over the stations themselves. Mark the lines with the name of the corresponding stations. Now, move the transparent sheet and try to intersect the lines again with the stations. An unlimited number of positions connect two of the three stations, but there's only one location that connects all three of them.

The problem in this approach is, currently, there is no known device in the NXT world that would emit beacons, so first you have to look for an appropriate device. Unless you're an electrical engineer and are able to design and build your own custom radio system, you better stick with something simple and easy to find. The source need not be necessarily based on radio waves—light is effective as well, and we already have such a detector (the light sensor) ready to interface to the NXT.

When you use light sources as small lighthouses, in theory you can make your robot find its way. But there are a few difficulties to overcome first:

- The light sensor isn't directional—you must shield it somehow to narrow its angle.
- Ambient light introduces interference, so you must operate in an almost lightless room.
- For the robot to be able to tell the difference between the beacons, you must customize each one; for example, making them blink at different rates (as real lighthouses do).

Laser light is probably a better choice. It travels with minimum diffusion, so when it hits the light sensor, it is read at almost 100 percent. Laser sources are now very common and very cheap. You can find small battery-powered pen laser pointers for just a few dollars.

WARNING

Laser light, even at low levels, is very damaging to eyes—never direct it toward people or animals.

If you have chosen laser as a source of light, you don't need to worry about ambient light interference. But how exactly would you use laser? Maybe by making some rotating laser lighthouses? Too complex. Let's see what happens if we revert the problem and put the laser source *on the robot*. Now you need just one laser, and can rotate it to hit the different stations. So, the next hurdle is to figure out how you know when you have hit one of those stations. If you place an NXT with a light sensor in every station, you can make it send back a message when it gets hit by the laser beam, and using different messages for every station, make your robot capable of distinguishing one from another.

The NXT light sensor is quite a small target to hit with a laser beam, and as a result, it was almost impossible to hit accurately. To stick with the concept but make things easier, we discovered you could build a sort of diffuser in front of it to have a wider detection area.

With this solution, you will still need several NXTs, at least three for the stations and one for your robot. Isn't there a cheaper option? A different approach involves employing the simple plastic reflectors or reflective tapes used on cars, bikes, and as cat's-eyes on the side of the road. You can find reflective tapes in your local art and craft stores. They have the property of reflecting any incoming light precisely back in the direction from which it came. Using those as passive "stations," when your robot hits them with its laser beam they reflect it back to the robot, where you have placed a light sensor to detect it.

This really seems the perfect solution, but it actually still has its weak spots. First, you have lost the ability to distinguish one station from the other. You also have to rely on dead reckoning to estimate the heading of each station. We explained that dead reckoning is not very accurate and tends to accumulate errors, but it can indeed provide you with a good *approximation* of the expected heading of each station, enough to allow you to distinguish between them. After having received the actual readings, you will adjust the estimated heading to the measured one. The second flaw to the solution is that most of the returning beam tends to go straight back to the laser beam. You must be able to very closely align the light sensor to the laser source to intercept the return beam, and even with that precaution, detecting the returning beam is not very easy.

To add to these difficulties, there is some math involved in deducing the position of the robot from the beacons, and it's the kind of math whose very name sends shivers down most students' spines: trigonometry! This leads to another problem: the standard NXT firmware has no native support for trig functions, but in theory you could implement NXT-G blocks for such functions. If you want to proceed with using beacons, you really have to switch to RobotC or NBC, which both provide much more computational power.

If you're not in the mood to face the complexity of trigonometry and alternative firmware, you can experiment with simpler projects that still involve laser pointers and reflectors. For example, you can make a robot capable of "going home." Place the reflector at the home base of the robot; that is, the location where you want it to return. Program the robot to turn in place with the laser active, until the light beam intercepts the reflector and the robot receives the light back, then go straight in that direction, checking the laser target from time to time to adjust the heading.

Map Matching Using Ultrasonic Sensor

Have you seen bats fluttering in evening twilight? Bats can get around just fine, doing their regular business in feeble light. Because, instead of light, they use sound to get around in the dark. In other words to "see" with their ears, they make sound in pulses. This sound reflects when it hits an object, and an echo bounces back to the bat. The time lapsed from making the sound and its echo tells the bat how far the object is. The ultrasonic sensor works on a similar principle.

You can use this technique to first detect your surroundings and recognize landmarks. Then you can use this information with a map to localize and establish your position.

In this method information acquired from the sensors is compared to a map or model of the environment. If features from the sensor's readings and the model map match, then the robot's absolute location can be determined. In the previous example, these landmarks can be incorporated into your program to localize the robot. When the robot approaches a door, the ultrasonic sensor returns longer obstacle readings. These readings continue until the robot moves across the door. As shown in Figure 13.7, this information can help deduce that the robot has crossed the first door (or second door) and can identify its precise location.

Figure 13.7 The Ultrasonic Sensor Can Detect Surroundings and Recognize Landmarks

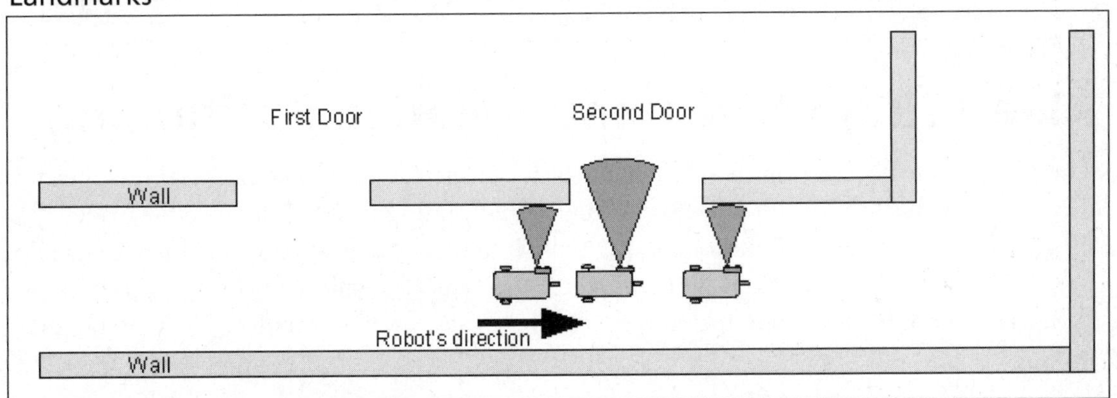

Combining Compass Sensor to Increase Precision

The inputs used in the preceding method included ultrasonic sensor readings, and a known map. Additional inputs can help improve accuracy. You could use distance from the wall, and orientation of the robot as additional inputs. The ultrasonic sensor would provide you the distance from the wall. These readings can be utilized with appropriate corrective action of wheel motors to maintain constant distance from the wall. However as shown in Figure 13.8, when unexpected drifts occur or objects appear in front of the ultrasonic sensor, the sensor will provide those readings too. You can either compare such readings with your previously known map information or ignore them. Better yet, you can use the additional input of a compass sensor to determine and maintain heading.

A compass sensor would give you the magnetic heading your compass is pointing to. As your robot starts to maintain constant distance from the wall, note the heading from the compass sensor, and during subsequent movement maintain that heading by applying corrective action of wheel motors.

Figure 13.8 A Compass Sensor on the Robot Can Help Avoid Losing Orientation

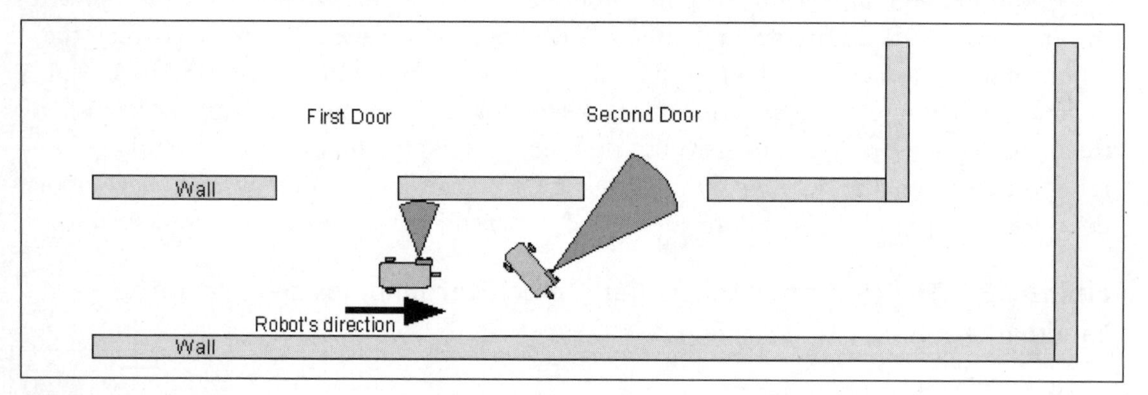

Measuring Movement: Relative Positioning

The technique of measuring the movement of the vehicle, called *odometry*, requires an *encoder* that translates the turn of the wheels into the corresponding traveled distance. To make things easier, the new NXT motors have a built-in encoder that is great for this purpose. To provide two degrees of freedom while moving, you typically will need two such motors, and as described in Chapter 3, you can also synchronize these two motors to get a straight-line motion.

The equations for computing the position from the decoded movements depends on the architecture of the robot. We will explain it here using the example of the differential drive, once again referring you to Appendix A for further resources on the math used.

Suppose that your robot has two wheels, each connected to a motor through gearing. Given D as the diameter of the wheel, R as the resolution of the built-in encoder in motors (the number of counts per turn), and G the gear ratio between the motor and the wheel, you can obtain the conversion factor F that translates each unit from the encoder into the corresponding traveled distance:

$$F = (D \times \pi) / (G \times R)$$

The numerator of the ratio, $D \times \pi$, expresses the circumference of the wheel, which corresponds to the distance that the wheel covers at each turn. The denominator of the ratio, $G \times R$, defines the increment in the count of the encoder (number of *ticks*) that corresponds to a turn of the wheel. F results in the unit distance covered for every tick.

Your robot uses the largest spoked wheels, which are 81.6mm in diameter. The built-in encoder of the motors has a resolution of 360 ticks per turn, and it is connected to the wheel with a 1:5 ratio (five turns of the sensor for one turn of the wheel). The resulting factor is:

$$F = 81.6 \text{ mm} \times 3.1416 / (5 \times 360 \text{ ticks}) \approx 0.14242 \text{ mm/tick}$$

This means that every time the sensor counts one unit, the wheel has traveled 0.14242mm. In any given interval of time, the distance T_L covered by the left wheel will correspond to the increment in the encoder count I_L multiplied by the factor F:

$$T_L = I_L \times F$$

and similarly, for the right wheel:

$$T_R = I_R \times F$$

The centerpoint of the robot, the one that's in the middle of the ideal line that connects the drive wheels, has covered the distance T_C:

$$T_C = (T_R + T_L) / 2$$

To compute the change of orientation ΔO you need to know another parameter of your robot, the distance between the wheels B, or to be more precise, between the two points of the wheels that touch the ground:

$$\Delta O = (T_R - T_L) / B$$

This formula returns ΔO in radians. You can convert radians to degrees using the relationship:

$$\Delta O_{Degrees} = \Delta O_{Radians} \times 180 / \pi$$

You can now calculate the new relative heading of the robot, the new orientation O at time i based on previous orientation at time $i - 1$ and change of orientation ΔO. O is the direction in which your robot is pointed, and results in the same unit (radians or degrees) you choose for ΔO.

$$O_i = O_{i-1} + \Delta O$$

Similarly, the new Cartesian coordinates of the centerpoint come from the previous ones incremented by the traveled distance:

$$x_i = x_{i-1} + T_C \times \cos O_i$$

$$y_i = y_{i-1} + T_C \times \sin O_i$$

The two trigonometric functions convert the vectored representation of the traveled distance into its Cartesian components.

Oh, ' this villainous trigonometry again! Unfortunately, you can't get rid of it when working with positioning. Thankfully, there are some special cases where you can avoid trig functions; for example, when you're able to make your robot turn in place precisely 90 degrees, and truly go straight when you expect it to. In this situation, either x or y remains constant, as well as the other increments (or decrements) of the traveled distance T_C.

Measuring Movement: Acceleration Sensor

The traveled distance can either be measured by the rotations of the motor that turns the wheels or the acceleration and velocity of the robot and the time it travels (the latter is called *Inertial Navigation*). You could use an acceleration sensor to measure acceleration of your robot.

The acceleration sensor measurements are influenced by the gravity of earth. So, if your robot is stationary on an inclined plane, you would still read a nonzero acceleration due to the gravitational influence. To correct that, you can use a gyroscopic sensor to know inclination of your robot.

If you are using the method of acceleration, in theory, the acceleration value A integrated over time T_0 to T_X will give you velocity V_X of your robot:

$$V_x = \int_{T_0}^{T_x} A$$

The velocity value V integrated over time T_0 to T_X will give you the distance traveled.

$$D_x = \int_{T_0}^{T_x} V$$

There are third-party gyroscopic and acceleration sensors with varying sensitivity levels from HiTechnic and Mindsensors. These sensors can be used to measure acceleration of your robot. This seems easy if you know the formulae, but in practice a small error can result in unbounded growth in integrated measurements. To minimize errors, you need to take readings at rapid and constant intervals. With standard NXT firmware, it is almost impossible to achieve the required sampling rate.

Summary

We promised in the introduction that this was a difficult topic, and it was. Nevertheless, making a robot that has even a rough estimate of its position is a rewarding experience.

There are two categories of methods for estimating position, one based on absolute positioning and the other on relative positioning. The first category usually requires landmarks or beacons as external references. We described some possible approaches based on laser beams and introduced you to the difficulties that they entail. With the powerful I^2C interface that NXT now offers, in future you can anticipate some advanced high-precision industrial positioning systems integrated with NXT. The Map Matching technique can make good use of Ultrasonic Sensor to read and verify map and determine position. It is possible to interface NXT robots with GPS systems, however outdoor GPS has granularity of 0.5 meters, which usually is inadequate for our robots.

Relative positioning is more readily applicable to NXT robots. We explained the math required to implement deduced reckoning in a differential drive robot, and suggested some alternative architectures that help in simplifying the involved calculations. We also explained how acceleration sensors could be utilized for measuring movements, and problems you may encounter. The real life navigation systems—like those used in cars, planes, and ships—usually rely on a combination of methods taken from both categories: dead reckoning isn't very computation intensive and provides good precision in a short range of distances, whereas absolute references are used to zero the errors it accumulates.

Chapter 14

Classic Projects

Solutions in this chapter:

- **Exploring Your Room**
- **Following a Line**

Introduction

From this chapter on, we will explore several example projects that could be the inspiration for many others of your own creation. As we already explained, the spirit of the book is not to provide you with step-by-step instructions, but rather to give you a foundation of information and let your imagination and creativity do the rest. For this reason, you will find some images of each model, some text that describes their distinguishing characteristics, and tricks that could be useful for other projects. Of course, we don't expect each detail to be visible in the pictures. It isn't important that your models look exactly like ours!

Another point we want to bring to your attention is that there is no reason to read Chapters 14-25 in order. Feel free to jump to the project that attracts you most, because they aren't ordered according to their level of difficulty.

In this chapter, we'll show you some projects that could be considered "classic," because almost everybody with an NXT kit tries them sooner or later. Though you might not find them exciting, working with them is a good way to build up some solid experience and learn tricks that will prove useful in more complex projects. If this is among your first forays into robotics, we strongly suggest you dedicate some time to them.

All the robots appearing in this chapter have been built primarily from the NXT kit. When describing some of the robots, a few extra parts may be necessary that do not exist in the NXT kit. However, most of these are pretty common and you can find them if you have other TECHNIC sets.

Exploring Your Room

Well, actually *exploring* your room is too strong a term for what we are proposing here; it's more like *surviving* your room—your robot and your furniture could take some hits! The task here is to build a robot with the basic capability to move around, detect obstacles, and change its route accordingly.

For simplicity of design, and for the robot's capability to turn in place, we suggest you make this robot from differential drive architecture, such as the one shown in Figure 14.1.

Figure 14.1 Start with a Simple Differential Drive

We deliberately chose a gear ratio that makes the robot rather slow: 1:9, obtained from two 1:3 stages (Figure 14.2). This ensures that if you make some error in the code and the robot fails to properly detect the obstacle, it won't collide with it at too high a speed. Never expect everything to go well on the first try—because it won't!

Figure 14.2 Detail of the Two-Stage Geartrain

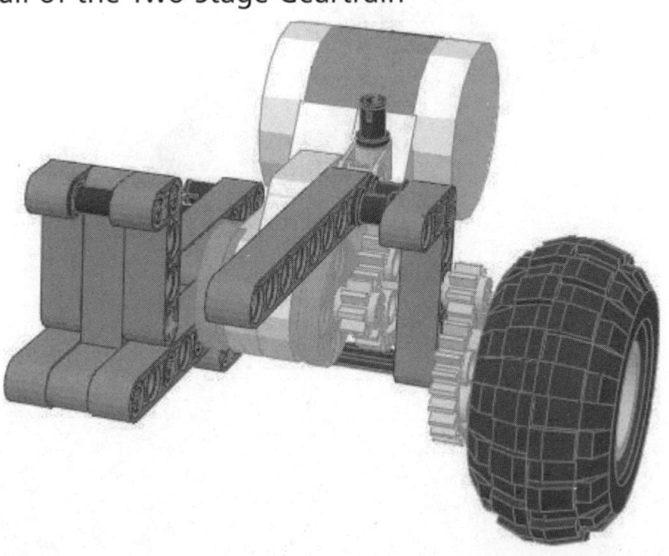

When you feel satisfied with your software and your robot runs safely around your room, you can always try a faster ratio. Substituting the second 1:3 gearing with two 16t gears will give you an overall 1:3 ratio, making your robot about three times faster.

The robot is rather large, keeping the main wheels far from the body of the robot. There's a reason for this too: In a differential drive, the distance between the drive wheels affects the turning speed of the robot, because the wheels have to cover a longer distance during turns. The farther the wheels are from the midpoint, the slower the turns. You could control turns through timing, which would make slow turns a desirable property that would provide finer movement control. However, with the NXT's built-in rotation sensors, navigation via rotations provides more consistent control of the turns.

The caster wheel is the same kind we showed in Chapter 9. Now add the NXT and a couple of bumpers that are normally closed (see Figure 14.3), and you're ready to go—well, ready to program the robot, anyway. Check out Figure 14.4 to see what the completed robot looks like.

Figure 14.3 Detail of the Bumper

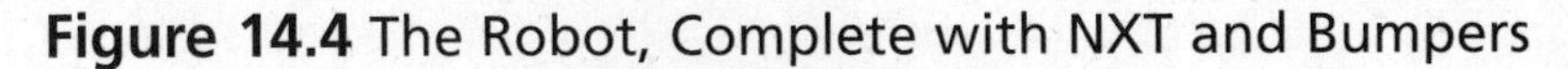

Figure 14.4 The Robot, Complete with NXT and Bumpers

The program itself is very simple: Go straight until one of the touch sensors opens. When that happens, reverse for a few fractions of a second (or rotations), then turn in place, right or left depending on which bumper found the obstacle. Finally, resume straight motion.

Experiment with different timing/rotations for turns, until you are happy with the result. You might also use some random values for turns to make the behavior of your robot a bit less predictable and thus more interesting. If you feel at ease with the programming, you can add more intelligence to your creature—for example, to make it capable of realizing when it's stuck in a cul-de-sac. You can achieve this by monitoring the number of collisions in a given time, or the average time elapsed between the last *n* collisions, and then adopting a more radical behavior (such as turning 180 degrees).

Detecting Edges

If your room has a flight of stairs going down, simply equipping the robot with an ultrasonic sensor facing the floor to sense the edge can avoid a bad fall. However, if you want to use your ultrasonic sensor for some other purpose, you could use a touch sensor instead, connecting the touch sensor to a feeler flush with the ground. When the feeler in front of the robot drops, you have detected an edge.

Unless you have a third touch sensor, you are forced to use the light sensor. It's time to look back at some of the tricks explained in Chapter 4 and see whether you find something useful. A light sensor can actually emulate a touch sensor: You have to place movable parts of different colors in front of it so that when contact is made, the parts move, and the color of the brick in front of the light sensor changes.

However, as your kit contains the ultrasonic sensor, use it, because this sensor is ideal for this purpose. We kept the ultrasonic sensor behind the bumpers so that in most cases it doesn't interfere in obstacle detection (Figure 14.5).

Figure 14.5 The Ultrasonic Sensor Installed on the Robot

Unfortunately, this system doesn't cover all possible scenarios, because your robot could approach the edge at an angle that allows a wheel over the edge before detection occurs. You can improve upon the design and avoid this by providing the robot with two left- and right-edge sensors, but you'll probably have to give up the double bumper and go with a single sensor bumper.

Using a different approach, you could write the software to make the robot very cautious, turning slightly left and right from time to time to see whether there's a dangerous precipice around.

Variations on Obstacle Detection

By using the NXT motor's built-in rotation sensor, you can experiment with indirect obstacle detection. Program the robot to monitor the rotation count while in motion. If both motors are on forward, but the count doesn't increase, the robot knows an obstacle has blocked it. As a positive side effect, the rotation count allows you to use the same platform for experimenting with navigation, applying some of the concepts about dead reckoning explained in Chapter 13.

You can also implement indirect obstacle detection using a "drag sensor." The idea requires that your robot keep a mobile part in touch with the ground, and that the friction this part exerts against the floor surface when the robot moves activates a touch sensor. For example, you can use the friction of a rubber tire to oppose the force of a rubber band that keeps a touch sensor closed. When the robot moves, the friction of the tire on the floor overcomes the force of the rubber band and opens the touch sensor; as soon as the robot stops—or has been blocked by an obstacle—the friction disappears and the touch sensor closes.

Following a Line

The line-following theme is often mentioned in this book, as we think it is a very useful indicator of how different techniques can improve the behavior of a robot. The time has come to give it an official place, and face the topic in its entirety.

Let's review what we have already said about line following:

- You must actually follow the edge between the tape and the floor, reading an average value between dark and bright, so that when you read too dark or too bright you know which direction to turn to find the route back.
- If you want to keep your software and robot as general as possible, you should use some kind of self-calibration process before the actual following begins. Calibration consists of taking readings of light and dark areas on the pad before actually starting line following. This lets your robot adjust its parameters to the actual lighting conditions at the time it runs; these conditions are almost certainly different from those for which your robot was designed.
- Some platforms can benefit from the introduction of a small quantity of hysteresis to reduce the number of corrections and get a higher efficiency. We explained in Chapter 12 that hysteresis "widens" the gray area between light and dark, which keeps the robot from spending too much time on course correction instead of moving forward!

To turn theory into practice and experiment with line following, you can use the same differential drive as in the previous project in this chapter. Remove the bumpers and mount a light sensor facing down, as shown in Figure 14.6.

Figure 14.6 The Differential Drive Equipped for Line Following

You can easily swap the ultrasonic sensor for a light sensor, which is attached to the structure with a long pin. Observing the result, the light sensor is now fairly far from the floor. The distance of the sensor from the pad is very important. Therefore, you should lower the attach point for the light sensor so that the sensor is much closer. Our experience teaches us that for the best results, this distance should fall in the range of 5 mm to 10 mm.

Now that you've finished, you're ready to program and test your robot for line following. If you want to, you can use the RobotC sample code from Chapter 7, or you can use some of the references in Appendix A. It shouldn't be difficult to adapt it to the language of your choice.

Suppose you have succeeded in the task of programming for line following, and you feel quite happy with the result. You have good reason to, but you should also wonder, as always, whether you can do anything else to make it better. What could "better" mean in this case? Probably "faster," along with little to no errors. This is what standard line-following competitions are about: going from one end of a line to the other in the shortest time, and at the highest speed.

Observe carefully your differential drive in action. When it turns in place to adjust the course, it makes no progress along the line. This approach gives your robot the capability to follow a winding line closely with very tight turns, but it isn't very efficient. A first step could involve changing the course adjustment algorithm to a better one—for example, making the robot turn with one wheel stopped and the other in motion, instead of turning with one going forward and the other reversed. Try this technique, and you'll see that your robot progresses much faster.

We haven't yet mentioned the most obvious improvement—increasing the speed of your robot! Try different gearings until you find the fastest setup that still allows your software to keep the course.

Nothing more you can think of? Imagine yourself standing over a line, straddling it with one leg on the left side and the other on the right. Now, gazing at the line, you advance one foot or the other, trying to keep your eyes centered above the line. Do you feel a bit stupid? We would. This isn't actually what you would do to follow a line in a real situation. You're not a differential drive. You would rather walk as usual, putting one foot in front of the other, simply changing your direction without changing your speed.

That's the key: You need something that changes its direction without affecting the speed. Looking back at Chapter 9 with a different eye, you will discover that the steering drive and the synchro drive actually share this property of allowing changes in direction without changing the speed of their driven wheels. The synchro is probably too complex in gearing to be really efficient, so use the traditional steering drive. If you exclude turns with a very short radius from your path, proceed with the steering drive architecture that is simpler to implement.

It's not imperative that a steering drive have four wheels, so ours will actually be a tricycle, because this makes the platform easier to build. Figure 14.7 shows our version of a steering line follower.

Figure 14.7 A Steering Line Follower

Let's dissect it to understand some of the choices we made, starting with the drive gearing; you might think this is a bit more complex than necessary (Figure 14.8).

Figure 14.8 Bottom View: The Gearing

Although it is possible to drive the differential from the motor without so many additional gears, we were looking for an easy way to change the gear ratio without having to take the robot apart. While you build the robot, you don't know yet at what maximum speed it will be able to keep following the line. Our solution brings a pair of gears on the outside of the model at a distance that allows at least three combinations, so it will be easy to experiment with different speeds (Figure 14.9, versions A, B, and C):

A. 24t 12t (2:1)

B. 20t 16t (5:4)

C. 16t 20t (1:2)

Figure 14.9 These Gears Are Easy to Replace with Different Combinations

Now turn your attention to the front assembly. It's very simple, and there are just a few things worth mentioning. There is a TECHNIC pin (3L double) with an axle hole that makes the fork drivable by the steering assembly, and several liftarms that hold the light sensor at some distance in front of the wheel (Figure 14.10).

Figure 14.10 The Front Wheel Fork Assembly

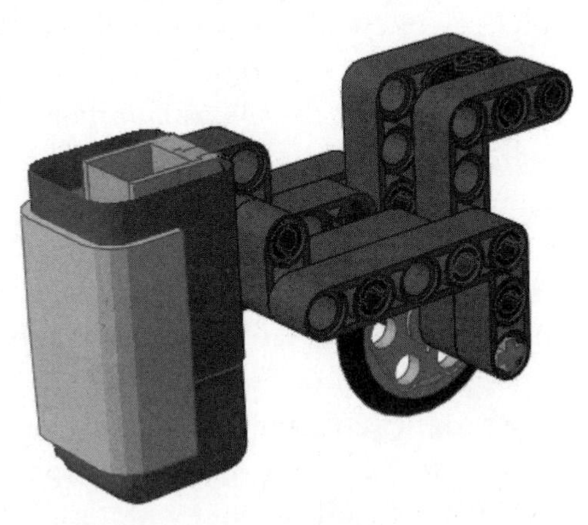

The front fork is driven by a steering assembly based on a worm gear, a 24t and two pulleys. As you learned in Chapter 2, the pulley-belt systems prevent any stall situation if the robot doesn't control the steering wheel properly when the software is not yet fully tested and debugged (Figure 14.11).

Figure 14.11 The Steering Assembly

Now it's time to program and test the robot. If you wrote the line-following program for the differential drive, it will work for this robot with a few minor modifications. During the run, the drive motor will always stay on, while the robot will adjust its course using only the steering motor, in either forward or reverse direction.

As for the calibration procedure, we suggest that with the robot still and placed on the borderline, you make it rotate the front wheel slightly left and right to read the minimum and maximum light values, then compute the average and position the light sensor onto that. This architecture needs very small hysteresis, or none at all.

Start at slow speed, mounting a 12t gear onto the motor shaft, and when your robot is able to follow the line properly, try to change the ratio and increase the speed.

The minimum radius that the robot can follow depends on a combination of the forward speed of the robot, how quickly the turn drive motor can move the steering wheel, and how far in front of the robot the light sensor is. We encourage you to experiment with these variables and see how the robot behaves when following lines with turns of different radii.

Further Optimization of Line Following

You should be happy with this line follower. It runs very smoothly compared to the differential drive configuration. Nevertheless, you might wonder whether you can do anything to make it run at an even higher speed.

Changing the gear ratio is not enough. There is a large margin for an increase in speed—it's not difficult to set up a LEGO NXT robot to run at about 2 m/s (6.5 ft/s). The problem is keeping the robot *on the line*.

When you are targeting excellence, the finer details are vital. There is a big difference between building a robot that "works" and a robot that is *optimized* for a given task. You already switched from a differential drive to a steering drive, and this change of architecture has proven to be a significant improvement, but now you have to dig into the particulars if you want to gain some additional speed. In working with line following, one of the key factors is to make the steering fast and accurate. To achieve this, you can reduce the *backlash* between the 24t and the worm gear, for example, keeping the steering gently pulled back from one side with a rubber band. Use a long and soft rubber band, because you don't want to introduce too much friction; you want to simply keep the teeth of the 24t in contact with the worm gear, and always from the same side (Figure 14.12).

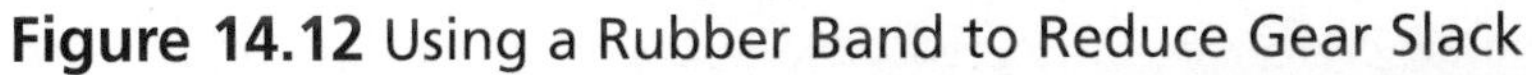

Figure 14.12 Using a Rubber Band to Reduce Gear Slack

At this point, speed limitations are probably due to the rotation speed of the steering assembly. This pulley, belt, and worm gear system is safe and accurate, but a bit slow. We leave you the task of developing a faster steering assembly. It's not very difficult: Try to connect different pulley sizes to the steering mechanism...

Now you will meet the last barrier: the reaction time of the software. If you used the standard firmware, either with NXT-G, RobotC, or other tools, the time it needs to interpret the instructions becomes relevant to the performance of your robot. Another very critical factor is the sampling frequency of the sensors, which is much higher in most replacement firmware than in the original LEGO one.

Summary

You may not have any interest in the topics covered in this chapter—you may think "What a bore. I'd like my robots to do more than just follow a black line or run around my room bouncing against obstacles..."

You're right; there *are* more interesting activities you can program your robots for, but these classic tasks help lay the foundation for more complex projects in the future. They reveal that even apparently trivial projects conceal unexpected difficulties, and we're sure the time you spend experimenting with line following and simple navigation will indeed pay off in the end. In this chapter, you also had the opportunity to review many of the tricks learned in this book and see them at work. We applied the concepts of Chapter 6 to make solid structures and build two of the most important types of mobile configurations described in Chapter 9: the differential drive and the steering drive. Naturally, to build these configurations we used the principles of the first two chapters concerning the geometric relationships of LEGO parts and the proper use of gears. We recalled the techniques of Chapter 4 about making good bumpers, about using light sensors for line following, and even about emulating a touch sensor with a light sensor, and we introduced an important application of the ultrasonic sensor. From Chapter 7, we took the idea that good code should be as general as possible, and for this reason we suggested using a self-calibration routine to make your line follower suitable under any light conditions. Even the math of Chapter 12 had its applications here: We recalled the idea that hysteresis can improve the efficiency of your robot in tasks such as line following.

Like a thread sewing together most of the topics that we covered in this book so far, the simple robots of this chapter demonstrated what we've stated on more than one occasion: Robotics involves many disciplines, and a good design cannot neglect any of them!

Another theme that pervades the chapter concerns the care you should put into the particulars: Building a robot that works roughly as expected, compared to looking for optimal performance, are two very different approaches. Obviously, the concept behind the *design* of a robot is an important element in regard to its functioning. But even after finding a satisfactory architecture, there is still much work to be done to optimize the subsystems of the robot. The details, as often happens, make the difference. You will see this prove true in Chapter 20 and 21, where we will explore the world of robotic *contests*. Contests are a great incentive to pursue optimization, the kind of motivation that makes you spend all night rebuilding a working robot from scratch in search of that little improvement that will make all the difference!

Chapter 15

Building Robots That Walk

Solutions in this chapter:

- **The Theory behind Walking**
- **Building Legs**
- **Building a Four-Legged Robot**
- **Building a Six-Legged Steering Robot**
- **Designing Bipeds**

Introduction

So far in this book, we have discussed in depth many mobility configurations, all of them based upon one of the most important inventions of mankind: the wheel. In this chapter, we will try to emulate what nature invented long before the wheel to provide mankind with a mode of transportation—legs!

Legged robots are rather impractical for all but a few special applications. However, there is much to learn in designing and building a walking robot, which is both challenging and fascinating.

This chapter owes a lot to the great designers who published their creations on the Internet, and patiently explained their choices through text and pictures, including Kevin Clague, Yoshihito Isogawa, Joe Nagata, Miguel Agullo, and many others.

The Theory behind Walking

How can one define walking? It's the process of lifting a leg from the ground while the other legs (one or more) support the body. When the leg has been lifted, it is advanced and lowered back to the ground. From there, the process continues with another leg, and so on.

The crucial point is this: What prevents a creature from falling down when it lifts its leg? To discover this, we need to introduce some basic concepts from a branch of physics called *statics*, which explains the laws of balance.

The weight of an object is the resulting effect of the force of gravity against the mass of the object. To describe a force, you need to determine three variables: its magnitude, direction, and point of application. For example, if you want to move a piece of furniture in your room, the magnitude is the amount of force you must apply to make it move, the direction is the bearing of the course on which you're pushing it, and the point of application is where you place your hands to apply the force. Returning to *gravity*, its magnitude is proportional to the mass of the object and its direction points vertically downward, but where is its application point? To answer this question, you should consider an object as being the sum of a very large number of very small particles, each one having its own mass. The gravity exerts a force upon every particle, and thus all of them can be considered a point of application. However, physics teaches that a combination of forces can be interpreted as a single force—called the *resultant*—that has its own magnitude, direction, and point of application. The resultant force of gravity has a magnitude that corresponds to the weight of the objects, a direction pointing downward, and a point of application called the *center of gravity* (COG) of the object (see Figure 15.1).

Figure 15.1 An Object's Center of Gravity

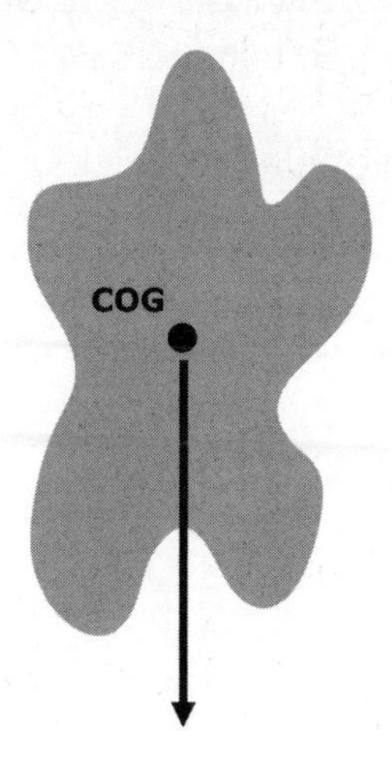

The force of gravity acts on an object and tries to move its COG as close to the ground as possible; this is why objects fall and shift until they reach a stable position. But what makes a position stable? Statics teaches that a body becomes stable when the vertical passing for its center of gravity falls inside its supporting base. The supporting base is the surface whose perimeter results from connecting the supporting points with straight lines. A supporting point is any point on the object that is in contact with the ground or with any other stable object (such as the floor of your room, or your desk). For example, a book placed on a table has the whole surface of its cover in touch with the table, and that defines its supporting base. A table has four legs, each one having a small surface in touch with the floor: Its supporting base is the area delimited by the legs, which includes points untouched by the table (see Figure 15.2).

Figure 15.2 The Supporting Base of a Table

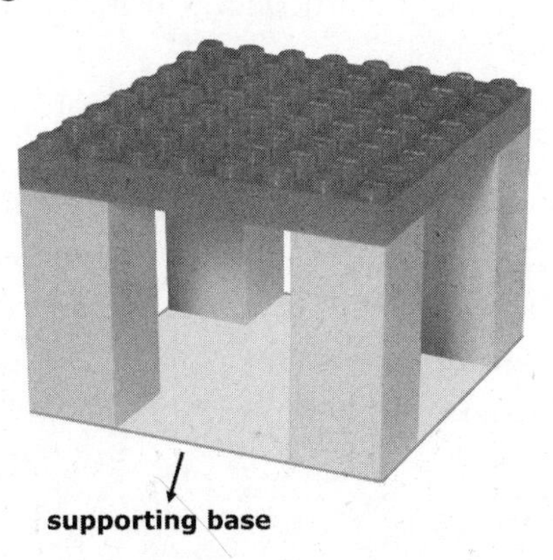

Every child learns this rule by experience when building towers of stacked blocks: While the COG remains within the supporting base, the tower is stable; as soon as it falls outside the base, the tower itself falls down (see Figure 15.3).

Figure 15.3 Stable and Unstable Piles of Bricks

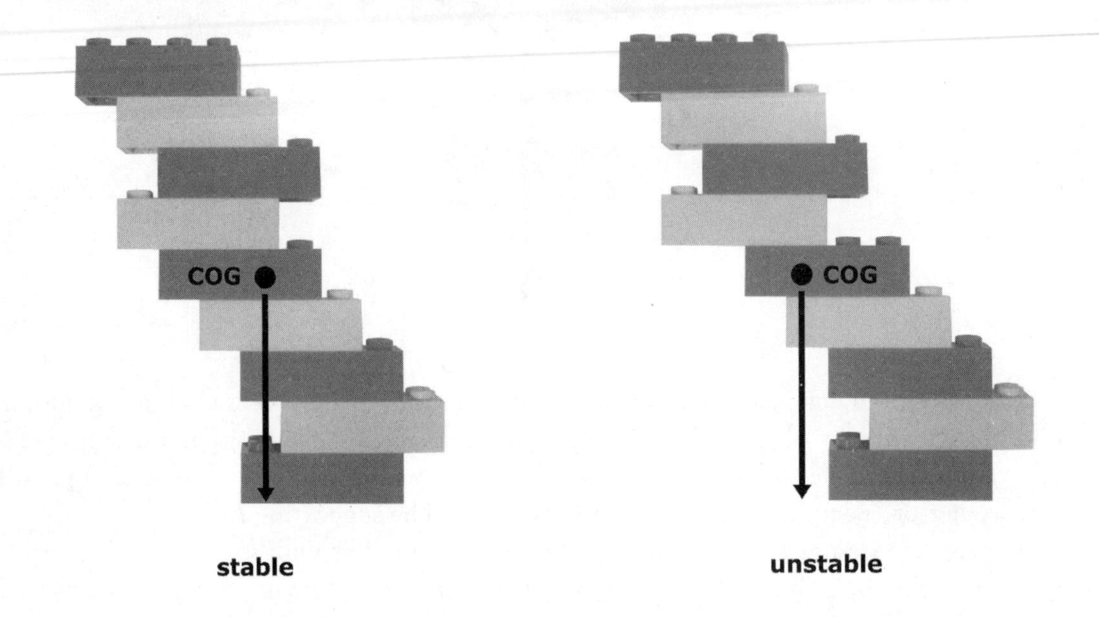

Okay, now that we know the rule, how do we find the COG of an object? For objects that are symmetrical in shape and density, the COG coincides with their geometrical center, but in more complex objects the COG is not very easy to find, and it is not guaranteed to be *inside* the object. A table is again a good example: The COG of a typical table lies somewhere below its top, as demonstrated by the fact that it has more than just one stable position (see Figure 15.4).

Fortunately, we do not need to find the actual position of the COG of our robot. We are actually interested in the position of the vertical line that passes through the COG, in order to see whether it falls inside the supporting base. This is easier to find: If our robot is mainly symmetrical, this line will pass very close to its geometrical center. Thus, what we actually need to do is to look at the robot from the top and determine whether the COG falls over the supporting base delimited by the legs.

Figure 15.4 The COG of an Object May Lie Outside It

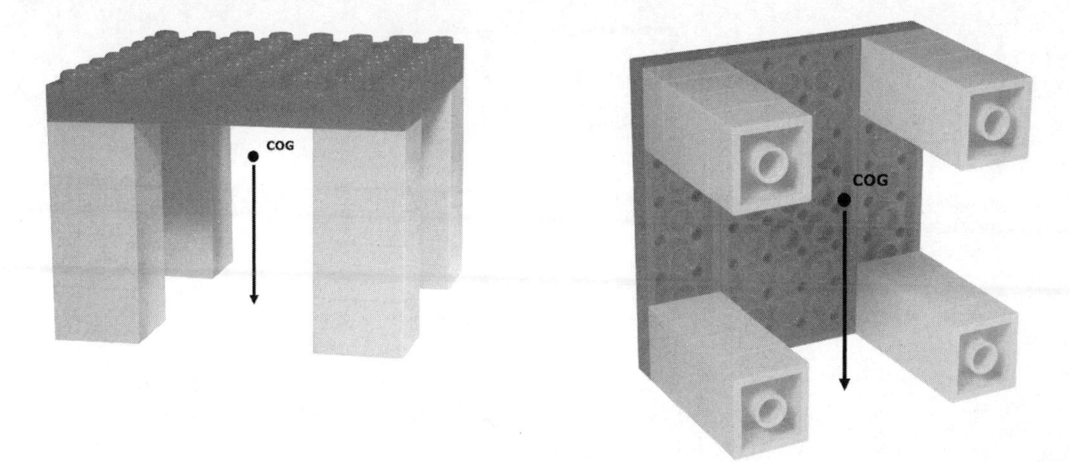

For example, in Figure 15.5, we see a scheme that represents a robot with four large legs (top view). One of the legs is lifted, and we see that the COG still falls inside the surface delimited by the other three legs. And thus, the robot remains stable.

Figure 15.5 A Four-Legged Robot with One Leg Lifted

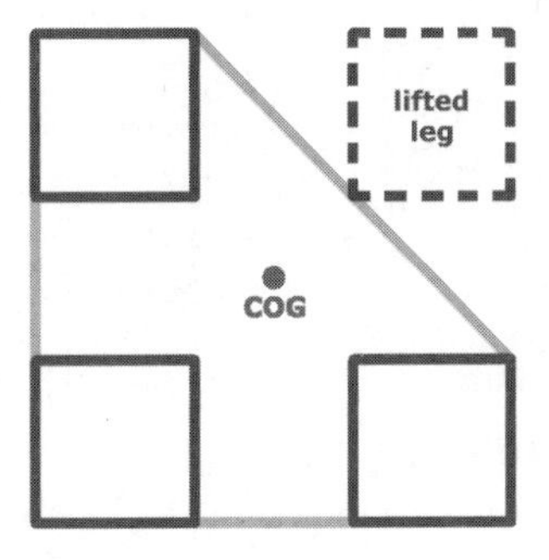

The same robot can stay balanced even with just two legs, because the COG still falls inside its supporting base (see Figure 15.6).

Figure 15.6 A Four-Legged Robot with Two Legs Lifted

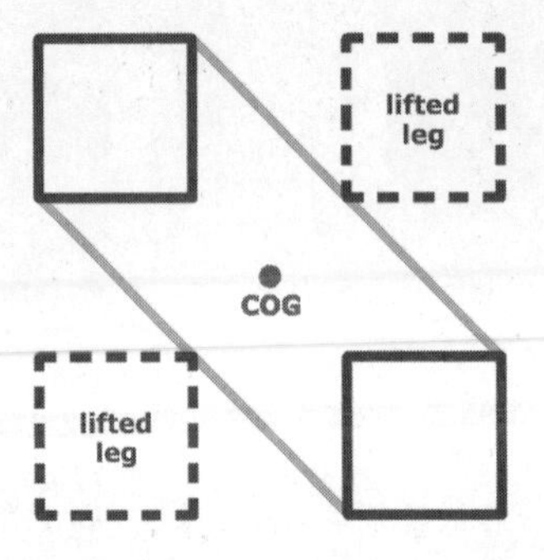

When the robot advances the two lifted legs, part of its mass moves forward, and the COG moves forward as well. And because the large contact surfaces of the legs delimit a zone wide enough to make the moving COG fall within the boundaries (see Figure 15.7), the robot, again, remains stable.

Figure 15.7 A Four-Legged Robot with Two Legs Lifted and Advanced

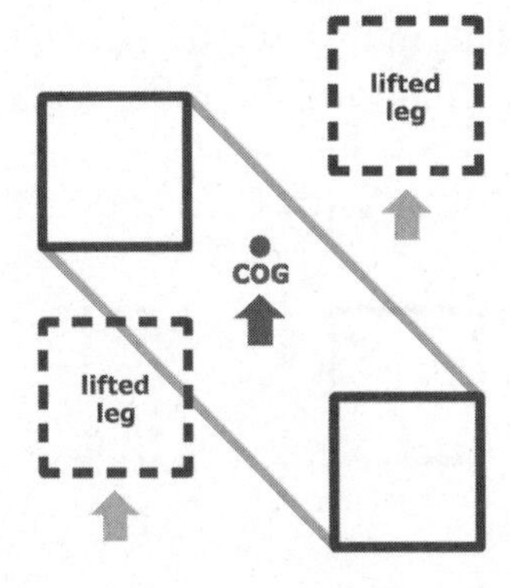

When we are using robots with more than four legs, we do not need to rely on their size anymore. A six-legged robot, for example, can walk with very thin feet provided it always has at least three of them touching the ground (see Figure 15.8).

On the contrary, when we start reducing the number of legs, things become more complicated. The making of two-legged (biped) robots requires a very careful design. A little trick is to build U-shaped legs that partly interlace, providing a large support for the robot (see Figure 15.9). LEGO suggested a similar approach in one of its Idea Books (8891, back in 1991).

Figure 15.8 A Six-Legged Robot with Three Legs Lifted

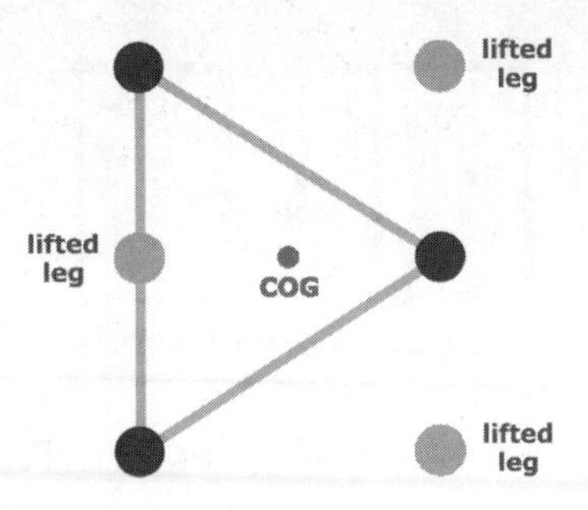

Figure 15.9 A Two-Legged Robot with Interlaced Legs

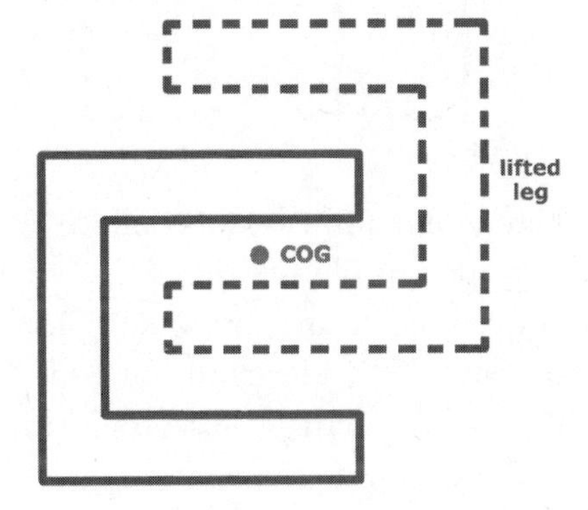

Though this works, it's a bit like cheating! If we want to emulate the way we human beings walk, we must understand what happens in the human body. Let's do a simple experiment. Stand still, being sure to distribute your weight evenly over your feet. Keep your arms at your side and keep all your muscles relaxed. Now slowly try to lift one leg: Your body tends to fall to that side. While walking under normal conditions, you unwittingly move your COG over one foot before lifting the other. This gives you balance and stability and prevents you from falling.

This is the behavior that we have to replicate to build a true biped robot. We have to shift its COG over one foot before lifting and advancing the other (see Figures 15.10 and 15.11).

Figure 15.10 A Biped Robot Standing

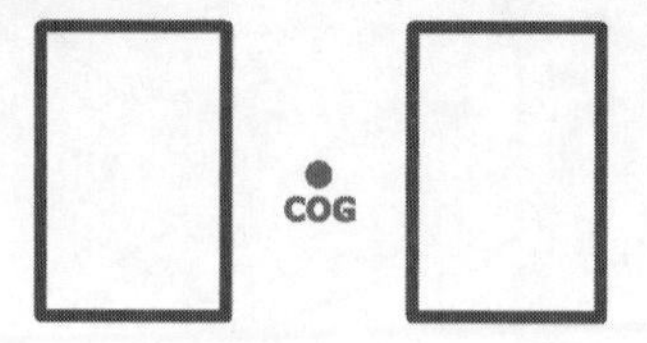

Figure 15.11 A Biped Robot Shifts Its COG over One Foot before Lifting the Other

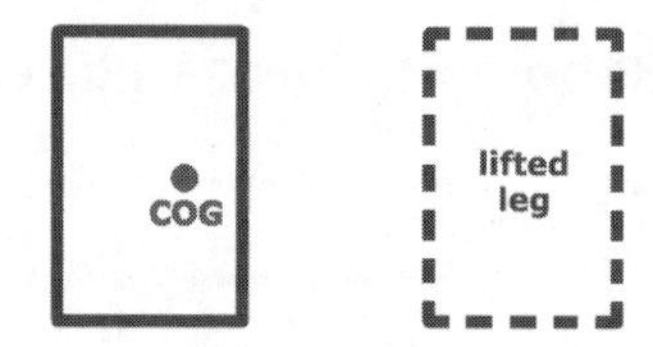

Actually, the way human beings and animals walk follows not only the rules of statics but also those of dynamics, the branch of physics that deals with matter in motion. When a man runs, for example, he is in dynamic balance, producing forces that oppose gravity and temporarily violate the rules of statics. To understand how this happens, you can study how *you* walk, and also look carefully at how animals phase their walking (bipeds, four-legged animals, insects, and arachnids). For example, elephants and other very large animals lift only one leg when walking slowly, to keep the static COG within the triangle bounded by their remaining legs. Once the pace picks up, the opposing gait takes over, which is similar to the sequence we described in Figures 15.6 and 15.7. Most four-legged animals use this scheme when trotting. At further increases of speed, such as in galloping, dynamic stability is more important than static: Only one leg needs to keep contact with the ground, and this allows the animal to cover more ground with every cycle.

Building a robot that walks or runs using dynamic balance is a very complicated task, and for this reason, in this chapter we will stay inside the comforting walls of statics.

Building Legs

Whatever kind of walking robot we're going to build, we must find a way to convert the rotary motion provided by the electric motors into the proper sequence of movements necessary for a leg to work. Animals and human beings use a very complex geometry operated by an impressive number of independent muscles. We must stick to the constraints imposed by the MINDSTORMS system, thus finding simpler solutions.

Figure 15.12 illustrates an initial idea: a leg mounted on two gear wheels of the same size, which are then connected in phase through a third gear. It's very important that the leg attaches to two corresponding holes of the gears; otherwise, it won't work because the holes will change their spacing as the gears turn.

Figure 15.12 This Leg Always Remains Vertical and Follows a Circumference

By driving any of the three gears, this simple leg will go up and down, forward and back, always in a circle. The leg always remains vertical. Figure 15.13 shows a slightly different approach, where only one point of the leg is attached to a wheel, and the leg itself slides freely into a rotating support (fulcrum).

In this assembly, the terminal point of the leg describes an ellipse—a flattened circle—whose height is equal to the distance between the uppermost and lowermost positions of the point where the leg is attached to the wheel, and whose length is a function of the distance between the fulcrum and the wheel. The closer the fulcrum is to the wheel, the longer the ellipse and, consequently, the stride of the leg will be. You can adjust this distance to make your robot take longer or shorter steps, affecting its speed. We invite you to experiment with this setup, changing the distance between the wheel and the fulcrum, to understand the effect on the stride. Later in the chapter, we'll use this feature to provide a legged robot with turning capability.

Figure 15.13 This Leg Describes an Ellipse

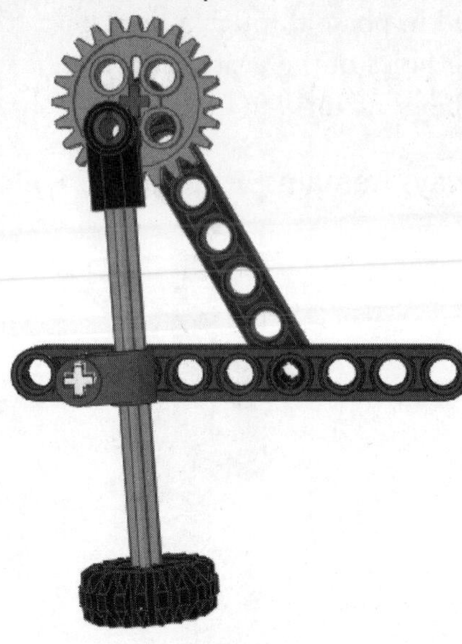

More complex leg geometries are also possible (see Figure 15.14). Designing legs is almost an art—it requires good intuition, and a lot of patience to test and improve your initial idea.

Figure 15.14 A Leg with a More Complex Geometry

Building a Four-Legged Robot

Let's start by building a robot with four legs in order to demonstrate the center of gravity principle explained in Figures 15.5–15.7. The architecture is very simple, and symmetrical:

Keep the COG as close as possible to the center (see Figure 15.15). We built it solely from NXT parts.

Figure 15.15 Our Four-Legged Robot

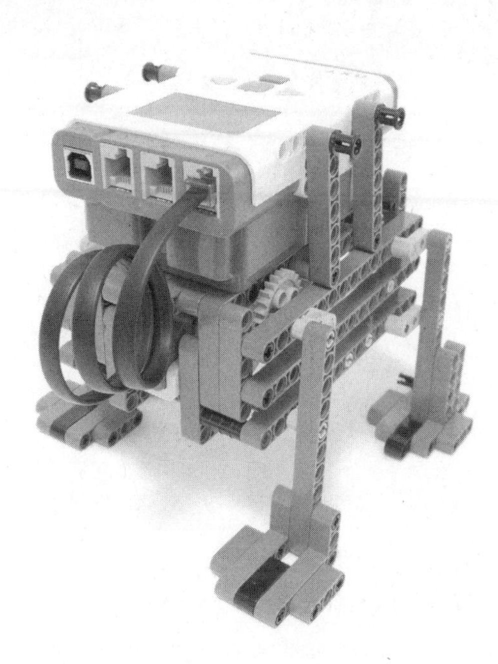

Removing the NXT, you'll notice there's a single motor, which through a series of gears provides motion to the front and rear leg assemblies (see Figures 15.16 and 15.17). However, notice the phase of the legs: They are diagonally paired. The front left goes together with the rear right, and the front right accompanies the rear left, which implements the walking scheme shown in Figures 15.6 and 15.7.

Figure 15.16 Top View (NXT Removed)

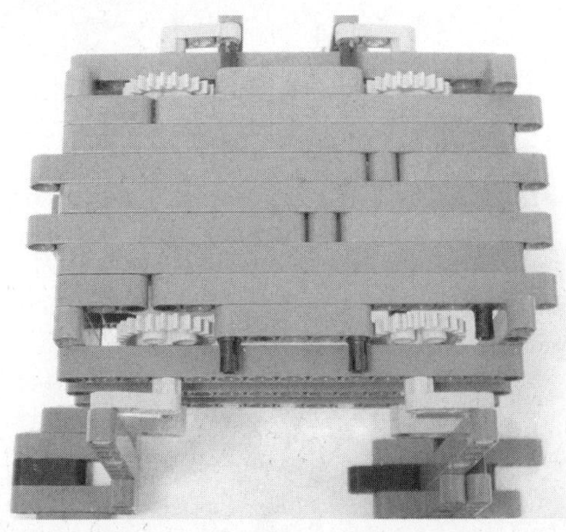

Figure 15.17 Bottom View

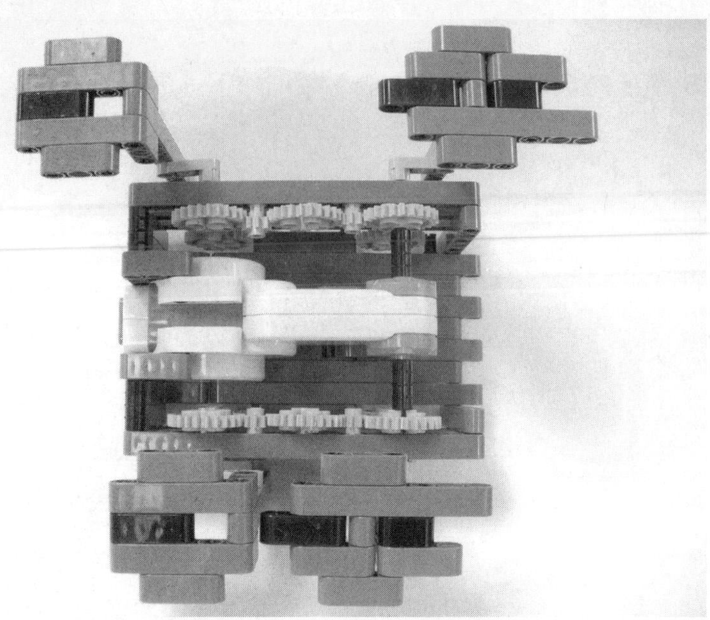

The legs follow the scheme of Figure 15.12, where just the gear wheels are inside the robot, their axles mounted on short 1 x 3 liftarms to which the legs are connected (see Figure 15.18).

Figure 15.18 The Front Left Leg

When the robot walks, it lifts two legs diagonally opposed, while standing on the other two (see Figure 15.19). Even when moving the legs, this robot always remains symmetrical; thus, its COG doesn't change position.

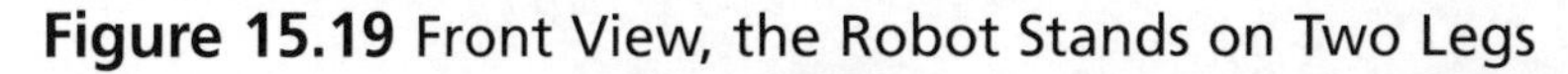

Figure 15.19 Front View, the Robot Stands on Two Legs

There's actually not much this robot can do. It's easy to build, and somewhat instructive, but it's able to go only forward in a straight line, and backward. You can mount two front and rear bumpers to make it reverse direction, but that's all you can expect from it. To provide your robot with directional control, unlocking all the opportunities that navigation affords, you need further sophistication. Let's move on and discuss some more challenging projects.

Building a Six-Legged Steering Robot

By increasing the number of legs, you can easily make a steering walker. The robot shown in Figure 15.20 has six legs similar to that in Figure 15.12.

Figure 15.20 A Simple Six-Legged Robot

The left and right leg groups are powered by two independent motors (see Figure 15.21), and in each group the wheels are phased so as to have the middle one raised when the front and rear legs are lowered (see Figure 15.22).

Figure 15.21 Top View (NXT Removed)

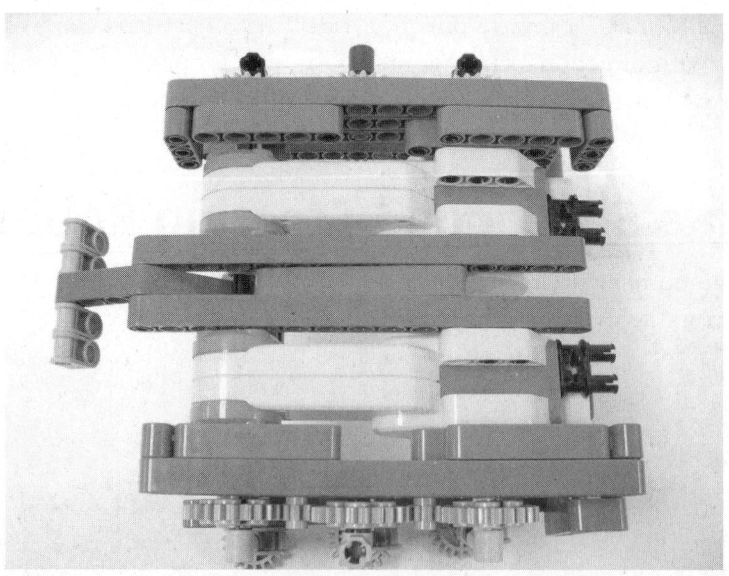

Figure 15.22 The Right Leg Group

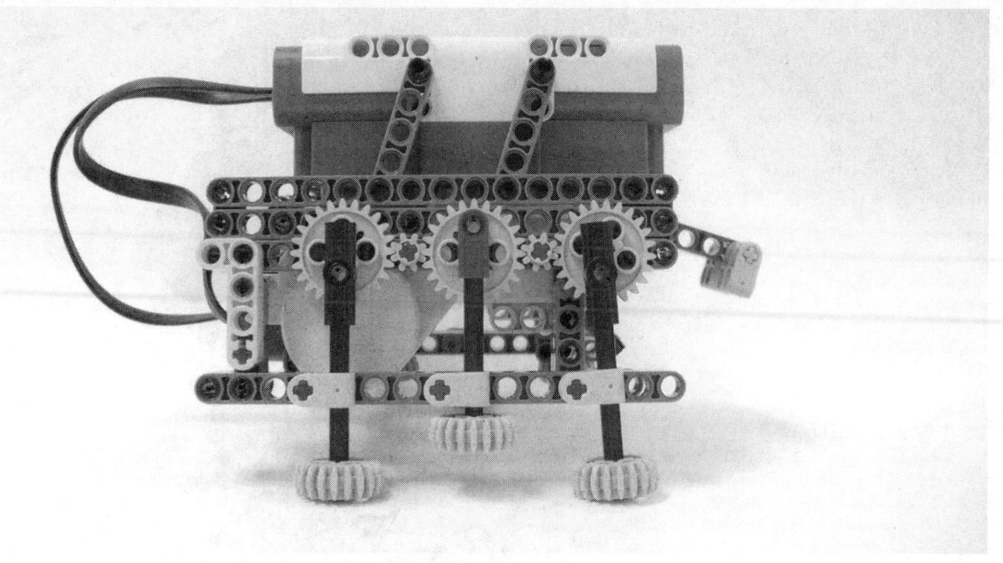

This robot turns and walks. You can make it turn by stopping or reversing one of the motors as though it were a skid steer drive. But it's affected by a serious problem: stability. The two groups of legs are not synchronized. Because of this, only the central legs are down at certain times. Because two legs are not enough for a stable balance, the robot tilts forward or backward a bit, ensuring that the additional legs make contact. As a result, its walking is rather irregular and jolting.

What could you do to smooth the walking motion? Using two sensors to detect the position of the legs, you could keep the two groups in sync so that one side goes on the middle leg only when the other one has two legs down.

There is another approach, more on the hardware side, that requires you to vary the geometry of the legs to make them change their stride. The left and right leg groups are connected together and they are powered by a single motor so that the robot is always supported by a triangle such as that shown in Figure 15.8. To change the stride of the legs, you have to change the distance of their fulcrums from the gear wheels. The robot in Figure 15.23 uses this technique.

All the legs are powered by a single motor, while the second motor controls the leg geometry (see Figure 15.24).

Figure 15.23 A More Sophisticated Steering Walker

Figure 15.24 Top View (NXT Removed)

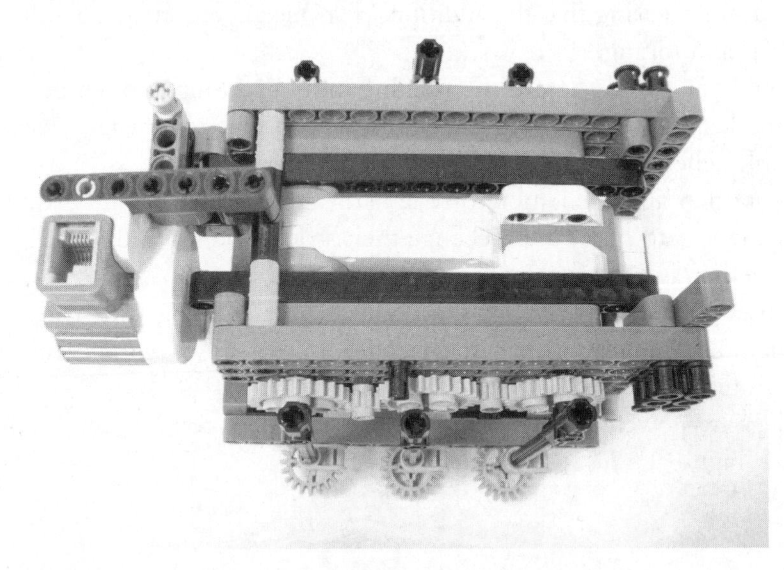

It's crucial that you connect the six legs in phase so that each side has the middle leg raised when the other two are down, and the left side has the middle leg down while the right one has its middle leg up (see Figure 15.25).

Figure 15.25 Three Legs Are Always in Contact with the Ground (Side View)

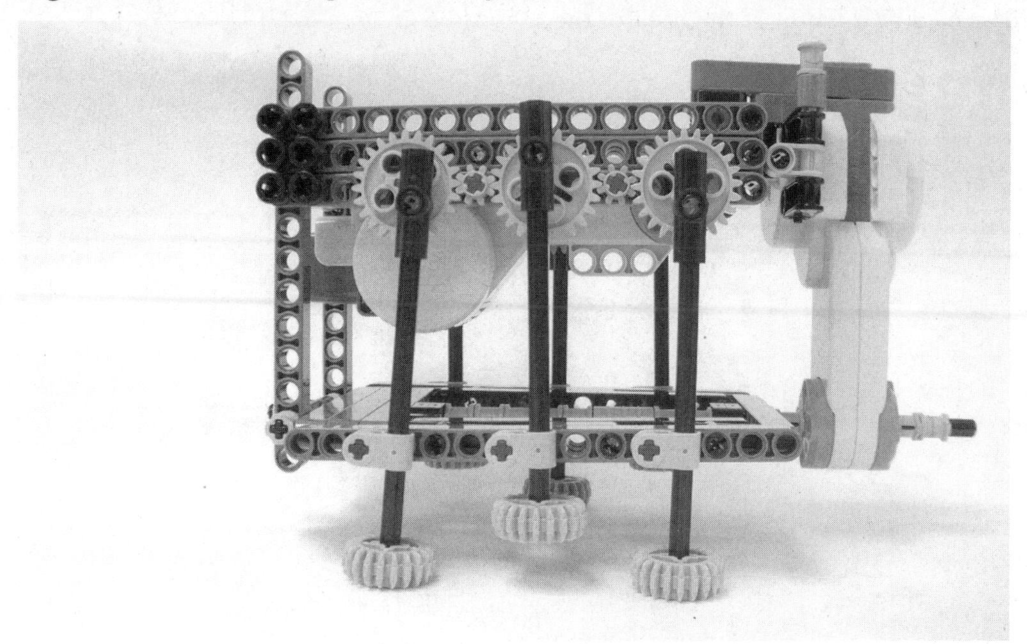

The fulcrums of the legs are attached to a swinging chassis that the second motor can incline on one side or the other (see Figure 15.26). The stride of the legs becomes shorter at the side where the fulcrums have been lowered, and longer at the other, thus making the robot turn.

Figure 15.26 Rear View

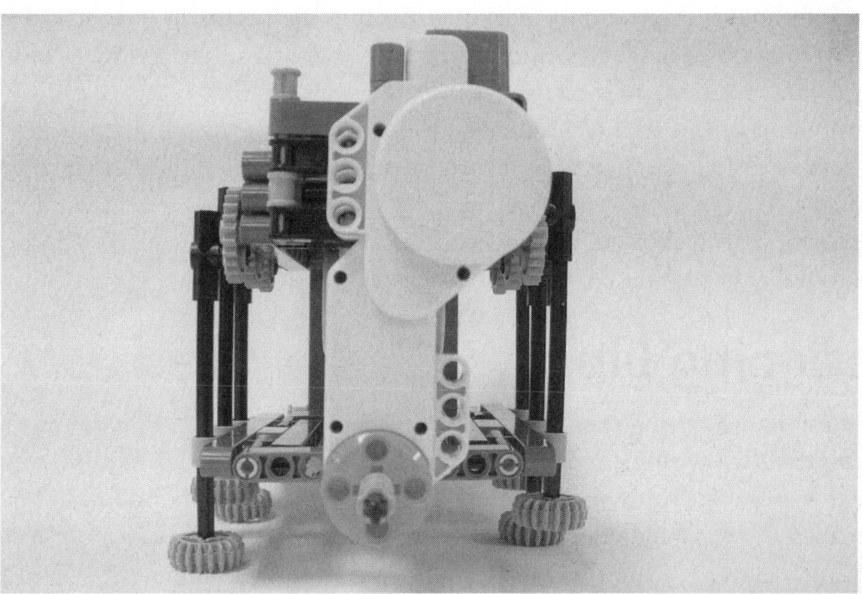

The front and rear sides of the swinging chassis are operated through a long joined axle that is connected by a motor at the back of the robot (see Figure 15.27).

Figure 15.27 Bottom View

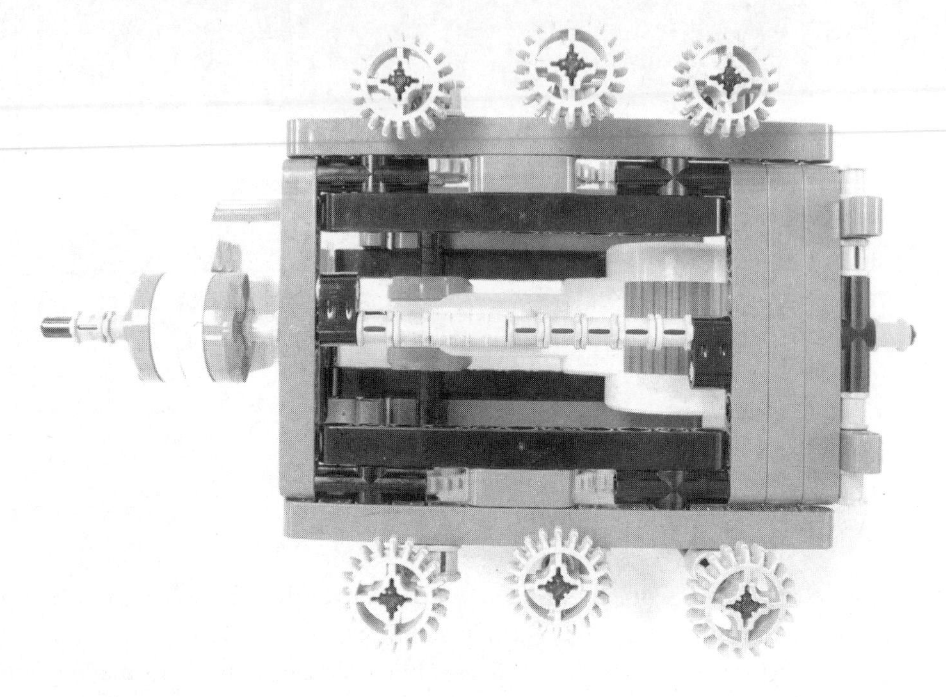

This robot needs no sensors to control its motion. When you want to make it turn, switch the motor on to the desired direction for a second to change the geometry, and then brake it to hold it in position. Recall that floating the motor might allow the swing chassis to turn. To resume straight motion, let the motor float and the swinging chassis will return to its central positions after a few steps.

The limit of this architecture is that the robot will turn only with a very large radius. It will be able to follow a line only if this doesn't make tight angles—a right angle, for example.

To give high maneuverability to your robot, you must remain with a skid-steer type drive, possibly increasing the number of legs to improve stability.

Designing Bipeds

Biped robots are among the most challenging projects we face. In a biped, the position of any single part, any single gram of mass, is critical to a stable balance. If you replicate the designs that follow, you will see that all of them walk quite smoothly, but you will also discover that you can't add additional parts anywhere in their body and not feel the pain!

We will go through the approaches described in the section "The Theory behind Walking," earlier in this chapter: interlacing legs (Figure 15.9) and COG shifting (Figure 15.11). For the latter category, we will explore a technique that requires that the whole body of the robot bend at the ankles to move the COG over a foot. In addition, there is also a technique (not described here) where an independent mechanism moves a mass from one side of the robot to the other to change the position of its COG.

At the end of the section, we will give you some tips about the next step in the challenge: the making of a biped robot capable of turning!

Interlacing Legs

Let's start with a biped based on the technique shown in Figure 15.9 using interlacing legs. The feet must be U-shaped and large enough to support the weight of the whole robot (see Figure 15.28, NXT MINDSTORMS parts only).

Figure 15.28 A Biped with Interlacing Legs

This robot uses a simple gearing, only an 8t and a 40t gear. The axle of the 40t connects to two opposing liftarms that operate the legs. The motor shaft has been prolonged with an axle (see Figure 15.29).

Figure 15.29 Top View (NXT Removed)

The leg geometry is very similar to the one in Figure 15.14, but here we used a double series of parallel beams in order to form two connected parallelograms, an upper (body to knee) and a lower (knee to foot) one (see Figure 15.30). Making this allows the foot to always remain parallel to the body, and thus the body always parallel to the ground.

Figure 15.30 The Left Leg

When you look at the robot from the bottom, it's easy to see how the feet interlace with each other (see Figure 15.31).

Figure 15.31 Bottom View

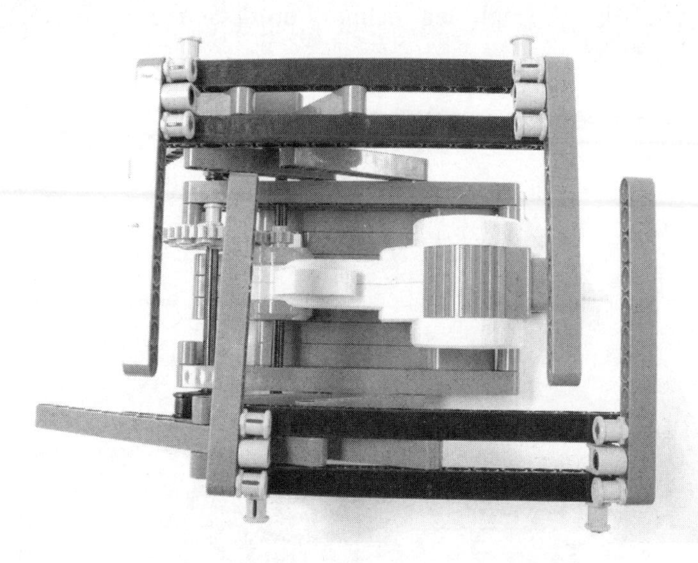

In our current design, the feet rest flat on the ground (see Figure 15.32). On other designs similar to ours, the feet are not flat. Nagata's Walker ND1 (see Appendix A), for example, has protrusions at the end of each inner tip of the feet (two on each feet). These tips compensate for the slackness of the leg that otherwise would make the robot lean at the side of the lifted leg, causing the COG to move beyond the base and make the robot fall. A similar feature is also observed on the interlacing-leg bipedal robot described in our RCX publication, although it only uses a thin plate instead.

Figure 15.32 Front View

COG Shifting

We will describe an approach to COG shifting of a walker by way of bending its ankle sideways in order to carry the COG over to the resting foot. The robot in Figure 15.33 uses this technique: You'll notice that the right leg inclines outside and that the NXT rests over the foot.

Figure 15.33 An Ankle-Bending Walker

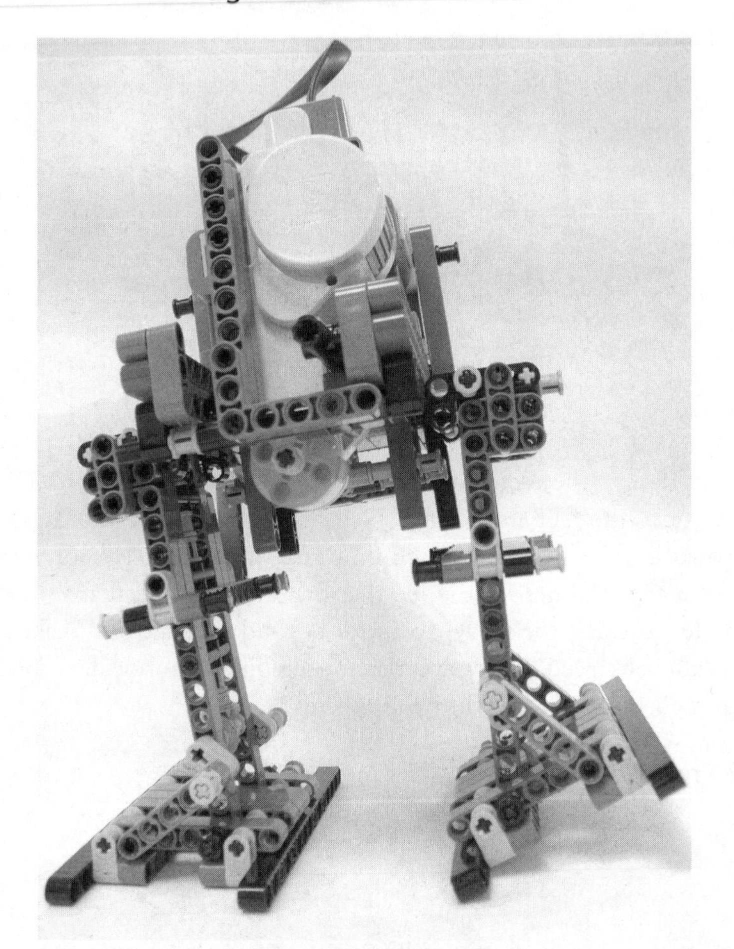

We used the leg designed by Miguel Agullo for his very nice Hammerhead (see Appendix A). The key component of the ankle is a *crankshaft* that manages the bending of the leg over the foot (see Figure 15.34). It looks pretty funny as it walks, lurching from side to side.

Figure 15.34 Detail of One Foot

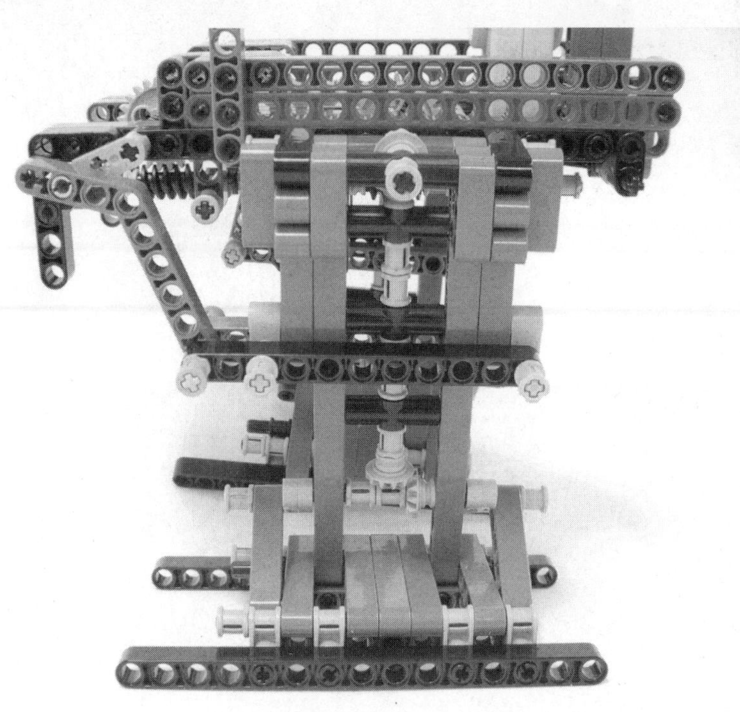

The hips are free to swing back and forth, and their supporting axle serves to also transfer motion to the ankle with a technique similar to what we described for the making of synchro drives or front wheel drive cars (see Figure 15.35). A second axle at the rear provides motion to the legs, through two crankshafts and two liftarms.

Figure 15.35 Top View (NXT Removed)

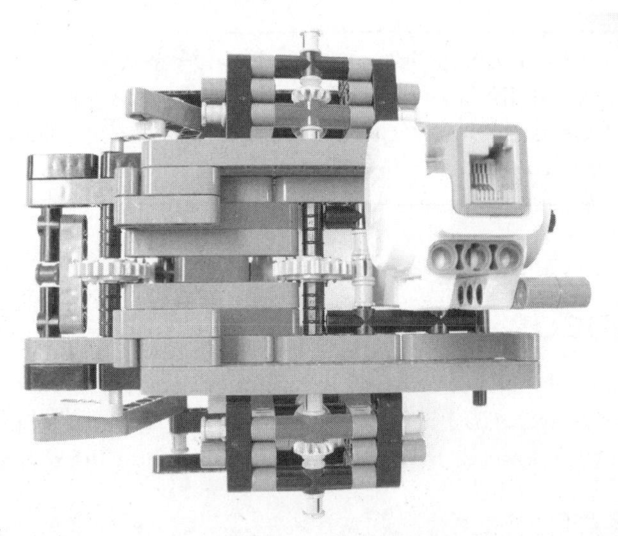

This walker uses a single motor and two worm gear and 24t pairs (see Figure 15.36).

Figure 15.36 Bottom View

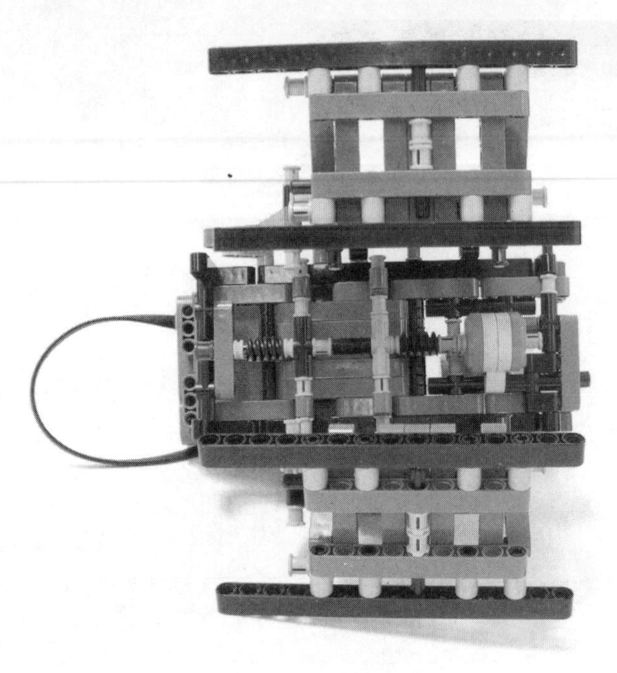

The difficulty of this model lies in finding the proper synchronization between ankle and leg movements. If you decide to give this technique a try, we suggest you follow Miguel's detailed instructions on his site (see Appendix A) to replicate his geometry.

WARNING

This ankle-bending model is not completely symmetrical, and it can walk forward only. Its COG lies in front of the hip joint, so the robot tends to lean forward and transfer its weight from one leg to the other as it is advancing. If you run it backward, it will fall.

Making Bipeds Turn

Is it possible to make a biped robot turn? It is, though it's definitely not an easy task. Once again, we invite you to experiment for yourself to find a working strategy. Observe your feet while you walk slowly, taking short steps and going straight: One foot is ahead of the other, but they remain parallel, as if they were running on tracks. Now try to change direction, but only take the very first step. If you look at your feet again, you notice that they are no longer parallel: The foot that's ahead is pointing in the new direction.

How can you emulate this behavior in your robot? We can do this by modifying the long legs utilized by either one of the bipedal architectures described here. The ankle-bending robot (Figure 15.34), for example, uses a 1 x 15 beam as its primary building piece for its leg construction. If, instead, it is replaced with two 1 x 7 beams connected with a joint that can vary the angle between the two beams, the legs can slightly converge or diverge. You will need a second motor to control the parallelism between the legs, and probably a third sensor to detect the straight position.

Summary

In this long chapter, we covered, hopefully, all of the most important aspects about the making of robotic walkers. Along the way, we discussed some important concepts, such as the *center of gravity*, that will prove useful in many other applications.

If you had the impression that we talked a lot about mechanics and not much about software, you're right. The task of balancing the weight over the legs is by itself so demanding that not much space remains to make your robot perform other actions. Although we showed some possible basic behaviors such as line following, more complex tasks such as grabbing objects are usually beyond the scope of walking robots due to the changes brought about regarding their delicate balance. Precise navigation is also not very suited to walkers, because the natural tendency of their legs to skid a bit on the floor makes them somewhat unreliable for positioning.

However, all this shouldn't keep you from experimenting with walkers. The pure reward of seeing them move compensates for all the effort put into building them. And who knows, in your enthusiasm you could develop some new solutions, or maybe design something as complex as a *running* robot!

Chapter 16

Robotic Animals

Solutions in this chapter:

- Creating a Monkey
- Creating a Mouse
- Creating Other Animals

Introduction

Trying to emulate animals in form and function with LEGO's MINDSTORMS NXT is a fun and instructive experience. You can approach the problem from different viewpoints—for example, concentrating your attention on the behavior of the animal, or on its shape; you can even create a pure fantasy animal or develop your own interpretation of some mythological creature.

In the following pages, we will discuss two robot projects: a monkey and a mouse. The robot monkey will try to emulate the swinging capability of real primates. Its function is to go hand over hand across a pole, dowel, or broom handle. The robot mouse is intended to be a roaming platform capable of performing multiple tasks. Like real mice, this robot has a playful nature as well.

Once again, the robots in this chapter will offer us the opportunity to revise and apply some of the concepts stated in the first part of the book. You will use a split gearbox to power the monkey's shoulders. The mouse, meanwhile, will use a large ball through which we will implement a sort of automatic steering feature.

The last section of the chapter contains a list of proposals intended as starting points for new projects inspired by the world of animals: squirrel, mole, ostrich, kangaroo, crab—take your pick!

Creating a Monkey

Our MINDSTORMS NXT monkey will need something to climb across. We recommend that you place a broom or similar object across two chairs. This will be the tree branch across which the monkey will be able to move back and forth. For the monkey to "see" his target point, you will have to drape a towel off the edge of the right chair, thereby providing the ultrasonic sensors a target that they can "see." Your monkey will start on the left end of the broom handle and cross to the right. When he gets to the right side he will see the chair, turn, and then, because he has been counting the number of swings required to get to that point, he will go back across the broom handle to the point where he began. Figure 16.1 shows a model of a fully assembled monkey. Now let's go through the step-by-step process of creating your monkey.

Figure 16.1 A MINDSTORMS NXT Monkey

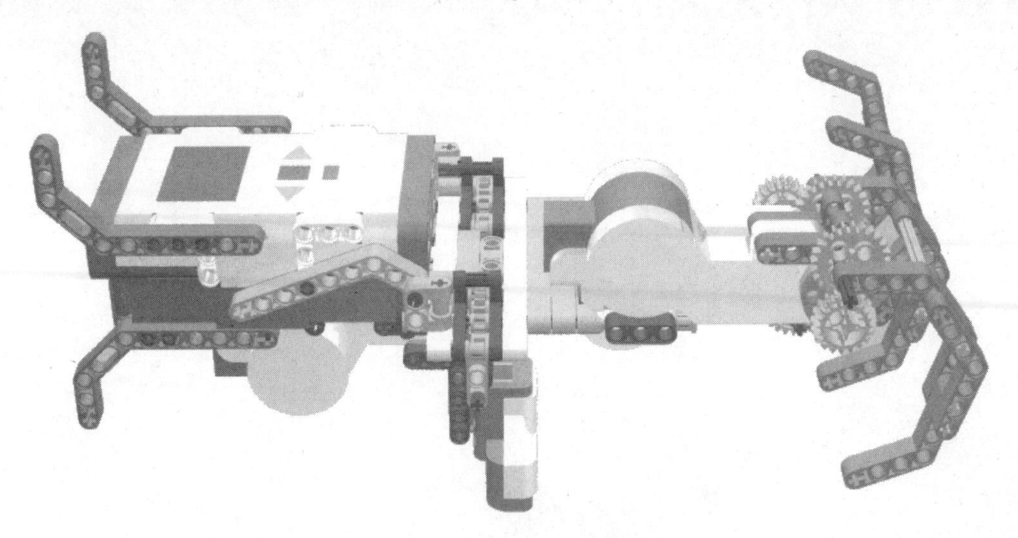

Step 1: Center Motor Assembly

The center motor (see Figure 16.2) will drive the two rotating arms of your monkey. You'll need the following parts for this step:

- 1x Motor
- 2x Pin with friction
- 2x Axle joiner perpendicular with two holes
- 4x Pin long with stop bush
- 2x Beam 1.15

Figure 16.2 The Center Motor

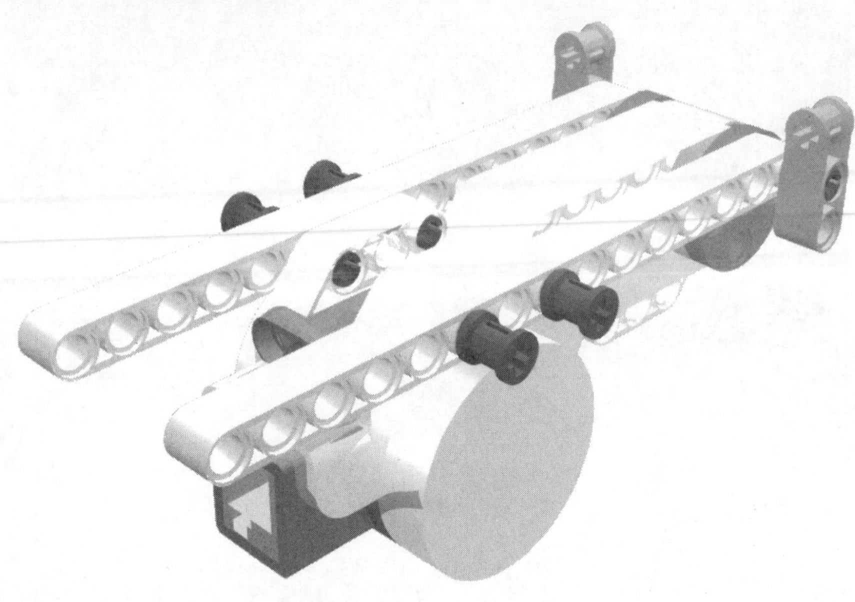

Step 2: Shoulder Assembly

When assembling the shoulder, you can use twin axles to power two functions. To transfer power through two 90-degree angles and to add torque we will use the worm gears and a pair of 24-tooth gears. This reduces load on the motor and gives the monkey the needed muscle to swing from its arms. You'll need the following parts for this step:

- 4x 1/2 Bushings
- 2x Worm screw gears
- 2x Gear 24 tooth
- 2x Axle joiner perpendicular with two holes
- 2x Axle 12
- 4x Triangles
- 4x Axle three with stud
- 2x Axle 8

Figure 16.3 is an example of what your monkey's shoulder will look like.

Figure 16.3 The Monkey's Shoulder

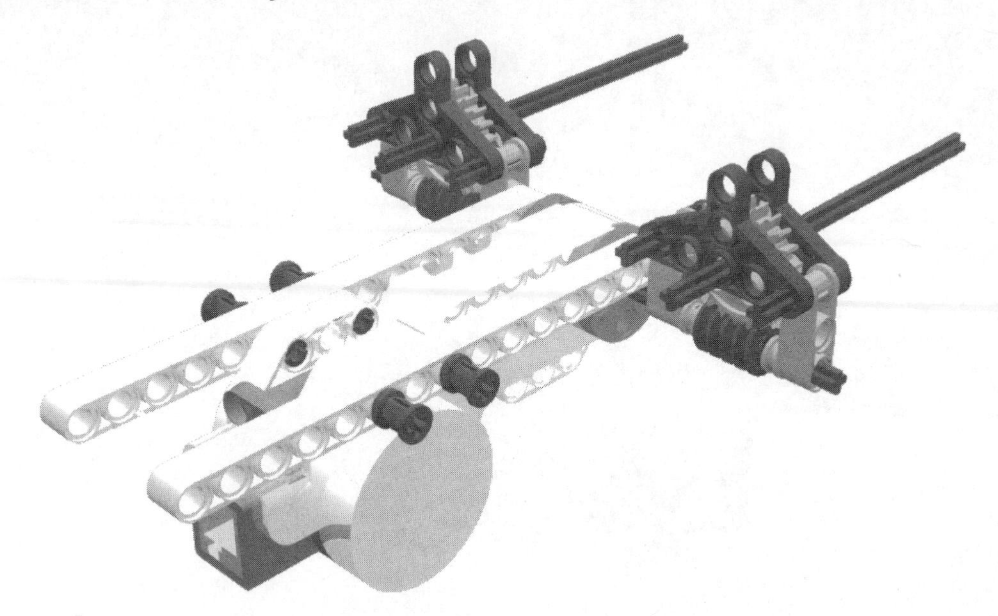

Step 3: Shoulder/NXT Brick Bracing

For this step, you'll need the following parts:

- 1x Beam 1 x 11
- 1x Joiner perpendicular 1 x 3 x 3 with four pins
- 2x Beam 1 x 15
- 4x Axle pin with friction
- 1x Beam 1 x 5
- 8x Pin long with friction
- 2x Axle joiner perpendicular with two holes
- 2x Joiner perpendicular 3L
- 2x Bush

Figure 16.4 shows the shoulder/NXT brick bracing.

Figure 16.4 Shoulder/NXT Brick Bracing

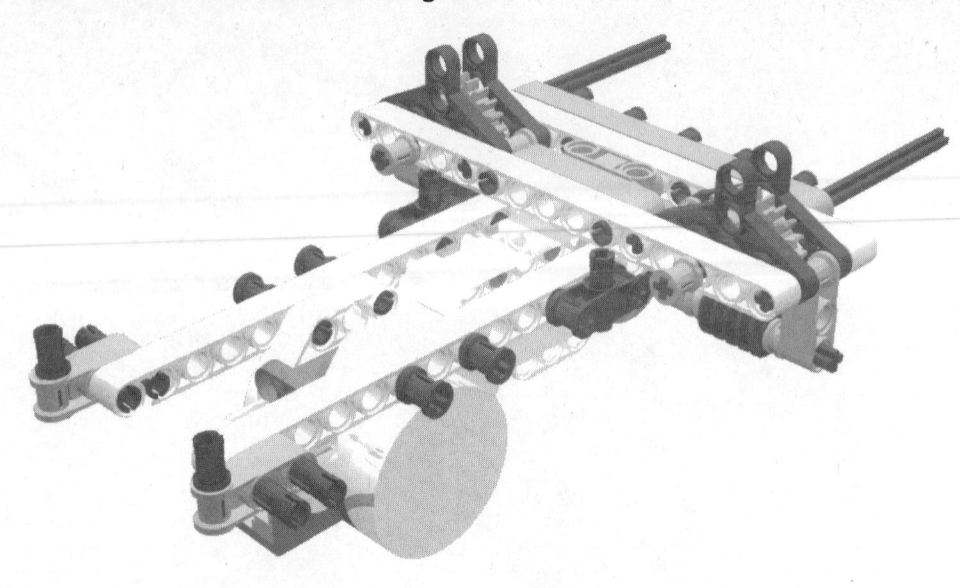

Step 4: Shoulder-to-Arms Support

For this step, you'll need the following parts:

- 3x Beam 1 x 11
- 2x Bush
- 9x Pin long with friction
- 4x Axle joiner perpendicular

Figure 16.5 shows the shoulder-to-arms support.

Figure 16.5 Shoulder-to-Arms Support

Step 5: Arm Motors

The arm motors (see Figure 16.6) will power your monkey's hands. You'll need the following parts for this step:

- 2x NXT motor
- 4x Pin long with friction

Figure 16.6 Arm Motors

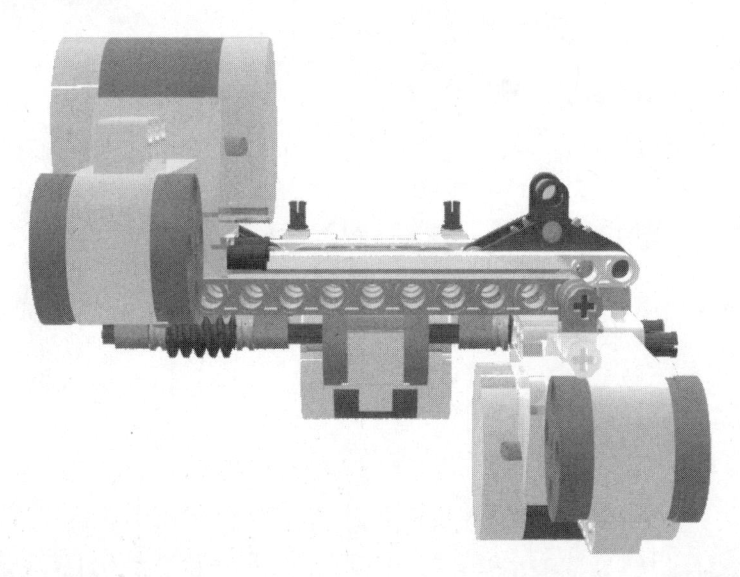

Step 6: Monkey Fingers

Monkey fingers are powered by your NXT motors. They use gear reduction for torque. With 3:1 reduction, you can give your monkey a grip so that he can hold on to the broom handle. You'll need the following parts to assemble your monkey's digits (see Figure 16.7):

- 2x Axle 10
- 2x Gear 20-tooth double-bevel
- 4x Beam 1 x 3
- 8x Pin long with friction
- 4x Axle 7
- 4x Gear 24-tooth
- 16x Liftarm 1 x 7 bent
- 14x Axle 3
- 4x Gear 8-tooth
- 4x Liftarm 2 x 4 L-shape

Figure 16.7 The Monkey's Fingers

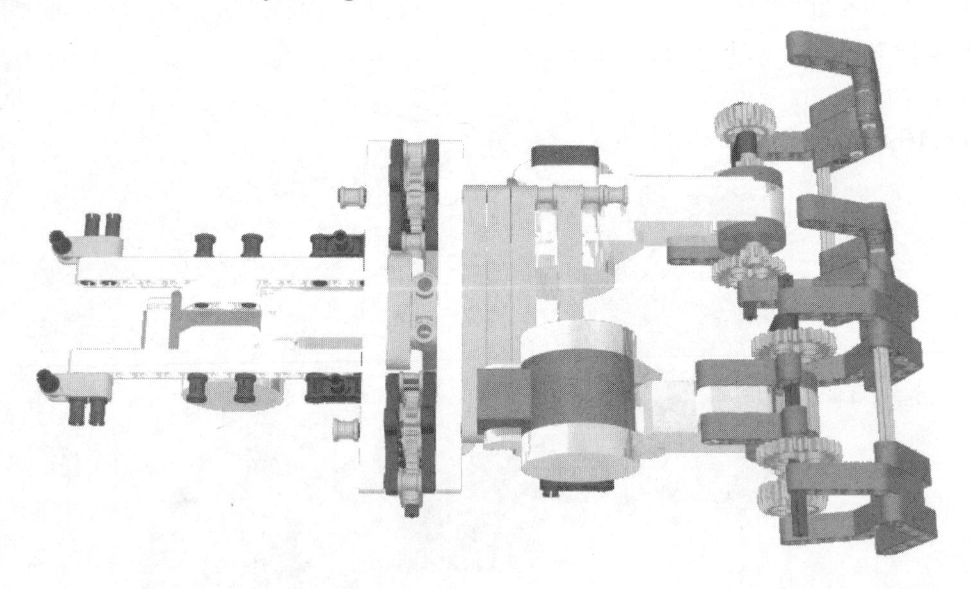

Step 7: NXT Brick Backbracing and Ultrasonic Sensor

You'll need the following parts for this step (see Figure 16.8):

- 1x NXT brick
- 2-Liftarm bent 2 x 7
- 2x Axle pin with friction
- 2x Axle joiner perpendicular double split
- 1x Utrasonic sensor
- 4x Liftarm double bent
- 2x Axle pin with friction
- 2x Axle joiner perpendicular double
- 1x Liftarm 3 x 5 L-shape
- 12x Pin with friction
- 2x Pin long with stop bush

Figure 16.8 NXT Brick Backbracing and Ultrasonic Sensor

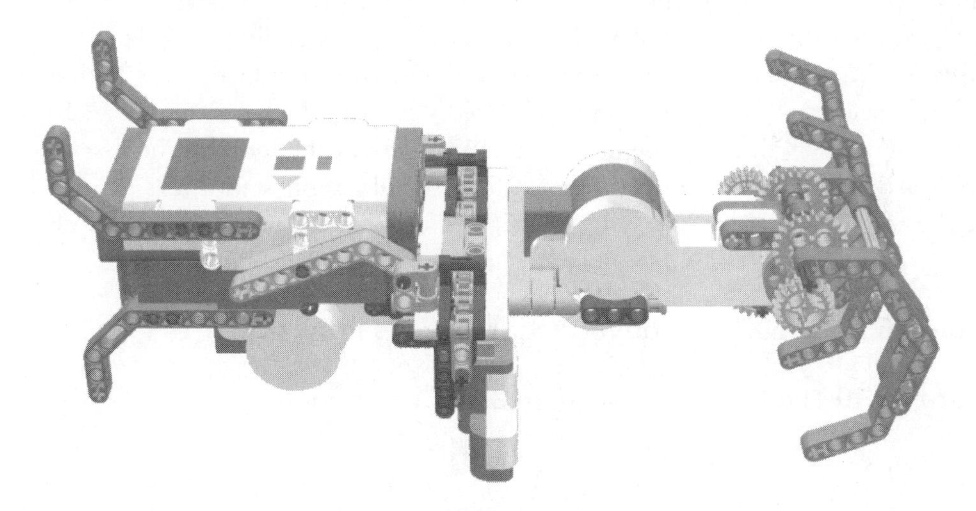

The Final Step: Wiring Your Monkey

Set up your monkey's wiring by facing the NXT brick toward you. The wire for the left arm is labeled "A", the one for the center shoulder motor is labeled "B", and the one for the right arm is labeled "C". Connect the ultrasonic sensor on port 4. Use the 20-tooth gears as tuning knobs to adjust the fingers when attaching them to the broom handle. Have both of your monkey's hands grip the handle to start. Tune the shoulders at the worm gear axle so that when you are facing the NXT brick, the right arm motor is on the side of the NXT brick that is facing you. Adjust this until the motor's 1 x 3 beam touches the bracing. Do not overtighten it. Adjust both hands so that they are closed over the handle. When the hands are in a gripped or closed possition, they should be at about 90 to 100 degrees from the vertical arm motors. If you have your monkey high up on something, protect him from acidental falls by using a tether line.

Programming Your Monkey

To program your monkey you will use a few of the more advanced features of the NXT-G: namely, *Logic* and *Compare* blocks, using data wires to send information. This will give your monkey the intellect to count how many swings he made, so he can go back to the start.

Figures 16.8 and 16.9 show the NXT-G code required for this application. Because NXT-G code is written in a horizontal fashion, it makes it difficult to display in publications such as this, so we have broken it into two chunks for ease of reading.

Most NXT programs are built within a loop, which keeps the program running. You can stop the program in a variety of ways. In this program, you will need to press the NXT **Stop button** (the gray square below the orange one). You can add a sound sensor and write some code to trigger the monkey to stop when you clap your hands.

Figure 16.9 shows our monkey program. We started by adding a *loop* block set to execute *forever*. We placed two more *loop* blocks inside the main loop, then set both loops using *logic* and a *counter*. In the second loop, we placed a *compare* block (see Figure 16.10), and set it to *equals*. What will this do? The *compare* block will count the loops of the first *loop* block. The *compare* block remembers the count. When the ultrasonic sensor tells the first loop to exit, the second *loop* block knows how far it has to go back to reach the point where the monkey started in the second *loop* block.

Figure 16.9 The NXT-G Code for Your Monkey's Program

Figure 16.10 The Monkey's Compare Block

Now let's look at the *ultrasonic sensor* blocks and the *logic* block. For this application, you

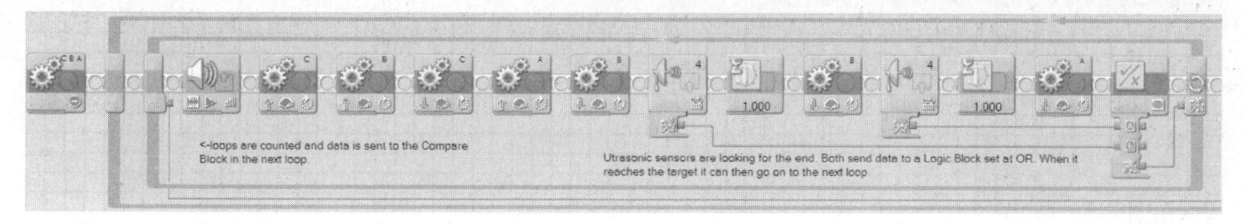

will use two *ultrasonic sensor* blocks. This technique could be used by different sensors or

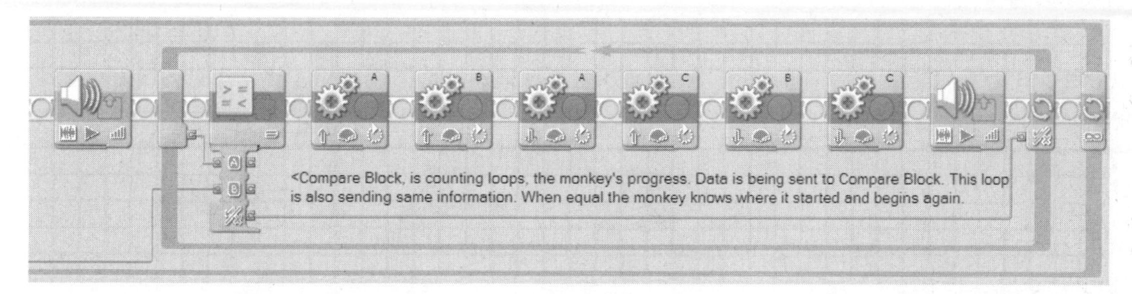

mixed sensors as well. You need to use two because the monkey makes a swinging motion in a right-to-left and left-to-right pattern for every loop. You don't want the monkey to miss an object and run into a tree! Connect both *ultrasonic sensor* blocks to the *logic* blocks' A and B hubs. Connect the *result hub* of the *logic* block to the end of the first loop. Set the *logic* block to *Or*. Now, if one or both of the *ultrasonic sensor* blocks detect an object in front of them, this information will tell the first loop to exit.

You will notice that there is one *motor move* block at the beginning of the program. This is telling all motors to brake on the A,B,C ports. This is so that when you place your monkey on the broom handle, it will hold on while things get going.

When you write a program it is a very good idea to test it before use. This robot has built-in stands (or legs) that facilitate the testing process. Test the functionality of all the motors at 50 percent power. A good test should work like this: Starting with both monkey paws closed, the right hand will open, the shoulders will rotate, the right hand will close, the left hand will open, the shoulders will rotate, and the left hand will close. This will repeat until you place your hand in front of the ultrasonic sensor, which will cause the monkey to perform the same motion, but in reverse. Test it and count the number of loops; notice whether the monkey goes back the same number of loops as it moved forward. If this works, the monkey is ready for action! Now set the motors to 100 percent and the monkey should be ready to swing.

Creating a Mouse

The mouse is a fast and timid creature. It is always on the lookout for danger. Our mouse has no legs, but it uses wheels and a ball castor for speedy turns and quick getaways. It has ultrasonic eyes that can see in the dark, and a microphone nose that hears the screams of startled people. His tail is a touch sensor; when you touch the tail, the mouse will scurry around (see Figure 16.11).

Figure 16.11 A Robotic Mouse

Step 1: Mouse Frame and Motor Assembly

For this step you'll need the following parts:

- 2x NXT motor
- 2x NXT rim
- 1x Beam 7
- 2x Beam 9
- 2x Beam 15
- 5x Axle 3
- 2x Axle 6
- 2x Pin long with stop bush
- 1x Touch sensor
- 14x Pin long
- 2x Axle pin with friction

- 2x Axle joiner perpendicular 3L
- 2x Beam 15
- 3x Pin with friction
- 4x Bush
- 2x Liftarm double-bent
- 2x NXT tire
- 2x Pincer Suukorak
- 5x Axle joiner perpendicular 1 x 3 x 3 with four pins
- 6x Liftarm bent 7
- 2x Axle joiner perpendicular
- 2x Axle 10
- 2x Axle 5

Figure 16.12 shows the assembled frame and motor.

Figure 16.12 The Mouse Frame and Motor

Step 2: Castor Bottom

For this step you'll need the following parts:

- 2x Long pin
- 4x Axle pine with friction
- 2x Pincer Suukorak
- 6x Pin with friction
- 2x Beam 7

Figure 16.13 shows the assembled castor bottom.

Figure 16.13 The Castor Bottom

Step 3: Tail Assembly

If the tail is too heavy for your touch sensor, shorten it by removing a few of the liftarms used in this assembly.You'll need the following parts for this step:

- 4x Liftarm 7 bent
- 1x Pin long with bush stop
- 2x Axle 2
- 1x Axle pin with friction
- 2x Liftarm 2 x 7 bent

- 1x Pin long with friction
- 1x Axle joiner perpendicular 3L
- 1x Double bend
- 8x Axle 3
- 2x Axle joiner perpendicular

Figure 16.14 shows the assembled tail.

Figure 16.14 The Mouse's Tail

Step 4: The Mouse Head Frame

We must add a special note here: The 3 x 5 liftarm should be connected to the NXT brick through the #3 and #15 beams. You will have to lift the #15 beam up one hole to set it with the long pin. This gives the frame a slight preload. For this step you'll need the following parts:

- 2x Beam 15
- 2x Liftarm 3 x 5
- 2x Pin long
- 4x Beam 3
- 28x Pin with friction
- 2x Gear 24-tooth

- 2x Liftarm 2 x 4
- 4x Axle pin with friction
- 2x Axle 3

Figure 16.15 shows what the assembled mouse head frame looks like.

Figure 16.15 The Mouse Head Frame

Step 5: Motorized Mouse Head Assembly

The mouse head is designed to move up and down; this feature mimics a mouse's inquisitive nature and adds to its form. You'll need the following parts for this step:

- 1x NXT motor
- 1x Axle joiner perpendicular
- double split
- 1x Pin joiner dual perpendicular
- 8x Angle connector #6
- 2x Gear 24-tooth
- 4x Gear 8-tooth
- 6x Axle pin with friction
- 1x Ultrasonic sensor
- 3x Beam 3
- 4x Liftarm 3 x 5
- 2x Axle 12

- 9x Axle 2
- 5x Pin long with friction
- 1x Sound sensor
- 1x Axle joiner perpendicular
- double
- 2x Angle connector #1
- 9x Axle 2
- 2x Gear 24-tooth
- 8x Pin with friction

Figure 16.16 shows what the motorized mouse head looks like.

Figure 16.16 The Motorized Mouse Head

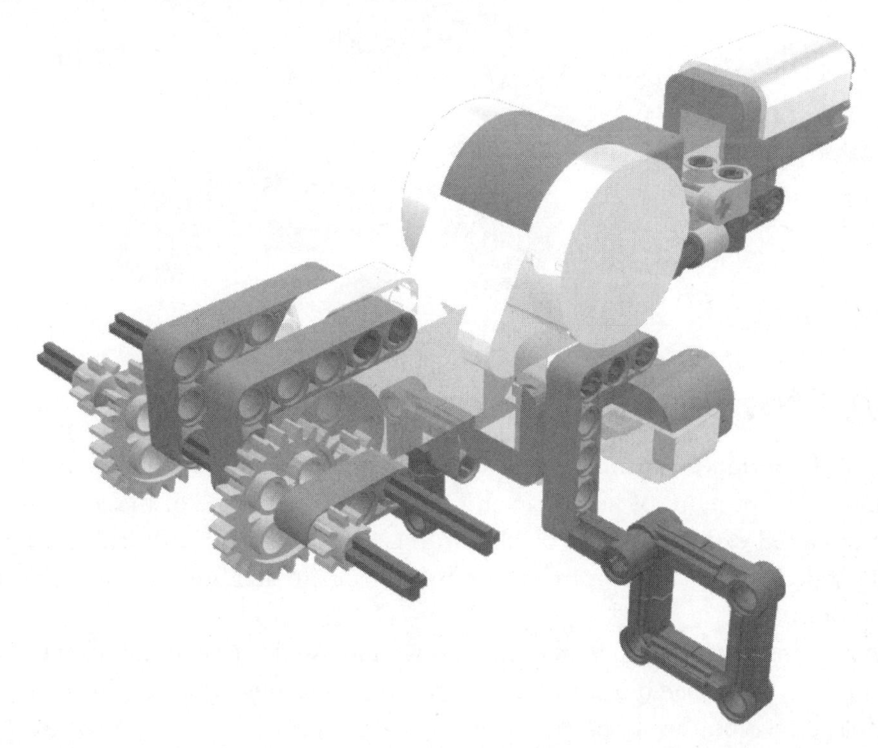

Step 6: Mounting the Mouse Head to the Body

When wring the mouse, port A is for the mouse head assembly motor; ports B and C are for the motors to wheel motors; port 4 is for the ultrasonic sensor; port 1 is for the touch sensor; and port 2 is for the sound sensor. Place one of the two large balls provided with the NXT set into the mouse's front paws for steering capability. You'll need the following parts for this step:

- 2x Beam 5
- 2x Bush
- 1x Large ball

Figure 16.17 shows what you'll see when you have mounted the mouse's head to its body.

Figure 16.17 The Mouse's Head Mounted to Its Body

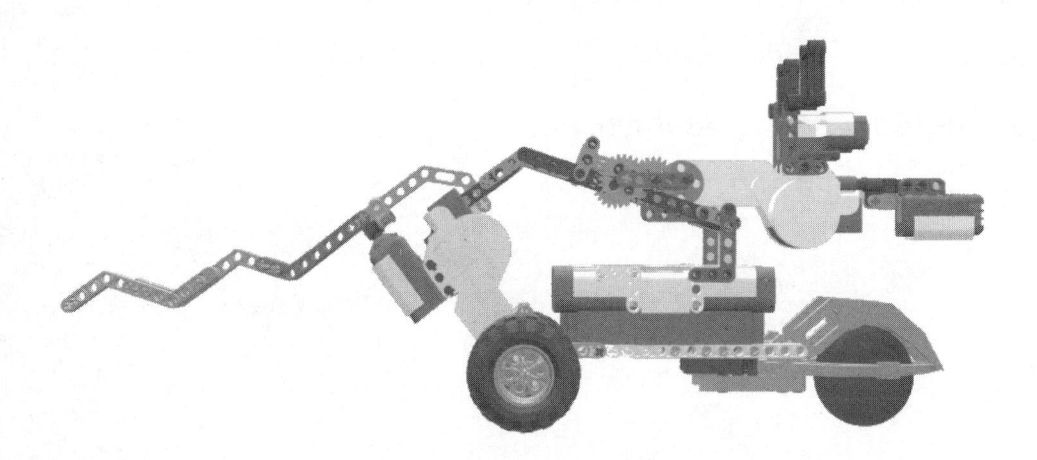

Step 7: A Programming Example

Because much of the form of our mouse is in its action, the program will have some actions that give the mouse robot more of its character. This example has some of the *logic* and *compare* blocks we discussed earlier. *Compare* blocks are very useful in terms of turning. Logic is needed to make a decision, or in this case, to use two sensors to activate the same operation or multiple functions.

This program is inside a *loop* block like the one we discussed earlier in the chapter. Here again, you start by placing a *loop* block that is first set to *Forever (infinity)*. All your code blocks will be placed within this loop.

Take a look at Figure 16.18. This code segment represents the "character" part of the mouse, but we will use it in other parts of the program as well. You see here that a touch sensor and a sound sensor are wired to a *logic* block. Set the *logic* block to *OR*. Wire the *logic* block to the *switch* block and set it to *value/logic*. (This could have been done just by making the *switch* block a touch or sound *switch* block. But then it could not do both. This is one reason why a *logic* block is so useful.) The *switch* block (Figure 16.18) is set to *flat view* because the bottom line is empty and is coded only in the top section. This makes the code easier to read and follow. As a result of this code, if the touch or sound sensor is activated, it will use the top code of the *switch* block. If not, nothing will happen because the bottom section of the block has no code.

In the next segment (Figure 16.19), start by placing a *switch* block in the ultrasonic sensor mode to the right of the last segment; use 12 inches for the setting here. The top code of this *switch* block is the drive-forward code within your character code. The character code makes the mouse's head move up and down and turn left and right. You can add sounds to give the mouse an "on the run" feeling.

The bottom section of this *switch* block is the turning segment with the character code as well. Start by placing a *motor move* block *(B/C=Stop)*. This will brake the mouse before it runs into a wall. Now place another *switch* block set to *Value/Logic/True*. Place the touch and sound sensors wired to the *logic* block and to the switch. The upper part of this switch has the same character code as before.

The bottom of this switch is the part of the program that directs the mouse to turn left and right. As the mouse approaches a wall, it stops, and then looks left and right. The ultrasonic sensor is measuring the distances and sending this data to the *compare* block. The larger value represents the direction of travel, as handled by the next *switch* block. Only the top section has code. If the value of one direction is greater than the value in the *switch* block, the mouse is already facing that direction. If the value of the other direction is greater than in the *switch* block, the mouse turns and the program starts again.

If you have a dog or cat, it probably will enjoy playing with the mouse robot. Be careful with dogs, as they tend to like mouse robots a little too much and may bite them. Also, note that you can program this robot to perform other tasks. For instance, you can make it follow lines, sounds, or light.

Figure 16.18 Programming Your Mouse's Character

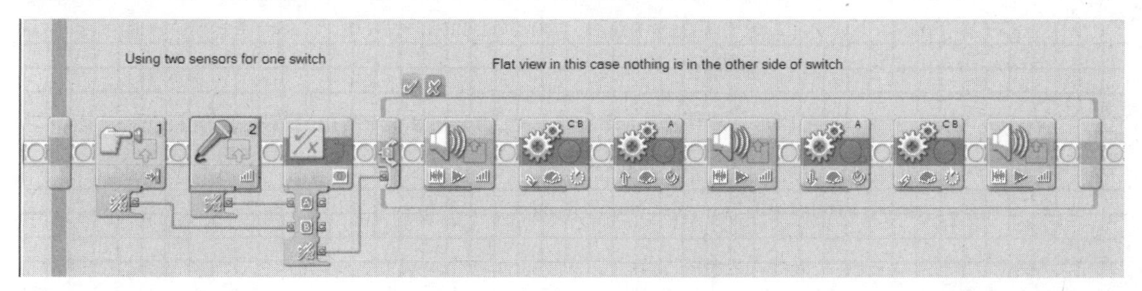

Figure 16.19 Programming the Switch Block for Your Mouse

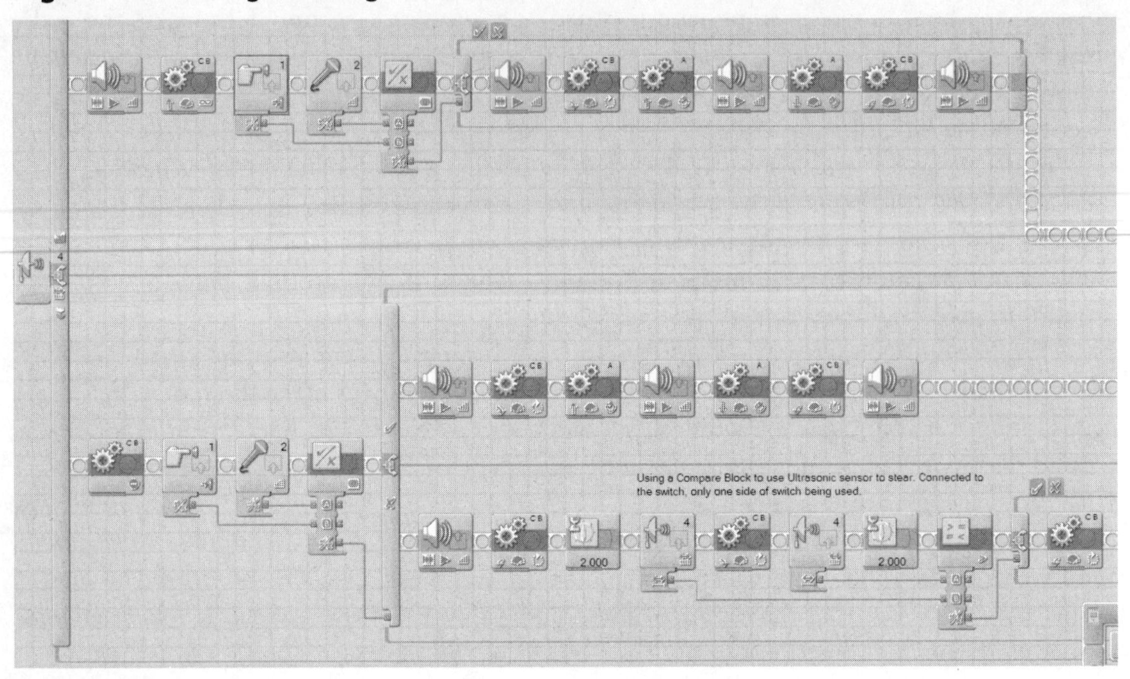

Creating Other Animals

Nature is a wonderful source of inspiration, and you can collect tons of great ideas just by browsing through books about animals. You can use insects and spiders as templates for multilegged walkers; other creatures higher on the evolutionary scale present an incredible range of shapes and behaviors. Take your pick.

Matching shape to function is almost an art. Even our simple monkey required many trials before we felt satisfied with its design. The following list provides some examples of what you can make. Keep in mind that many other creatures which are at least as interesting and challenging as these are just waiting for you:

- **A squirrel** This robot would collect 2 x 2 bricks as though they were nuts, and perhaps bring them back to its hole (difficult).
- **An ostrich** It won't bury its head in the sand, but it can hide its head between its legs.
- **A kangaroo** We like the challenge of designing a jumping robot, but so far we haven't succeeded. The idea was to implement a sort of spring or rubber band mechanism that, when slowly loaded by a motor, could release its energy all at once, thus making the kangaroo jump. We conducted a few experiments, and

although our prototype actually skipped and advanced a bit, we didn't consider it fully successful. It's still an open challenge!

- **An armadillo** This would roll up like a ball when disturbed.
- **An oyster** Even a simple animal such as this offers some ideas. For example, you can build one that closes its shell very quickly when someone steals its pearl, and the game could be to try to remove the pearl without being touched by the shell.
- **A dinosaur** This category includes such a broad variety of creatures that you shouldn't have a problem finding one to emulate.
- **A porcupine or hedgehog** This one could raise its quills (axles) when detecting a stimulus.
- **A crab or lobster** This type of robot would clamp down on everything that touches its claws.
- **A koala** This would climb a tree (you'd probably need a specially shaped LEGO tree for this one!).
- **A mole** This would be a dark-seeking robot that looks for the darkest places in your room, presumably under a piece of furniture, and rests there until light disturbs it.

With additional MINDSTORMS NXT sets you could model two or more animals of the same type so that they can cooperate with each other as they perform certain tasks; for example, ants. Or you could model animals of different types so that one hunts the other, which tries to escape. In addition, you could build a multi-NXT animal robot that could have many more functions.

Summary

In this chapter, we discussed how to design a robot based on existing creatures. In addition to technical issues, you have to face the difficulties that come from the need to match shape to function. In fact, in previous chapters, we concentrated on solving technical problems without introducing concerns about the size or appearance of our robots. However, you cannot emulate animals without carefully studying the shape of the robot; actually, that is the most important factor, and it's what makes your robot look like an animal instead of a vehicle. Decisions about the appearance of a robotic animal usually come before any other mechanical choice, and they will push you to look for technical solutions that suit your desired structure. Sometimes you will find them and will be able to carry out your original design; other times you will have to introduce some adaptations into the structure to make a mechanical solution possible.

The monkey and mouse described in this chapter are good examples of this approach. One of our goals in building the monkey was to build it with the strength it needed to hang on to a branch (represented by a broom handle). This led us to use various gear systems to drive its shoulders and hands.

The mouse project started with a different premise: The robot had to be mobile and fast. When making shapes from beams, we are confined to angles that may not emulate the real shape of a mouse. But it can have a degree of similar appearance. Much of the character of the mouse, then, is evident more in its programming.

Having stated the importance of shape when emulating animals, we don't want you to think that shape is the only thing that counts. You should take into account what you learned in previous chapters, paying particular attention to the strength of the structure, the gear ratio of the mechanisms, and the effectiveness of the sensors.

As for programming, start with a simple program and test it for function. It is best if your first programs have power settings that are as low as possible. This way, you can be sure the program will function the way you intend. If you have a problem that causes binding, or a mechanical issue that results in limited travel, you can stop the program and change certain parameters. Once you have a simple program that functions correctly, you can add more complexity one step at a time, which will make the program much easier to debug later.

The last lesson to remember from this chapter is that by studying and observing animals, you can learn many tricks that are useful to robotics. Remember: Form and function are a balancing act of engineering. Animals provide endless inspiration when it comes to challenging robotic projects!

Chapter 17

Solving a Maze

Solutions in this chapter:

- **Finding the Way Out**
- **Building a Maze Runner**
- **Building a Maze Solver**

Introduction

Humankind has always been fascinated by labyrinths, and mythology is crowded with heroes busy finding their way out of mysterious buildings. It was not unusual for large European 18th- and 19th-century villas to have a hedge labyrinth in their gardens. Indeed, mazes of different varieties are still common in the amusement parks and games of our era.

The ability to find your way through a maze is considered a good test of intelligence and has been used with mice and other animals to measure their capacities. Now the time has come to test your robots too!

Before building robots capable of solving a maze, we must understand what "solving a maze" means. In other words, we must understand what knowledge and skills are necessary to find the way out. If you ask anybody to solve a simple maze drawn on a sheet of paper, he will probably do it very quickly. But if you ask him to *describe* the procedure he used, you will likely receive some very generic explanations. This happens because human beings tend to ignore the details of what they do: They employ the knowledge and experience accumulated throughout their lives—especially during their childhood—without realizing that such a simple action actually hides a multitude of operations. If somebody were to stop you on the street to ask for directions, would you explain to her what "turn" and "left" mean? Surely not. However, in regard to robotics, there's no background knowledge you can take for granted. We explained in Chapter 14 that even an apparently easy task such as moving around the inside of a room or detecting obstacles requires a thoughtful analysis of the environment and of its interactions with your robot.

This is also the kind of analysis necessary to implement maze solving: You need a strategy, and it has to be detailed enough to be translated into program instructions for your robot. For this reason, we will begin exploring some theories about maze solving, which will lay the foundations for the projects that follow.

On the hardware side, the robots that you will come across in this chapter don't require more parts than what you find in your MINDSTORMS NXT box. As well as teaching some concepts about maze solving, this chapter will also strengthen your skills in working with touch and light sensors, consolidating ideas that appeared in Chapter 4.

Finding the Way Out

Even a simple maze, the kind you can solve in a few seconds with a pencil if you see it printed on a sheet of paper, assumes a completely different perspective when you are inside it. If you don't have any external reference point and are not allowed to take note of your moves, well, be prepared to spend a few hours!

How can external references or note taking help you find your way out of a maze? Because they help you understand where you are. To introduce this concept, we invite you to perform an experiment: You need a friend who will play the role of the robot inside the maze, while you simulate the sensors that return information about the environment around him. Your friend must find the exit from the maze of Figure 17.1 without actually seeing the picture, and only by using your verbal feedback. He can use only four commands inside the maze to direct himself: forward, back, right, and left. You track his position in the maze with a pencil, and if his command is acceptable—that is, if the desired direction doesn't come up against a wall—you move the pencil to the specified adjacent square, answering "OK"; otherwise, you keep the pencil stationary and answer "wall."

Figure 17.1 The Test Maze

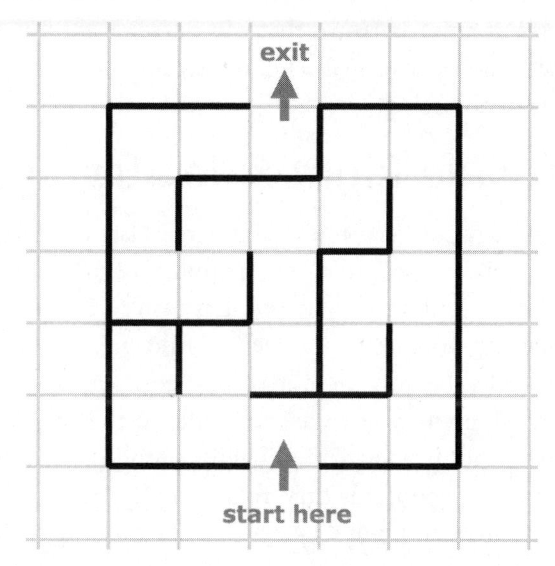

Will your friend be able to exit the maze under these conditions? Probably yes, but only after a long time, and with an effort that seems enormous when compared to the simplicity of the maze. In the second phase of the experiment, provide your friend with a square sheet and a pencil so that he can log his movements. When you answer "OK," he will move his pencil to the adjacent square and mark it as "visited," and when you answer "wall" he will remain in the same square, but will mark the specified side of his square with a line which represents the wall. Now things will go much more smoothly for your friend: Looking at his map, he can avoid visiting the same location more than once, sparing himself many "collisions" and exploring all possible routes until he finds the way out.

Some of you may have noticed that the aids mentioned pertain to the two basic categories described in Chapter 13 regarding knowing your position: absolute and relative positioning. In fact, the use of external reference points represents an application of absolute positioning—you use landmarks to locate yourself—while note taking has many similarities with relative positioning: You deduce your new location knowing the direction and the distance you covered from the previous location.

Finding one's way in a labyrinth is, in fact, a special case of navigation and requires similar abilities, with the addition of some memory to remember which branches have already been visited. In our previous experiment, the memory was symbolized by the sheet of paper where your friend logged his moves.

Thus, generally speaking, to solve a labyrinth, your robot should be equipped with a navigation system and a map in its memory. There are some notable exceptions, such as labyrinths that simply require slavish application of a rule to lead you to the exit, which could be handled by robots with less demanding equipment.

The strategies we are going to explain work with flat mazes—not just the ones you can draw on a piece of paper, but any labyrinths that can be *represented* on a piece of paper. For example, hedge and crystal labyrinths usually belong to this category provided that they don't contain any bridges or tunnels.

Using the Left Side-Right Side Strategy

The left side-right side strategy solves an incredibly large class of mazes, its rule being quite simple to remember and apply. It states that, when applicable, if you follow the left wall and turn left whenever possible, you will find the exit. Easy, isn't it? You're not guaranteed to cover the shortest distance, but you are guaranteed to find the way out. Actually you can just as easily keep to the right side, the two methods being complementary and leading to the exit along different paths. We invite you to test the rule on the simple maze of Figure 17.1. Imagine physically entering the maze and then trying to follow the left wall—eventually, you arrive at the exit. Now try again, this time following the right wall. Again you reach the exit, but from a different route (see Figure 17.2).

Figure 17.2 Following the Right and Left Walls

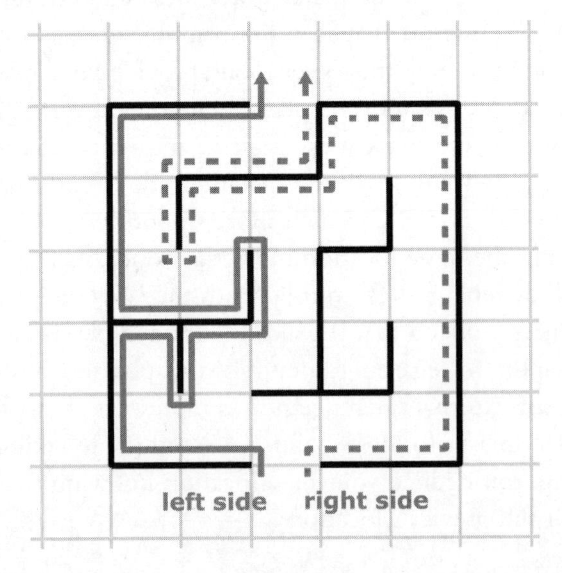

To be more precise, if you follow the right wall, you use the same route you would if you followed the left wall from the exit to the entrance.

This strategy has a great advantage in that you need not know anything about your position and orientation. The only capabilities required are that your robot can follow a wall and that it can recognize the exit when it's there.

At this point, the crucial question is, when can you apply this rule? There are essentially two cases in which you can do this:

- When the maze is flat, and has both the entrance and the exit placed along its perimeter (as in Figure 17.2).
- When the maze is flat, and the entrance and exit are points arbitrarily chosen anywhere in the maze, where the latter doesn't contain any loops. That is, it doesn't contain multiple paths that connect any two points (see Figure 17.3).

Figure 17.3 The Exit Is Inside a Maze with No Loops

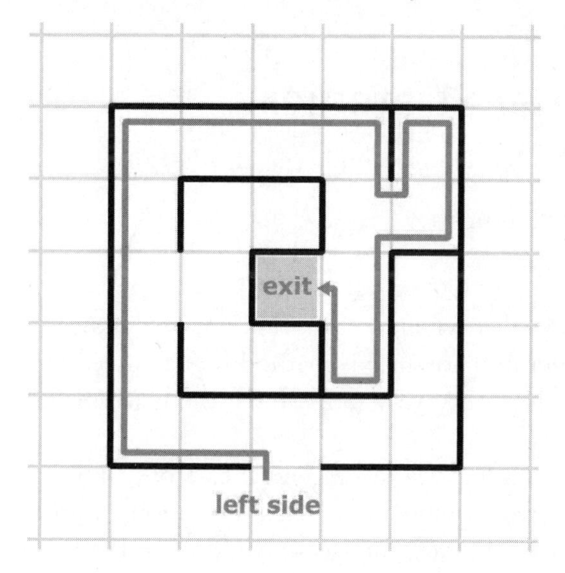

This rule covers many practical cases. However, it doesn't work when the entrance and exit are not along the perimeter *and* the maze contains loops, as in Figure 17.4. Notice that the route covered following the left wall brings you back to the entrance without reaching the exit point.

Figure 17.4 The Exit Is Inside a Maze with Loops

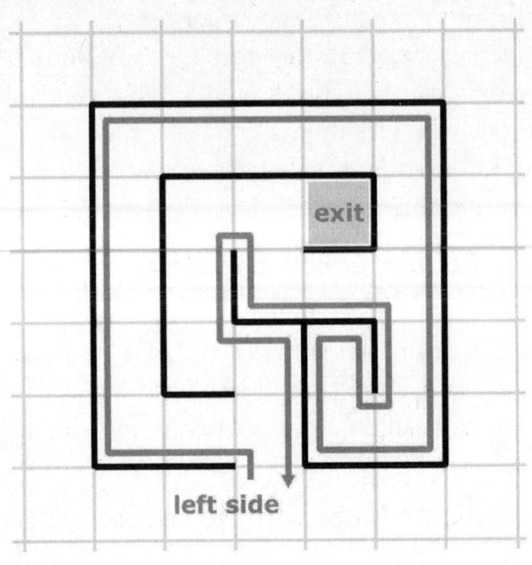

Applying Other Strategies

When you cannot apply the rule previously stated, you can rely on two strategies:

- Executing random turns
- Tracking your route

The first strategy says that whenever you find yourself at an intersection, you decide which way to go at random. Though this method is guaranteed to find the solution sooner or later, that "later" can be a very, very long time if the maze includes more than a handful of intersections!

The second approach solves the more general case of mazes with more than a few intersections, but it requires two valuable ingredients: a position control system and a memory. You must be able to recognize each intersection and mark the branches already explored so as not to explore them again. The right-side rule can still be useful as a basic rule, but when you find yourself in a place you've already been, you must be able to backtrack to the first intersection with unvisited branches and take one of those.

We imagine you already see the difficulties in this: You must provide your robot with an affordable navigational aid and an inner map to represent the maze so that you can mark the visited corridors. Fortunately, with the new MINDSTORMS NXT servo motors and RobotC, this mission is not so difficult.

But let's start with something simpler. We designed the first robot of this chapter, the Maze Runner, to apply the left-side rule inside a maze.

Building a Maze Runner

Our Maze Runner applies the left-side rule and follows the left wall of the maze toward the exit. It has no intelligence, only the capability to follow a wall.

Constructing the Maze Runner

To construct the Maze Runner, we used two servo motors, a ball caster, and one ultrasonic (US) sensor. You can replicate the whole robot with parts solely contained in the MINDSTORMS NXT set (see Figure 17.5).

Figure 17.5 The Maze Runner

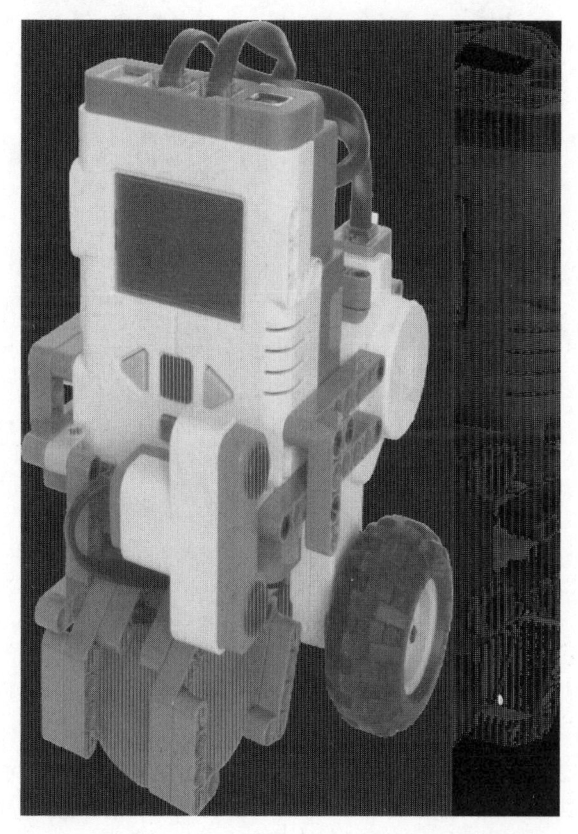

It works on a simple principle: The US sensor keeps the robot at a fixed distance from the left wall. When the distance changes abruptly, it considers this to be an opening to the left. If the gap is large enough for the robot to pass, it turns left. This covers the case of straight walls and left turns, but the robot will also have to face situations in which it hits a wall in front of it and must turn right. For this the robot monitors another encoder, and detects whether it is stalled by comparing the previous encoder reading to the current encoder reading.

We designed the Maze Runner to be as small as possible in the planar dimensions so that it can move through narrower mazes. We mounted the US sensor in front vertically to measure the distance to the left wall, while keeping the robot design compact. Figure 17.6 shows the front view of the Maze Runner, and Figure 17.7 shows the left-side view.

Figure 17.6 Front View

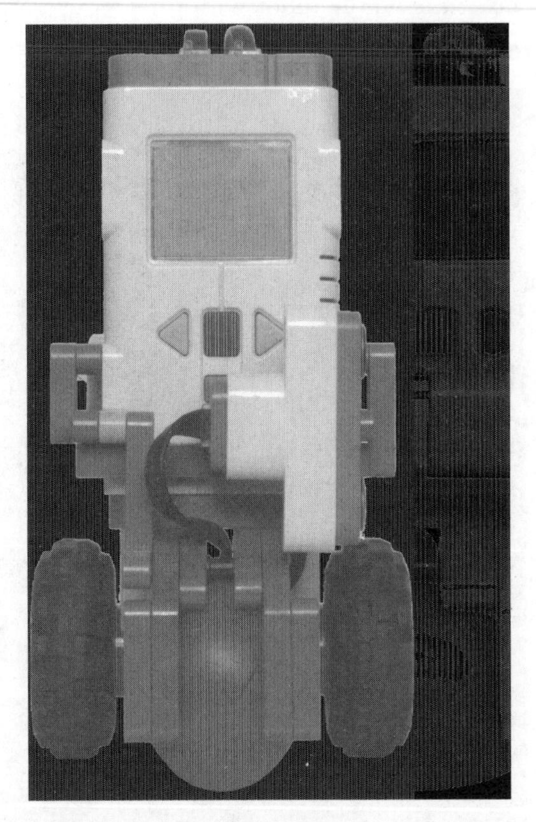

As we discussed in Chapter 6, the motors are an inherent part of the robot's "chassis." In our Maze Runner, the robot is divided into three "modules": the motors, the ball caster, and the NXT brick with the US sensor. As we emphasized in Chapter 6, building a robot in a modular fashion allows you to disassemble and fix each part without reconstructing the whole structure.

Figure 17.7 Left-Side View

The front ball caster was designed to be as small as possible, and to provide clearance for the US sensor (see Figure 17.8). Now, you may ask why the two L-shaped five-stud pieces are in the back. Well, when the robot moves forward, these two liftarms are in the air. However, when the robot hits a wall, due to the vertical design we found it can easily fall back. These two liftarms prevent this from happening.

Figure 17.8 Ball Caster Design

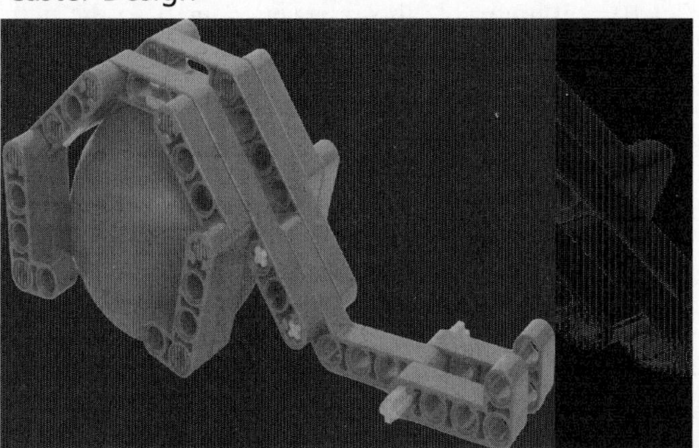

Programming the Runner

"Playing robot" is always a great exercise for devising or testing the strategy you are going to implement in your program. Even before you actually write any code, imagine running the program in your head, and try to explore the test maze of Figure 17.1 following the instructions step by step.

You will discover that this robot is relatively easy to program. The main program is composed of two nested loops (see Figure 17.9). One handles moving forward and turning right and the other handles the task of turning left and resuming forward movement.

Figure 17.9 Main Loop

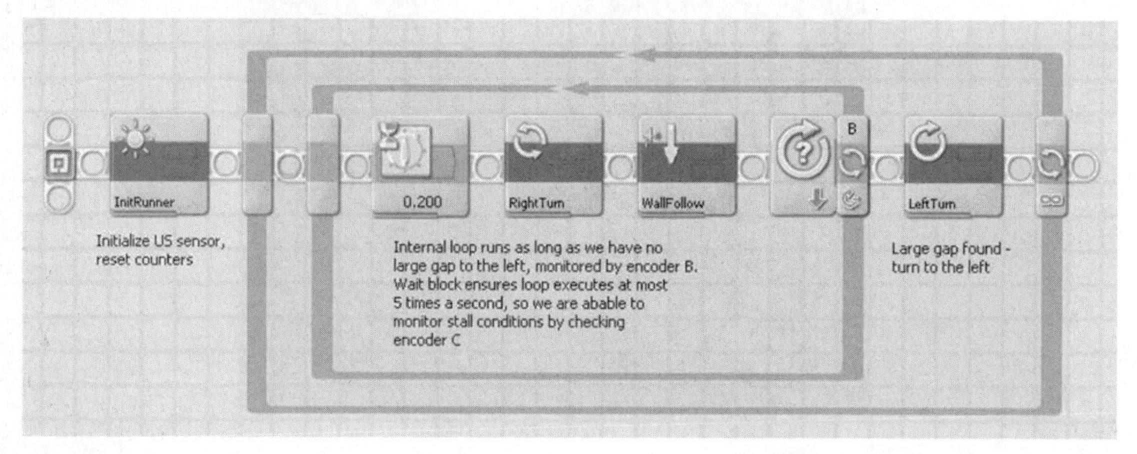

We wrote the program in NXT-G using MyBlocks to ease programming, improve code readability, and allow reuse of program pieces in the future. As you can see in Figure 17.9, this makes the diagram easy to understand. The wall-following MyBlock (see Figure 17.10) measures the distance to the left wall. If the distance is less than 20 centimeters, it assumes there is a wall to the left and provides a steering ratio to the *Move* block to correct the distance to the target of 10 centimeters. It also resets encoder B, so the nested loop in the main program exits only if there was a large enough gap to the left. Finally, it resets encoder C.

Figure 17.10 Wall-Following MyBlock

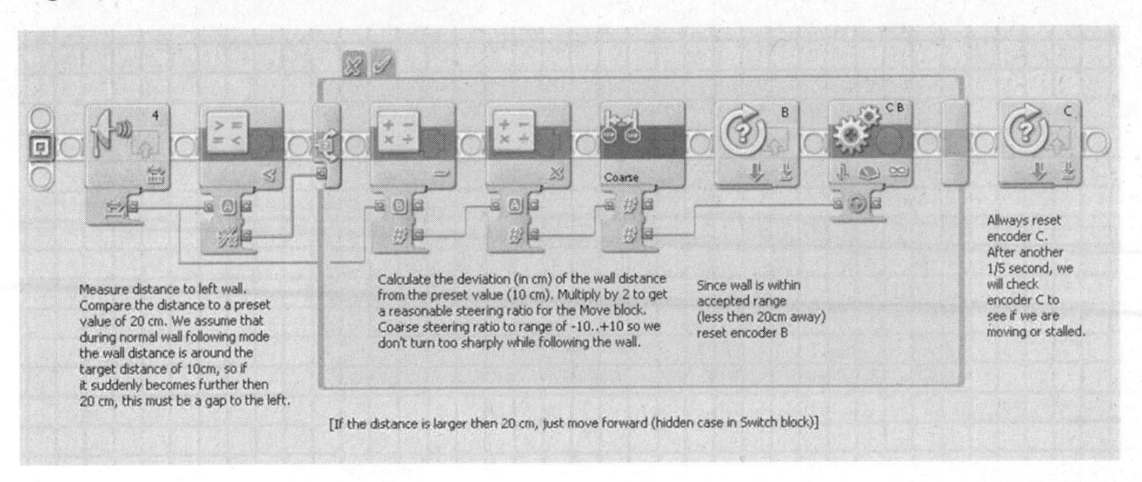

As the robot moves forward, the encoder C value will increment unless we are stuck at a wall. This is checked in the *RightTurn* MyBlock (see Figure 17.11). If encoder C did not increment at least 5 degrees from the previous loop iteration, we know we got stuck, and we need to retract and rotate to the right before we move forward again.

Finally, if encoder B is larger than some preset value, we know there was a large gap in the wall to the left, and the *LeftTurn* MyBlock (see Figure 17.12) turns the Maze Runner to the left.

Figure 17.11 Stall Detection and RightTurn MyBlock

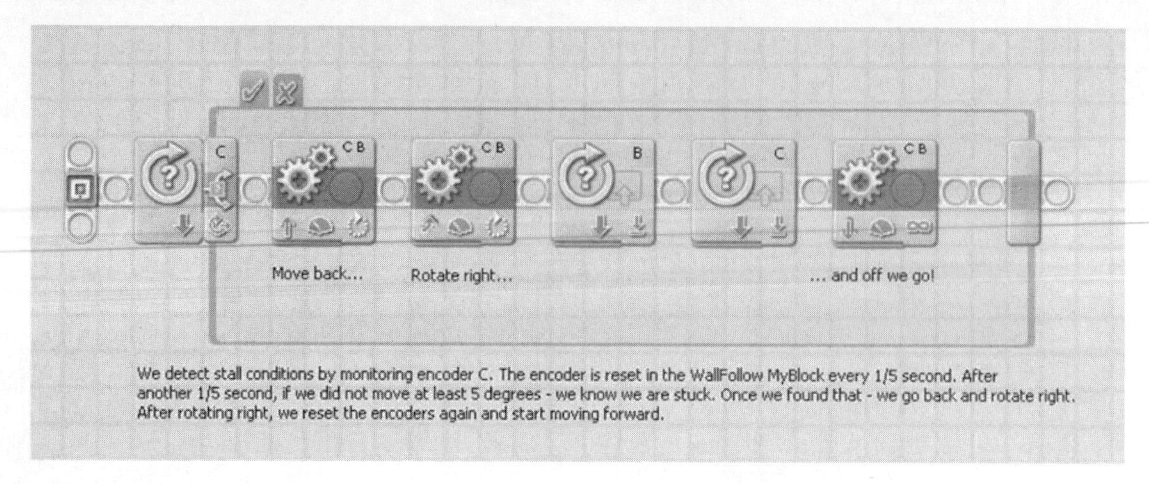

Figure 17.12 LeftTurn MyBlock

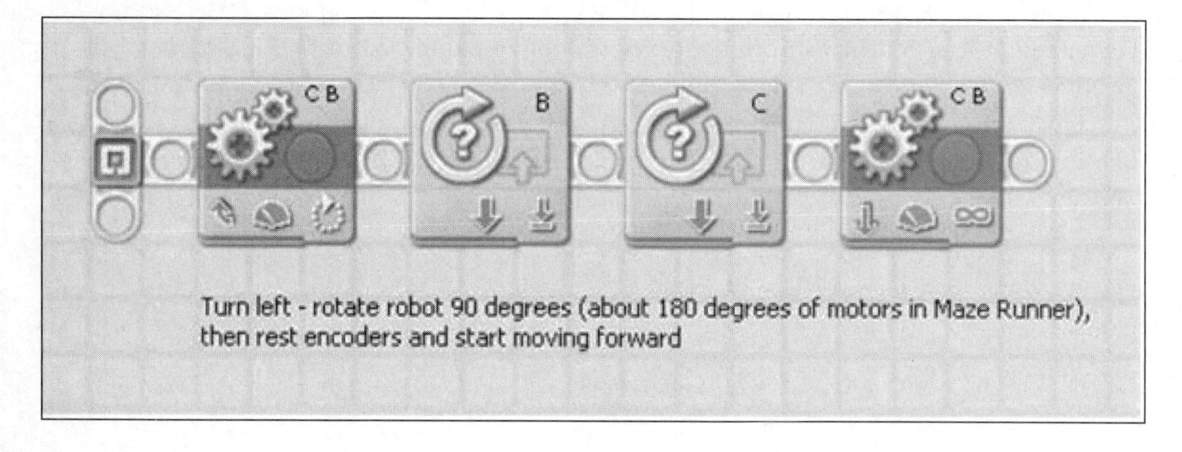

Creating the Maze

Now that you have a maze runner, you presumably would like a maze too! Unless you want to show off your robot at some sort of public exhibit, it's not necessary to build a lovely structure made from wood or other materials. You can test your creature against a makeshift labyrinth made with small pieces of furniture, piles of books, large boxes, cardboard, or anything else your imagination will suggest. Because we're using a US sensor to follow the walls, we avoided using soft materials or cloth-coated pieces as these do not reflect ultrasonic waves. Apart from that, the only thing we needed to ensure was that all the "walls" have a reasonably smooth surface at the height of the US sensor, so the sensor measures the distance to the walls correctly. You can find a video of the Maze Runner solving a simple maze (made of piles of books) at www.nxtasy.org (see Appendix A).

Building a Maze Solver

To overcome the limitations of the Maze Runner and its "left-side rule" tactic, and to solve the more general case of a labyrinth with an entrance and exit at two arbitrary points, we designed our own Maze Solver.

Constructing the Maze Solver

In Figure 17.13, you can see we changed our approach somewhat. We added a third motor to allow rotation of the US sensor so that we can scan in all three directions. The construction is again modular—the three motors and the back wheel caster comprise one module, the sensor with its rotating gear is another, and the third is the NXT brick mounted on top.

Figure 17.13 The Maze Solver

Figure 17.14 shows the rear side of the Maze Solver. The two driving motors are held together with an 11-stud beam and a 10-stud axle, also holding the third motor and the wheel caster.

Figure 17.14 Rear View

Figure 17.15 shows the front view with the NXT brick and the US sensor removed. The double bent beams connect the sensor motor to the other two motors, setting the angles between the three motors.

Figure 17.15 Front View (NXT Brick and US Sensor Removed)

Programming the Solver

Obviously, the hard part in making a maze-solving robot which uses memory is the programming. Our Maze Solver uses the algorithm we briefly described in the "Applying Other Strategies" section for tracking your path. This method is called Tremaux's algorithm. Before we get into the actual RobotC code, we will describe some concepts and assumptions needed for the program.

First, we define four "directions"—north, west, south, and east. These directions do not have to coincide with the actual four winds. Let's call the direction of our robot at the entrance "North," and let the other directions follow the usual convention. The current "direction" will be held in a variable called *MyDir*. The following code snippet declares the directions and some useful macros to find which direction is to the left/right of another direction:

```
typedef enum { // Define the directions in space
  North = 0,
  West  = 1,
```

```
  South = 2,
  East  = 3,
} tDirection;

tDirection MyDir = North; // My present direction

// What direction is to the left/right of 'Dir'
#define LeftOf(Dir) (tDirection)(((int)Dir+1) % 4)
#define RightOf(Dir) (Dir == North ? East : (tDirection)((int)Dir-1))
```

Next, we assume that the maze is flat and has a typical "length scale"—a characteristic length of which all maze dimensions are multiples. We call this length scale "maze units," and we shall use them to map physical space into a two-dimensional array holding the stored maze map in the robot's memory. Moving one maze unit forward corresponds to moving one "position"—in other words, changing our position in the map by 1. The map is stored in a variable called *Map* which holds a value between 0 and 3 for each maze "position":

- **Value = 0** This corresponds to a map position which we did not visit, or those which we cannot reach at all.
- **Value = 1** This corresponds to a position which we've passed already; in addition, we have already tried all possible routes emerging from it.
- **Value = 2** This corresponds to a position we have already visited, but from which we haven't tested one emerging route. For example, coming at a T-intersection, we need to choose one path. The other path is still unexplored, so we mark this position in the map with a value of 2.
- **Value = 3** This corresponds to a position in which two emerging routes have been left unexplored. This occurs when we first come to a four-way junction. We'll pick the right path, leaving the other two for the future.

Why do we bother with these different values? Well, when we move around the maze, we will probably encounter cases in which either there's no way to proceed (we've reached a dead end), or all positions accessible from our present position have already been visited in the past. In such cases, we need to backtrack through our path until we reach a position in which we left an unexplored path. To facilitate the backtracking procedure, we mark these positions on the map with a value of 2 or 3, making it faster to go back to these. The current position in *Map* is stored in two variables, *MyPosX* and *MyPosY*, defined in the following code snippet:

```
// size of maze map
#define SIZE_X 22
#define SIZE_Y 22
```

```
// Hold map of maze.
// 0 - haven't visited this position
// 1 - been here, no unvisited paths from here
// 2, 3 - there are more ways to go from here!
int Map[SIZE_X][SIZE_Y];

int MyPosX = 10, MyPosY = 1; // Keep position in map, start at (10,1)

int MyPath[SIZE_X*SIZE_Y]; // Hold path traveled
int PathCounter = 0;
```

As we said earlier in this chapter, we use *Dir* to describe our robot's orientation. We define two more useful macros that return a relative map position at a particular "direction":

```
// Convert directions to relative position in map
#define RelPosX(Dir) (Dir == West ? -1 : (Dir == East ? 1 : 0))
#define RelPosY(Dir) (Dir == North ? 1 : (Dir == South ? -1 : 0))
```

Now we can easily find out what's around us in the maze map! Our current position is *Map[MyPosX][MyPosY]*, in front of us is *Map[MyPosX+RelPosX(MyDir)] [MyPosY+RelPosY(MyDir)]*, and to the right (similarly to the left) is *Map[MyPosX+RelPosX(RightOf(MyDir))][MyPosY+RelPosY(RightOf(MyDir))]*. There is, however, a small problem with the maze boundaries. We do not want to refer to positions outside the array bounds. We solve this with a very widely used "trick." We define the maze array to be two maze units bigger in each dimension than the real maze. Then, at the start of the program, we set the outer "frame" of the array to a value of 1, and the rest to 0:

```
// Set initial map
void Initialize_Map() {
  int j, k;
  for (j = 0; j < SIZE_X; j++)
    Map[j][0] = Map[j][SIZE_Y-1] = 1;
  for (k = 0; k < SIZE_Y; k++)
    Map[0][k] = Map[SIZE_X-1][k] = 1;
  for (j = 1; j < SIZE_X-1; j++)
     for (k = 1; k < SIZE_Y-1; k++)
       Map[j][k] = 0;
}
```

Because we consider a value of 1 to be an "exhausted" position, we will never try to reach it as we move around the maze. This will also prevent us from trying to move out of the entrance if we encounter it again!

Finally, we will define some subroutines that do the actual motion; in other words, rotate the US sensor in all directions; rotate the whole body left, right, or 180 degrees back; move forward one maze unit; and check whether we are at the maze exit:

// These functions perform robot actions - you should write these!

```
void rotate_US_left();
void rotate_US_front();
void rotate_US_right();
void rotate_robot_left();
void rotate_robot_right();
void rotate_robot_back();
void move_forward();
bool is_in_exit();
```

As you can see, we intentionally left these functions empty. You must complete these for your robot to be capable of moving a single inch!

We now get to describe the main loop. At each iteration, the code will check which of the three possible ways ahead (left/front/right) is accessible (the Boolean array *Walls* will contain those which are not accessible) and decide which direction to go next in a variable called *NextMove*:

```
#define DecreaseOne(Var) { Var -= 1; Var = ((Var) < 1) ? 1 : (Var); }
bool Walls[3]; // which directions have a wall near? (0 - left, 1 - front, 2 -
right)
int NextMove; // which move should we do next?

#define WALL_NEAR 20

// This subroutine does the actual maze solving using Tremaux's algorithm
void Solve_Maze() {
while (!is_in_exit()) { // loop until we are in the maze exit
  // check in which directions we see walls
  rotate_US_left();
  Walls[0] = SensorValue(sonarSensor) <= WALL_NEAR;
  rotate_US_front();
  Walls[1] = SensorValue(sonarSensor) <= WALL_NEAR;
  rotate_US_right();
  Walls[2] = SensorValue(sonarSensor) <= WALL_NEAR;

  // first, assume all possible paths are unvisited
  Map[MyPosX][MyPosY] = 3-(int)Walls[0]+(int)Walls[1]+(int)Walls[2];
```

```
  NextMove = -1; // This value will tell that all paths are visited...

  // check left path
  if (!Walls[0]) {
    // Have we visited the position to our left?
    if
(Map[MyPosX+RelPosX(LeftOf(MyDir))][MyPosY+RelPosY(LeftOf(MyDir))]==0)
{
      // No. Let's go there now!
      NextMove = 0;
    } else {
      // Yes. Decrease number of new paths here and at the position to the left
      DecreaseOne(Map[MyPosX][MyPosY]);

DecreaseOne(Map[MyPosX+RelPosX(LeftOf(MyDir))][MyPosY+RelPosY(LeftOf(MyDir))]);
    }
  }

  // check forward path
  if (!Walls[1]) {
    // Have we visited the position in front of us?
    if (Map[MyPosX+RelPosX(MyDir)][MyPosY+RelPosY(MyDir)]==0) {
      // No. Let's go there now!
      NextMove = 1;
    } else {
      // Yes. Decrease number of new paths here and at the position in front
      DecreaseOne(Map[MyPosX][MyPosY]);
      DecreaseOne(Map[MyPosX+RelPosX(MyDir)][MyPosY+RelPosY(MyDir)]);
    }
  }

  // check right path
  if (!Walls[2]) {
    // Have we visited the position to our right?
    if
(Map[MyPosX+RelPosX(RightOf(MyDir))][MyPosY+RelPosY(RightOf(MyDir))]==0)
{
      // No. Let's go there now!
      NextMove = 2;
    } else {
```

```
        // Yes. Decrease number of new paths here and at the position to the right
        DecreaseOne(Map[MyPosX][MyPosY]);

DecreaseOne(Map[MyPosX+RelPosX(RightOf(MyDir))][MyPosY+RelPosY(RightOf(MyDir))]);
      }
    }

    .
    . // next code snippets come here!
    .
}
```

The first lines of the code check the distance to the wall at each of the four winds. If the distance is smaller than some predefined value, we decide there's a nearby wall in this direction. We then assume that all accessible directions lead to an unexplored position (i.e., a position in which *Map* equals 0), and we check the map at each of these directions. If we indeed find it unexplored, we decide to go there next. Otherwise, we decrease the number of "waiting" paths from our present position, as well as from the position we look at. Why the latter? Because the only way we will encounter a visited position while traveling in an unexplored route is if we're closing a loop, so the previously traveled position must have had more than one path available when we were there. Notice that this code is a "right-hand rule" code: If all paths are accessible and unexplored, the program preference is to first turn right; if not, the next option is to move forward, and the last option is to turn left.

When we decide on the next step (i.e., *NextStep* is 0, 1, or 2), we rotate the robot toward that direction and move forward:

```
  switch (NextMove) {
    case 0: // go left
      rotate_robot_left(); // rotate the whole robot left
      MyDir=LeftOf(MyDir); // change to new direction
      break;
    case 2: // go right
      rotate_robot_right(); // rotate the whole robot right
      MyDir=RightOf(MyDir); // change to new direction
      break;
  }

  if (NextMove != -1) { // Should we move forward?
    move_forward();
    MyPath[PathCounter++] = NextMove; // keep track of the move we did
    MyPosX += RelPosX(MyDir); MyPosY += RelPosY(MyDir); // change position
    continue; // go back to loop start
```

```
}
.
. // next code snippet comes here
.
```

As you can see, we keep track of the steps we performed: left turn, forward, right turn. If we did not find a path to use, we start backtracking using this path until we reach a position in which there are still unexplored routes:

```
// If we got here, either we are in a dead end or we visited all possible
// directions from this position. We should backtrack our way out!
rotate_robot_back(); // rotate robot 180 degrees

// go back until you find a position with unexplored paths
while ((Map[MyPosX][MyPosY] < 2) && (PathCounter-- > 0)) {
  switch (MyPath[PathCounter]) {
    case 0: // go back right
      rotate_robot_right(); // rotate the whole robot left
      MyDir = RightOf(MyDir); // change to new direction
      break;
    case 2: // go back left
      rotate_robot_left(); // rotate the whole robot right
      MyDir = LeftOf(MyDir); // change to new direction
      break;
  }
  move_forward();
  MyPosX += RelPosX(MyDir); MyPosY += RelPosY(MyDir); // change position
}
```

Once we reach a new emerging route position, we return to the start of the *while* loop and look again for a path to take.

Improving the Program

We can still improve our Tremaux's algorithm program in several ways. Test your strength in RobotC by trying to implement these improvements:

- When you decide to backtrack, first check whether backtracking will not lead you past one of the positions around you. This would occur, for example, if you moved in a circle and returned to the starting point. In this case, turning 180 degrees and going back is wasteful; it's better to move one step and remove the whole loop from *MyPath*.

- When you find that a nearby position is accessible and its *Map* value is 3, this means it contains an unexplored path; go for it!

Bricks & Chips...

Struggling with Limited Memory

The RobotC firmware gives you more than 15,000 bytes for program memory. For Tremaux's program, this allows you to solve a maze that is 80 x 80 maze units in size, which is probably bigger than you'll ever encounter. Yet, if you encounter memory issues with this or other programs, you may try to improve memory usage by using bit manipulations.

Consider the values we kept for each map position: 0, 1, 2, or 3. Thinking in terms of bits, you know that two of them are enough for a position (represented by the binary values 00, 01, 10, 11). Thus, you can store four maze units in one byte of memory, reducing memory usage by a factor of four!

The technique of addressing single bits of a variable requires that you have a bit of programming skill, and that you're familiar with bit masks and bitwise binary operations in particular. Any good programming text will help you understand how this mechanism works.

Summary

If you want to test your skills in maze solving, the first step you have to take is to understand the details involved in the process of finding your way out of a maze. We encourage you to draw a simple maze on a sheet of paper and to "play robot" with it: Take a pencil which represents the position of the robot in the maze and move it according to the "program" you execute in your head. This preliminary study will provide you with the necessary knowledge to successfully build and program your robot.

The robots we discussed in this chapter prove that maze solving is in the range of MINDSTORMS NXT robotics. In discussing the theory, we explained that maze solving requires a robot with both an accurate navigation system and a memory to store a map of the labyrinth. The navigation system is the more demanding of the two requirements (recall Chapter 13 and the problems involved in finding the robot's location).

We discovered in this chapter that maze solving may be no more complex than wall following. This means your robot needs only minimal intelligence—a trait reflected in our Maze Runner robot. If the maze's entrance and exit are not placed along its perimeter, a cleverer algorithm is required. We designed our Maze Solver to solve such mazes, using Tremaux's algorithm.

Chapter 18

Drawing and Writing

Solutions in this chapter:

- Creating a Logo Turtle
- Creating a Tape Writer
- Further Suggestions

Introduction

Can a MINDSTORMS robot be made to draw or write? Sure. Believe it or not, that's not even a very difficult thing to implement. In the following pages, we will show you two projects, the first mainly meant for drawing and the second for writing. Both of them require some additional parts, but both have wide margins for modifications and allow for less demanding variants.

Creating a Logo Turtle

Many of you may already know that Logo is a programming language specifically targeted to education. Born in the late 1960s at the Massachusetts Institute of Technology (MIT), Logo is derived from Lisp (with far fewer parentheses!) and features interactivity, modularity, and extensibility. More than a programming language, Logo is a learning tool which has gone through a number of changes and improvements over the years.

The most known characteristic creation of Logo is the *Turtle*, a symbolic turtle that moves across the computer screen according to the instructions it receives. With simple instructions such as *forward 10*, the turtle moves straight 10 units, and with *right 90* it turns clockwise 90 degrees. The statements *penup* and *pendown* specify whether the turtle leaves a track behind it, thus producing drawings, or rather just moving to a different location. Obviously, the language includes many other commands, but these are enough to understand the principles of the Turtle graphics that made Logo so famous.

What many people don't know is that in its first version, the Logo program controlled a small robot that actually drew lines on the floor. In subsequent releases, the turtle became just a virtual animal on the screen. Our interest here, however, is in replicating the first robotic version.

NOTE

Dr. Seymour Papert was one of the early promoters of Logo, and he designed the original Turtle. Under his guidance, the Epistemology and Learning Group at MIT devised the first *programmable brick*, whose concepts led to the development of the LEGO MINDSTORMS line.

Building the Turtle

The idea is quite simple: Build a small robotic platform that's able to go forward and backward, turn in place, and lower and raise a pen. Despite this apparent simplicity, if you want a turtle that works as expected, the task has many stringent requirements that must be adhered to. For instance:

1. The robot must go absolutely straight.
2. The pen must be exactly in the pivoting point of the robot, because it must stay in the same place on the floor while the robot turns (otherwise, it would trace a curve).
3. You need a tracking system to measure both traveled distances and angles.

If you remember the driving architectures described in Chapter 9, you already know the solution to the first point: Use a dual differential drive. The simple differential drive is suitable for this project only if you apply an active control to the wheels to be sure they travel exactly the same distance, whereas the synchro drive would work as well but at the price of greater complexity and a not so evident change in orientation during action. Another advantage of the dual differential drive is that it requires a single encoder to comply with point 3: When the robot goes straight it measures the covered distance, and when turning it measures the angle.

Okay, so we have requirements 1 and 3 covered, but there's still the matter of the pen being the center of rotation, which is at the midpoint of the imaginary line that connects the wheels. Conceptually, it sounds easy, but you have to build your robot with this point in mind.

The original turtle—a differential drive—featured a transparent plastic dome to cover the gears. We provided our turtle with a triangular shape (see Figure 18.1), because we wanted to mimic the screen turtle of some widespread Logo systems. Anyway, those V-shaped beams are definitely not necessary and you can shape your own turtle according to your wishes.

Figure 18.1 The Logo Turtle

Our differential drive does not use a caster wheel, because they tend to affect the direction of the robot slightly when resuming straight motion after a turn. With casters, the straight lines would have a short wiggly segment, so we preferred to use a simple tile as the third supporting point. To keep the friction on the floor to a minimum, we placed the NXT suspended behind the drive wheels, like a sort of counterweight, bringing the COG of the robot very close to the drive axles, and thus, most of the weight upon the drive wheels.

Let's start exploring the dual differential drive chassis that drives the robot (see Figure 18.2). The gearing is more compact than those shown in Chapter 9, but it works exactly the same way: One motor makes the differential gears and the wheels rotate in sync (motor C), and the other rotates them in the opposite direction (motor B in this case). The dark gray 16t gear right in the middle of the photo is an idler gear which connects the other two 16t gears; its center hole is not cross-shaped and thus it doesn't couple with the long joined axle that crosses the base of the robot.

Figure 18.2 The Turtle Dual Differential Drive Platform (Top View)

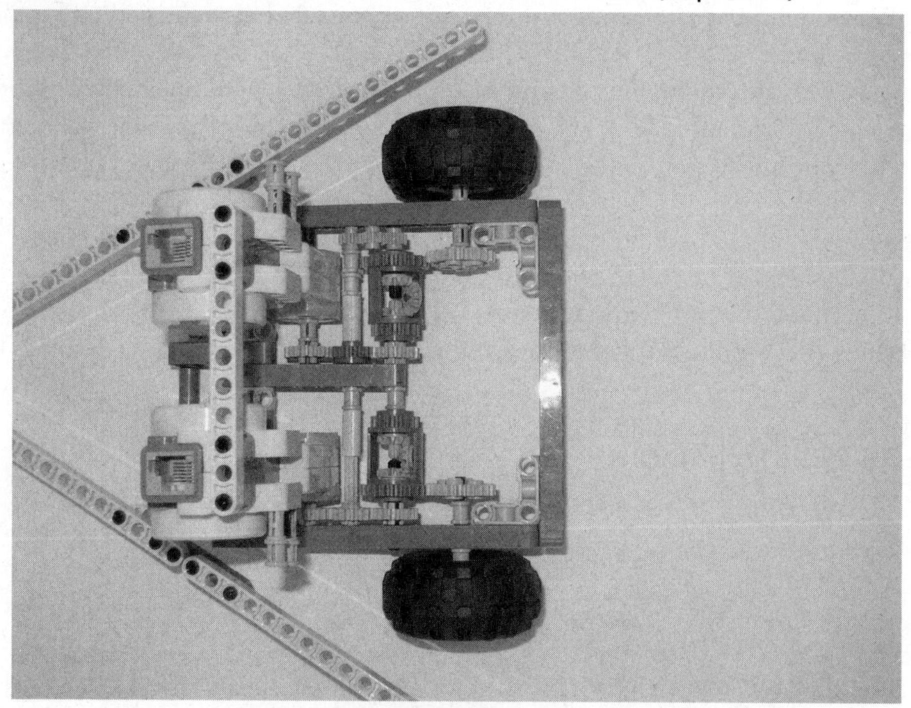

Looking at the bottom, you can see the front skid roller (see Figure 18.3). Figure 18.4 shows a side view of the turtle pen mechanism.

Figure 18.3 The Turtle Dual Differential Drive Platform (Bottom View)

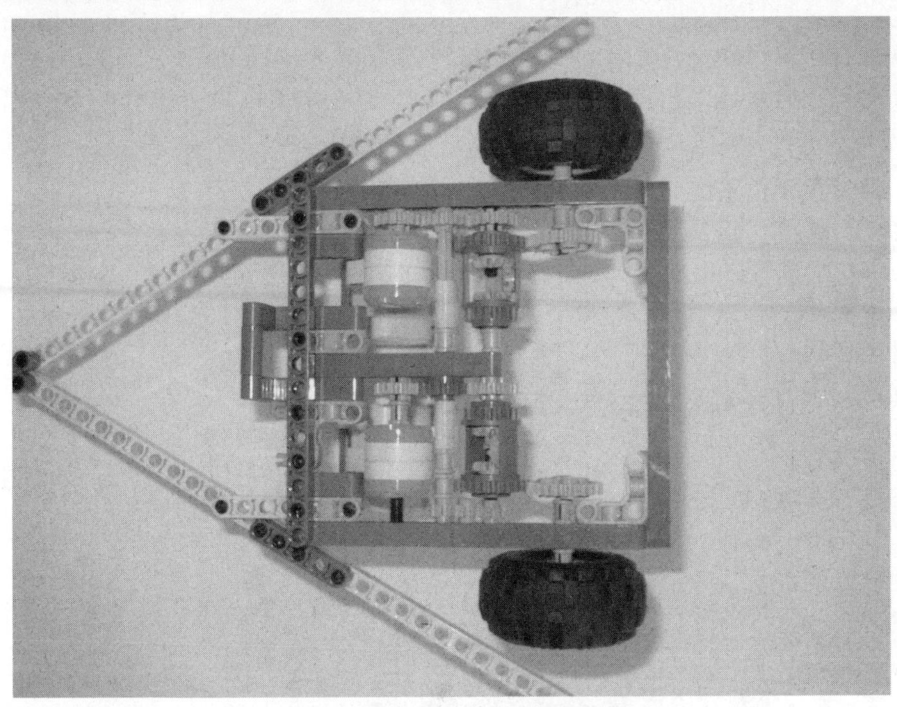

Figure 18.4 Side View of the Turtle Pen Mechanism

The pen is a non-LEGO part, a common marker with its body wrapped in adhesive tape so as to make it fit tightly into the 2 x 2 stud's squared hole reserved for the purpose. It stays there with nothing but friction. Two rubber bands are placed to ensure this friction.

The pen control mechanism is a swinging assembly operated by a third motor (in this case, motor A) (see Figure 18.5).

Figure 18.5 Turtle Top View

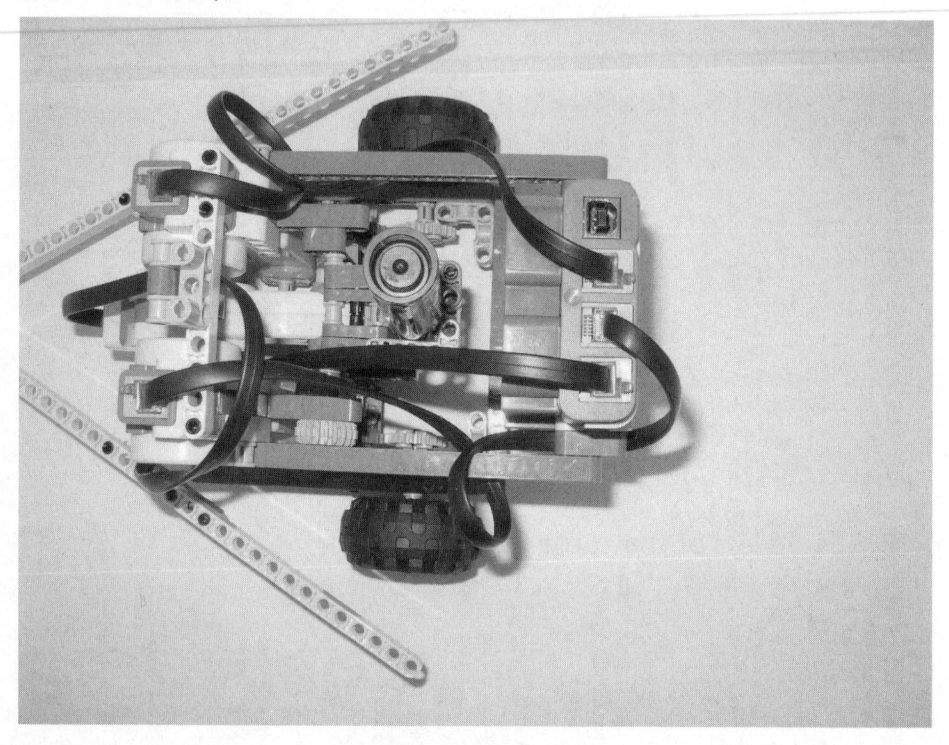

Now the turtle is ready. Place a large piece of paper on the floor, uncap the pen, and adjust its height so that it touches the paper gently when it is in the down position (see Figure 18.6). We strongly discourage you from writing *directly* on the floor. We're sure *somebody* won't like it!

Figure 18.6 Side View of the Turtle Ready for Operation

Programming the Turtle

The first task in programming the Turtle is to create the primitives that control the basic actions. Let's start with the easiest ones: the commands to move the marker *up* and *down*. A short degree rotation to the marker motor (motor A) does the trick—nothing more is required. In this case, it is easiest to set the marker down position on the ground and use the view rotation function on motor A to see the number of degrees necessary to move the marker adequately above ground. For this design, 75 degrees did the trick.

The *forward* and *back* commands, meanwhile, are not very difficult to implement, but require that you dig into the physical properties of your robot. You must discover what distance it covers for any increment of the motor rotation. Programming the distance you want the robot to travel affects the size of the image you are intending to create. If you have a desired size object—for example, a square with 30cm sides—you can calculate the amount of motor rotation required for the turtle to create this length. All you need to know is the wheel circumference (in this case, 17 cm to 17.5 cm) and the gear ratio from the motor wheel (in this case, a 12- to 24-tooth ratio [1:2]). If you have a fabric measuring tape or other flexible measuring tape, it is easy to measure the wheel circumference. If you do not have this type of measuring tape you can just as easily measure the distance you want by rolling the wheel on a ruler one rotation. For the wheels used in this design, the wheel circumference varied from 17 cm to 17.5 cm, compressed or uncompressed. Because these wheels were designed with a large amount of built-in shock absorption, they will vary in diameter with the amount of weight placed on them.

To get our turtle to travel exactly 30 cm, for example, all we have to do is multiply the gear ratio by the ratio of the distance we want the turtle to travel, and divide by the wheel circumference. This is displayed in the following equation:

$$Rotation\ forDrive\ Motor = \frac{GearRatio * DesiredLength}{WheelCircumference}$$

For our purposes:

$$\frac{2*30cm}{17cm} = 3.529 \text{ rotation or } 1{,}271 \text{ degrees.}$$

You should use this rotation for the forward drive motor; in this case, motor C (see Figure 18.7).

Figure 18.7 Programming for 30cm Forward Movement

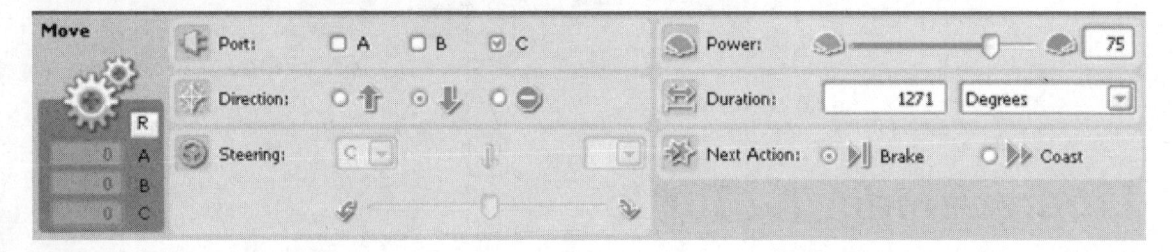

This is the theory. The actual robot will probably require some in-the-field tuning, because the distance covered by the wheels is affected by other factors: The weight compresses the tires and reduces their diameter. There might be some slippage too. We suggest you proceed by experimentation, making your turtle draw a line, measuring it, and then correcting the factor until you're happy with the result. All this process is meaningful only if you care about having your turtle use units that correspond to some common length unit. If you don't care, simply program the robot to go forward two or three rotations, for example.

Now the last part: the turning primitives *right* and *left*. If you remember from before, you find that the change in orientation □O_R (in radians) depends on the distance covered by the wheels ($T_R - T_L$) and the distance between the wheels (B). When the dual differential drive turns in place, both wheels travel the same distance (T) in opposite directions, so we can express the equation in simplified terms (see Figure 18.8):

□O_R = 2 x T / B

Figure 18.8 Computing Changes in Orientation

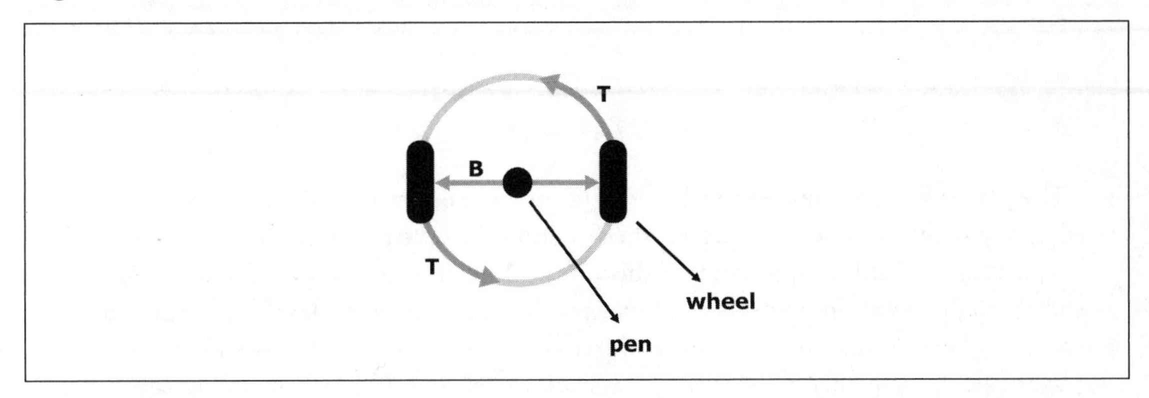

Actually, you know the □O_R you want to get; it corresponds to the desired turning angle of the turtle, and it's the input to your subroutine. What you're looking for is the *Count* of the rotation sensor that produces that □O_R. The first step is to obtain T from the previous equation:

For our robot to make a star, we need the robot to make a very sharp turn. The angles of the points of a star are 36 degrees. For our robot to make this point, it has to turn 180 – 36 degrees, or 144 degrees. To make any turn we need to consider the *outer* turn angle for the robot. This may seem a little confusing at first, but if you stop to think about the way the robot travels, it makes sense.

Now that we know our robot should turn 144 degrees to make a star, we need to figure out how to program it to do so. The following equation will explain this process:

$$\textit{Rotation for Turn Motor} = \frac{GearRatio * DegreesToTurn * \pi * WheelBase}{WheelCircumference}$$

For our purposes:

$$\frac{2 * 144 * \pi * 15.5cm}{17cm} = 825 \text{ degrees or } 2.29 \text{ rotations}$$

You should use this rotation for the turning drive motor; in this case, motor B (see Figure 18.9).

Figure 18.9 Programming for 144 Degrees Turning Movement

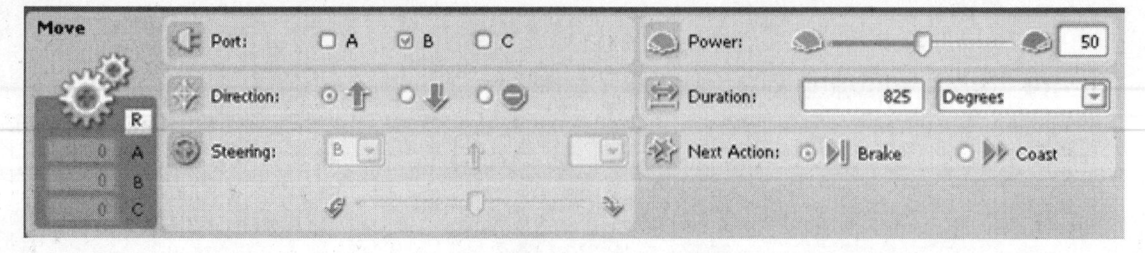

The power level is reduced to allow the robot to make more accurate angles. You may need to adjust this more depending on the friction of the surface on which you are writing.

As for the forward motion control, this one will need some adjustment too. Your turtle is not likely to draw proper angles on the first try. We suggest you make it draw a simple polygon, such as a square or an equilateral triangle, and check that it closes the path properly. For example, the sequence of *four forward 1 rotation right 518 degrees* should draw a square; if the last segment intersects the first one but not at its starting point, the count is too high, and you have to increase the rotation (e.g., 520 degrees instead of 518 degrees) and vice versa: If the square doesn't close at all, you should decrease the rotation (see Figure 18.10).

Figure 18.10 Tune Calculations by Testing Your Turtle in Drawing a Square

Instead of working on the software, you can often change the geometry of the robot. Altering the distance between the wheels by moving them in or out along the axles is a very effective way to tune the robot. Make small adjustments until your square comes out perfect.

Once you have the correct rotation, it is easy to program the turtle to make a figure with a count loop. Figures 18.11–18.14 show an example program for creating the star.

Figure 18.11 Example Program for a Star

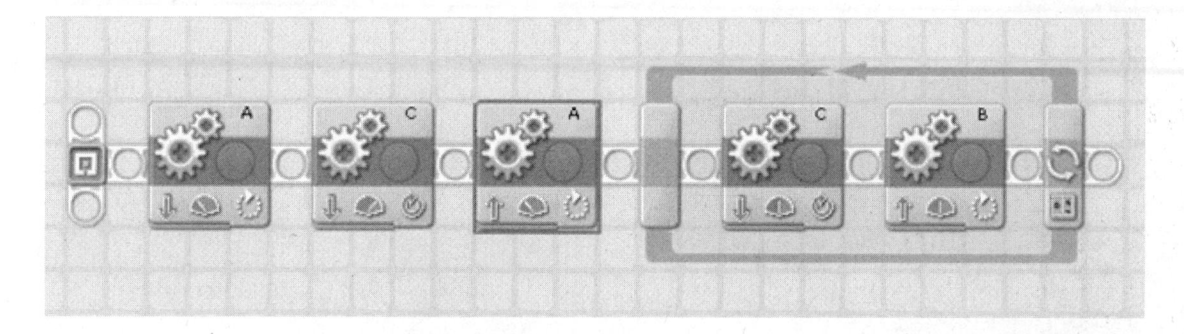

The marker is lifted and then the turtle travels to a specific location. Then the pen is placed down on the surface and the turtle enters a five-count loop that creates a star.

Figure 18.12 Example Program for a Star: Drive Motor

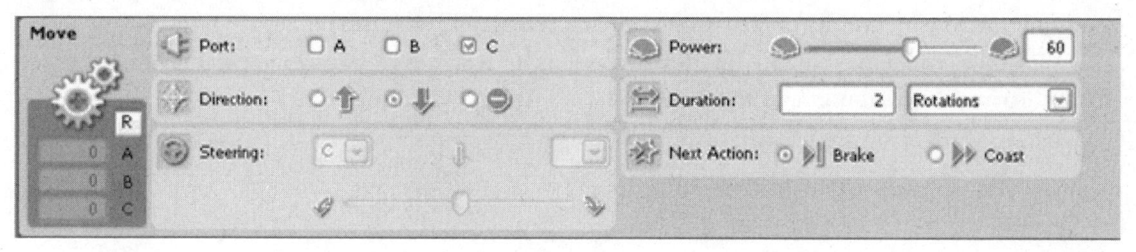

In the star, the turtle moves forward two rotations at 60 percent power before each turn.

Figure 18.13 Example Program for a Star: Turn Motor

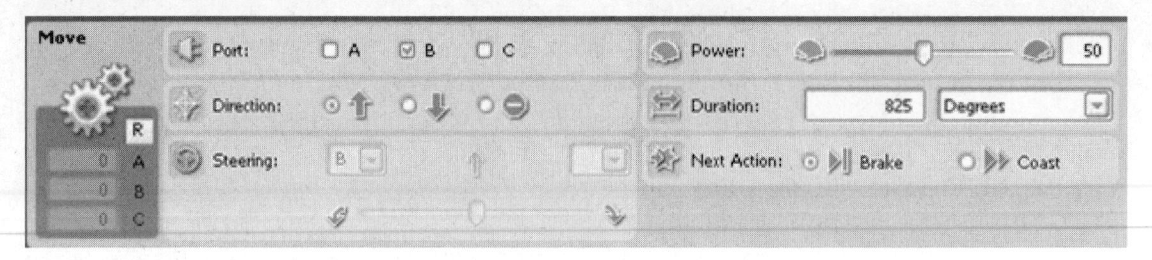

The turtle turns 825 degrees at 50 percent power during each turn.

Figure 18.14 Example Program for a Star: Loop Count

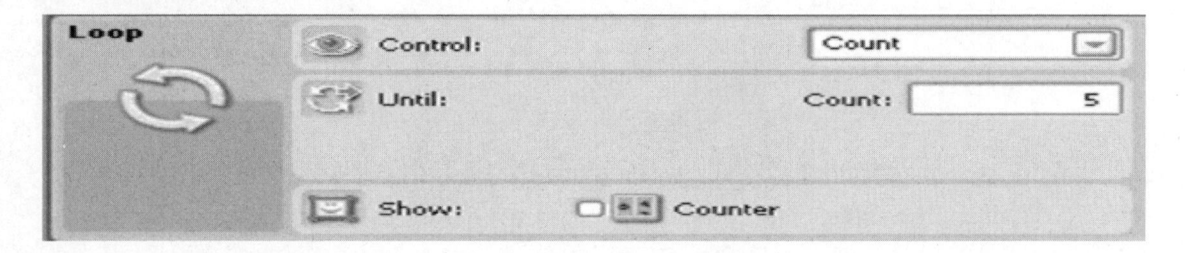

The turtle executes a total of five counts, creating the five lines of the star.

Work patiently on your turtle and its code. The result will astound you! Figure 18.15 shows our turtle drawing an almost perfect five-pointed star and a pentagon.

Figure 18.15 The Logo Turtle in Action

Tape Writer

The second project in this chapter uses an approach somewhat opposite to that of the Logo Turtle: Here it's the paper that moves, while the robot stays still. The principle is similar to the one used in ink-jet printers: A mechanism feeds the paper under a writing head, which by itself moves perpendicularly to the direction in which the sheet advances. From what you learned in the previous chapters, you can tell that such a system has two degrees of freedom (DOF), controlled respectively by a paper-feeding motor and the writing head motor (actually, our robot implements a third DOF, needed to move the pen up or down over the paper). This Tape Writer is also a Cartesian system, because the movements of the mechanisms are linear and perpendicular to one another.

This robot requires some extra parts: gear racks, beams, and a large tile; however, if you don't have the needed parts, there are many things you can do to downsize the project to keep within your inventory (we'll describe some of them).

Building the Writer

What we have in mind is a robot that writes on one of those common paper tapes made for printing calculators or cash registers. One motor moves the paper strip forward and backward, and a second moves the pen in a perpendicular (side-to-side) direction. The third motor controls the up/down pen movements.

Starting from the end, here's our finished robot that writes the word *LEGO* (see Figure 18.16).

Figure 18.16 The Writer Composes Its First Word

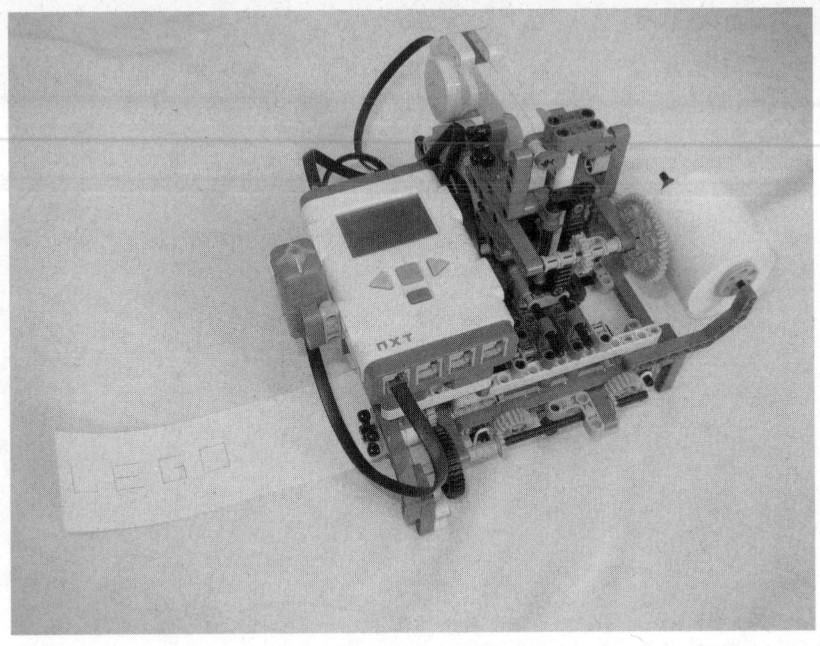

Analyzing the Tape Writer in detail, you can see that it's made of a body and three sub-systems, all of them with one degree of freedom:

- The body provides the main structures and hosts the paper transport system.
- There's a movable carriage over the body, which transports the pen in a direction perpendicular to the tape.
- Over the carriage, the pen assembly moves up and down.
- At the bottom of the body, there's the writing surface, a smooth surface that presses the paper against the wheels.

Looking inside the main body, you catch a glimpse of the transport wheels and the pen assembly (see Figure 18.17).

Figure 18.17 Writer Side View

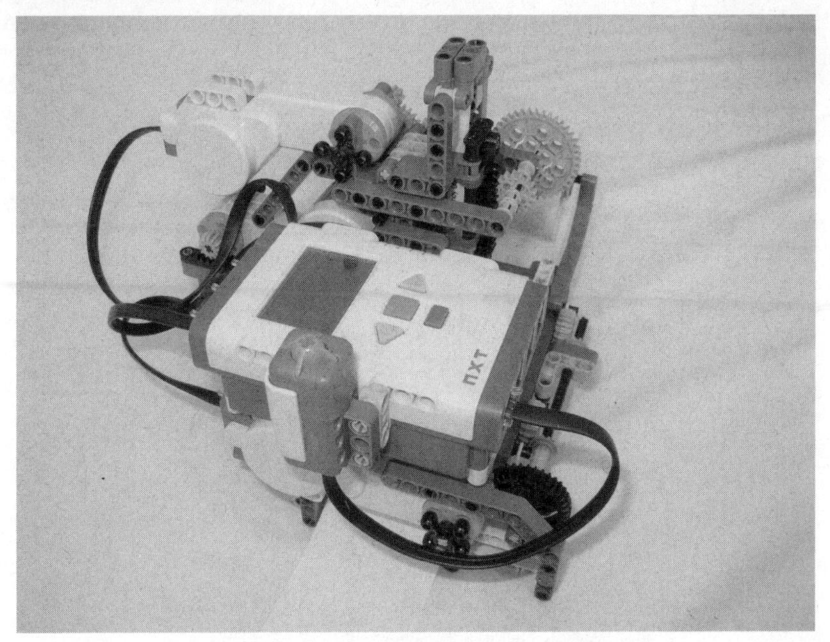

The wheels are operated by a motor through a 16t gear, a 36t gear, and two connected 12t and 24t gears (see Figure 18.18). This latter geartrain is necessary to keep the two groups of dragging wheels turning in the same direction. You need the paper to go back to shape some letters, and this is why there are wheels both before and after the pen.

Figure 18.18 Writer Rear View

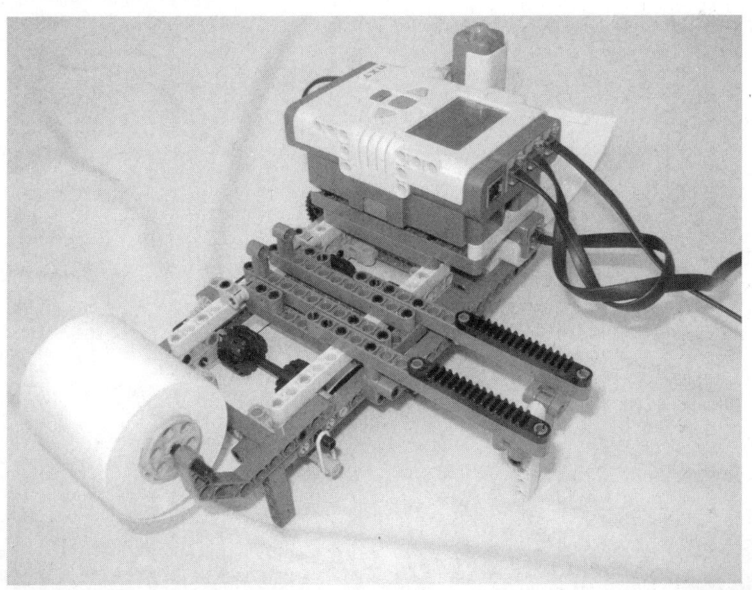

Removing the pen carriage, you see the wheels and the paper tape down below (see Figure 18.19). The carriage is translated using a rack and pinion assembly, powered by a second motor on the body.

Figure 18.19 Writer Top View, Pen Carriage Removed

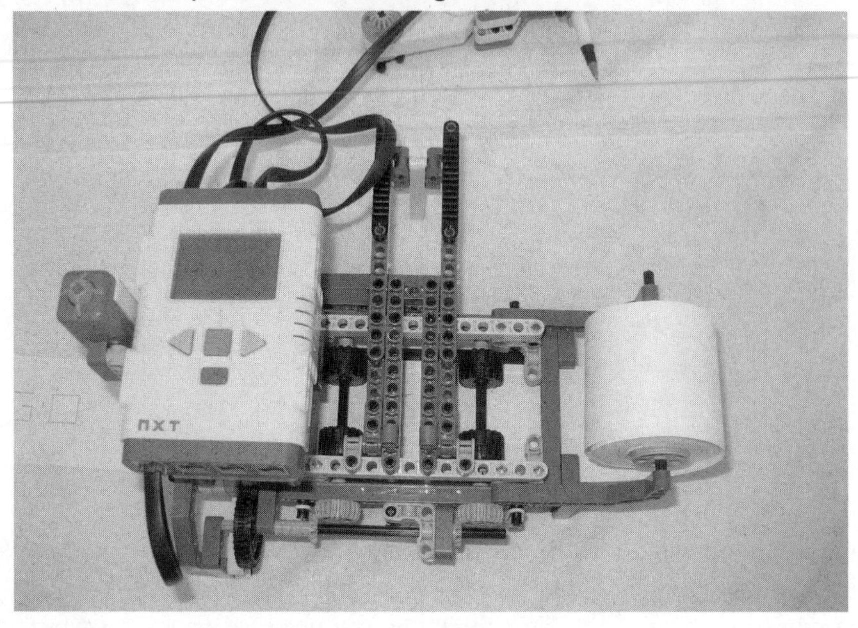

A second rack and pinion system, operated by the third motor, controls the vertical movement of the pen (see Figure 18.20). Figure 18.21 shows a close-up image of the pen.

Figure 18.20 The Writer's Pen Assembly

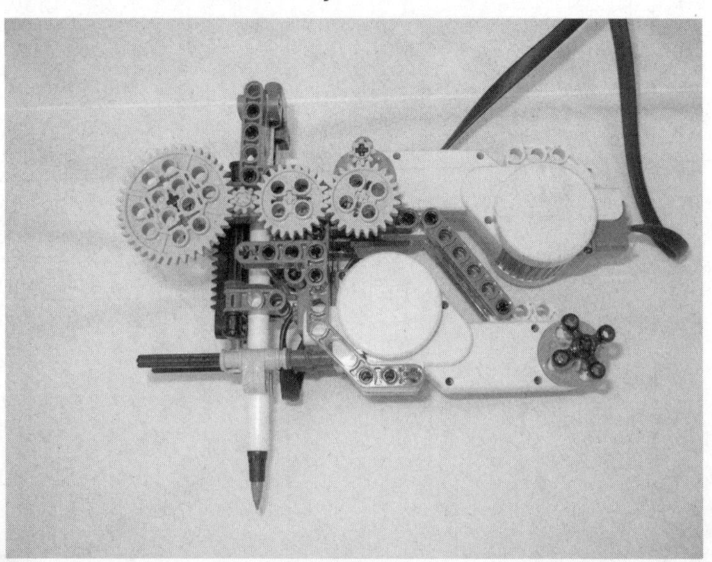

Figure 18.21 Close-Up of the Pen

Bricks & Chips...

Washing LEGO Parts

LEGO bricks are, generally speaking, not very difficult to clean. You can remove small ink spots using a cotton ball soaked in some alcohol. For large-scale cleaning, hand-wash your bricks in a tub of warm water with some dish detergent. Machine washing in a clothes washer on a warm setting is possible too, provided that you put your parts in a canvas bag.

Some warnings apply:

- Do not use any solvent, unless you're sure of what you are doing. Test it on a brick you don't care about.
- You should not clean transparent and printed bricks with alcohol or any other solvent. You should wash them only by hand.
- Never use hot water, and never put your bricks in a *dishwasher* because the settings will be too hot and will warp the parts.

The writing pad is one large 8 x 16 stud tile piece (see Figure 18.22). If you do not have this piece you could use a bunch of little tiles covering a plate as the writing pad. The irregular surface covered with studs wouldn't work. In case you don't have tiles, or you do not have enough of them, cover the plates with a smooth, thin support, such as a glossy cardboard, an aluminum or plastic sheet, or anything else similar that comes to mind. You can also build a top out of standard LEGO bricks laid on their side, which should provide an even more regular surface than tiles.

Figure 18.22 The Writer's Writing Pad Taken Apart

The writing surface is an independent part linked to the main body through short rubber bands (see Figure 18.23). Those bands pull the surface up against the pen and against the wheels of the feeding mechanism.

Programming the Writer

Programming the writer may seem difficult, but it is much easier to divide the program into My Blocks for each letter and then bring them together in one large program. Our robot uses a loop program to write the word *LEGO* and a touch sensor to order more copies. Figure 18.24 displays this program.

Figure 18.23 Writer Front View

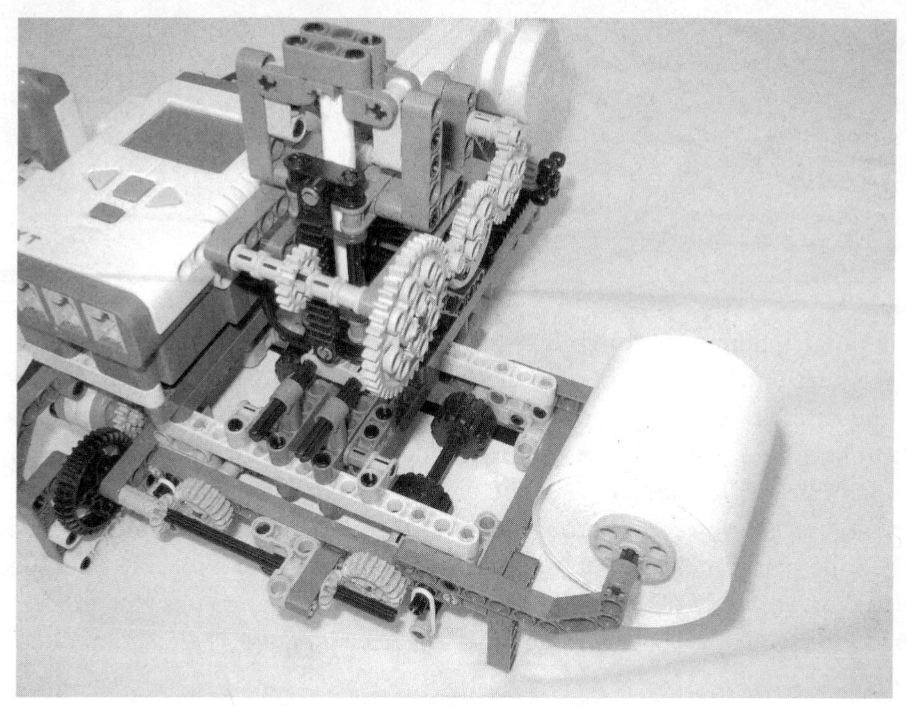

Figure 18.24 Program for Writing the Word LEGO

As you can see, the My Blocks are labeled with the letter they write in the program. The program writes the four letters and then lifts the pen up, advances the paper roll so that you can view the words, lowers the pen to the original starting position, and then waits for the touch sensor to order another copy. The My Blocks can be small or large programs depending on the complexity of the letter. For example, the program for the letter "L" is a very short program displayed in Figure 18.25. However, the program for the letter "E" requires much more movement.

Figure 18.25 Program for Writing the Word LEGO: My Block for the Letter "L"

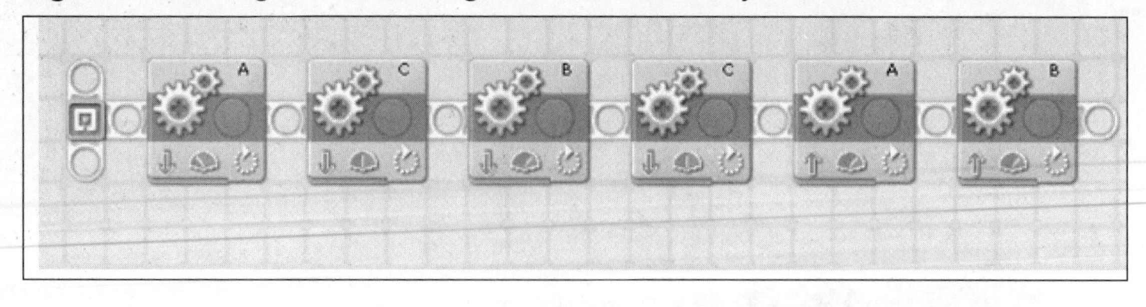

In this program, the A motor controls the vertical movement of the letters, the C motor controls the horizontal movement of the letters, and the B motor lifts the pen up and down. To draw the letter "L" the A motor moves down the entire length of the vertical movement from top to bottom. For this design, that is 130 degrees of motor rotation. The power level is set to 35 percent to give a crisper line. The pen then feeds the paper running C motor 300 degrees at 50 percent power level. You can adjust this value depending on the size font you desire. Then the pen is moved up to the starting height and over a spacing of 200 degrees, and is placed down, ready to write the next letter. You can change this spacing depending on how much space you want between the letter characters. Figure 18.26 shows the program for writing the letter "O."

Figure 18.26 Program for Writing the Word LEGO: My Block for the Letter "O"

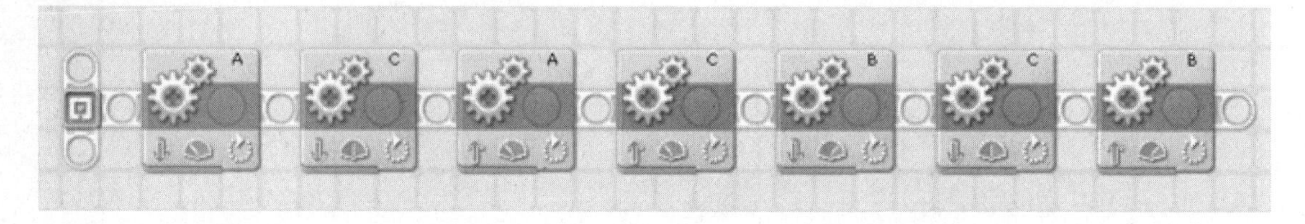

Once you create a My Block for any letter you want, you should save them so that you can recall that letter later. Who knows; maybe you can program the entire alphabet!

What to Write

We had the idea of making this robot an Automatic Haiku Writer, but the truth is that you can make it write whatever you want. In the last part of this chapter, we will give you some hints about other possible uses of this robot: a label-writing machine, a graphing system, and more.

Now, what's a *haiku*? It's an ancient Japanese poem with a formal structure. Though not everybody agrees on all the rules involved, the most accepted form is a three-line verse where each line is composed, respectively, of five, seven, and five syllables. It usually contains a reference (even indirectly) to time, and it's broken into two parts, such as an introduction and a theme, or an action and its consequence. Here's our own example of a haiku contemplating the theme of this book (we ask your forgiveness in advance!):

The robot is on

Feel a bit worried, about

The things it could do

You can program your robot to produce a written haiku on request, generating it randomly from a "database" of predefined sentences. Or using a more sophisticated approach, you can pluck random words from a (small) inner dictionary, combining them according to simple predefined grammatical structures (see Appendix A for some links to useful Internet resources in this regard).

Further Suggestions

The writing and drawing theme offers many other ideas. The following suggestions are far from being exhaustive. Consider them starting points for your own creations.

Copying

This has become an almost classic project, but it's still interesting, instructive, and challenging. You need a feeding mechanism similar to those of our Tape Writer, but duplicated for two pieces of paper. It must be able to drag two sheets of paper: the one being copied and the blank one. Obviously, the whole machine will be much larger than the Tape Writer if you plan to use standard letter or A4 sheets.

The copying system is made of a translating assembly that moves a light sensor back and forth across the original sheet and a pen in the corresponding position over the copy sheet. With the paper feeding motor stopped, the software scans a row with the light sensor, and depending on what intensity it detects, it puts the pen up or down. After each row, the paper will feed a bit for the next scan.

The requirements for this project are not very high: three motors, one light sensor, and one or two touch sensors for the carriage movements; but we suspect that in creating a standard sheet copier, you will likely need many additional beams and plates, because the structure will be rather large.

Emulating Handwriting

Using a completely different technique, you can build a robotic arm that writes with movements similar to those of a human arm. You need an arm with two degrees of freedom: Two levers move on a horizontal plane, the first attached to the body of the robot and the second to the end of the first. At the end of the second lever, there's the pen with its lifting mechanism. We suggest you keep it very lightweight, using pneumatics, the flex system, or a micromotor.

The software to control this beast is not very simple. Converting the angles of the arm into Cartesian coordinates on the sheet requires some trigonometry, and all that that implies.

Learning by Example

To make the preceding project even more interesting, and at the same time get rid of all the trig (yeah!), you can design your robot to learn from your movements. In this case, your robot will have a training phase, where you guide its arm to write or draw what you want, and a production phase where it replicates your movements.

Make the motors easy to decouple, so you don't have to move them while driving the arm during the training phase. The program will save the sensor readings at small intervals in order to reproduce those positions later in the production phase.

For the pen up/down movements, you can keep the motor connected and controlled by a touch sensor that you press when you want to flip from one state to the other.

The most challenging part of this project is storing the data collected during the learning process. You have basically two options: using a language that allows large memory structures such as arrays, or doing the dirty work on the PC, leaving all the "intelligence" and data there and using the NXT as merely an executor.

Summary

In this chapter, we explored some techniques described in Part I that had not yet been applied to robots in this book. The Logo Turtle offers a good opportunity to find a use for the sophisticated dual differential drive of Chapter 9, which is capable of turning in place like a simple differential drive, but also of going perfectly straight. In fact, at the price of some mechanical complexity, it provides a way to separate straightforward motion and turning capabilities using two independent motors. Its advantages include the fact that you can monitor both kinds of movement with a single rotation sensor attached to one of the wheels. Using the dead reckoning math you can precisely control your Turtle. We went through those equations again, providing a concrete example of how to implement them in an NXT program.

Though conceptually simpler, even the Tape Writer showed some construction tips. It is a Cartesian system not too different from those used in the robots of previous chapters (the Maze Solver and the Tic-Tac-Toe machine), but it does demonstrate once more that by reevaluating the terms of a problem, you can find an easier solution. For example, a Tape Writer built with a technique similar to the Maze Solver would have required very long rails; so, moving the paper instead of the robot, your construction results in a more compact design that is also capable of writing texts of unlimited length.

In the suggestions we provided at the end of this chapter, we described the possibility of emulating handwriting using an arm. This includes a glimpse at how robots can learn by example too; a feature used in many real-life robots, including industrial robots. In a case where you want your robot to perform handwriting, you can guide the movements of the robotic arm to copy the shape of any written character; the robot "remembers" your movements, and then is able to replicate them and write by itself.

Chapter 19

Racing Against Time

Solutions in this chapter:

- **Hosting and Participating in Contests**
- **Optimizing Speed**
- **Combining Speed with Precision**

Introduction

This chapter explores the world of MINDSTORMS robotics contests and challenges. The information in this chapter is mainly based on direct experience, accumulated while participating in competitions organized by various organizations. Some of the competitions referenced in the following pages are routinely run in various locations. They include the following:

- **The Northeast Indiana Robot Games (NEIRG)** February and August
- **Chicago Area Robotics (Chibots) competitions** May and November
- **Central Illinois Robotic Club (CIRC)** March
- **Lafayette LEGO Robotics Club (LafLRC) competitions** May (typically)
- **HiTechnic sponsored competitions** August (typically)
- **Brickworld** June
- **RTL Toronto** year-round

This chapter won't be discussing the specific details of the contests; instead, it provides you with a good starting point for more general considerations. But we do recommend that you look up the competition rules on the Web sites of the competitions listed and learn more about the variations for competitions that may sound simple, such as "line following."

The first section of this chapter is about robotics contests in general. It explains what robotics contests are all about, from the definition of the rules to the course of competition. For those of you interested in participating in LEGO robotics contests, the chapter will give you some hints about how to find a LEGO Users Group not far from where you live.

In the later sections of the chapter, contests related to pure speed, as well as those demanding great amounts of mechanical and programming acumen, are introduced. There are many different kinds of contests and challenges. Because of this, they are grouped into three categories: contests based on speed, contests based on strength, and contests based on ability. These categories are not absolute, because most of the competitions require a mix of these capabilities. For example, a line-following contest is mainly about speed, but each robot is also required to run without veering too far from the line. Nevertheless, we tried to sort a few typical contests into the categories previously mentioned because in our opinion, this helps in focusing on their key points.

Hosting and Participating in Contests

A contest offers many opportunities to learn new concepts and build some experience. There are at least four main phases of participating in a contest, each one requiring extensive usage of your know-how while contributing to your knowledge base. They are:

1. **Defining the rules** Participating in this phase depends on whether you are the one who organizes the contest, or you are part of a group that does. Unless you're deciding on your own, this will prove to be a very creative moment, where the group develops a list of rules, adjusting them until it feels they are meaningful and consistent. A set of rules always has a specific purpose (whether declared or not), which has been chosen to test the ability of the competitors on a specific field. The "legislator" should take care to close any possible loopholes that might allow a contestant to escape the main difficulties of the contest, which requires that he or she imagine all the possible approaches to the problem. The rules should also try to ensure that the contest is fair to all competitors, regardless of their monetary resources. Luckily, most of the time, you don't have to worry about defining the rules because you are a competitor, not a sponsor.
2. **Studying the rules and deciding on a strategy** From this moment on, you are in the competitive arena, and you must find a strategy to beat your competitors. Don't limit your choices to what the organizing committee expects you to do. In our experience, most contests have been won by people who found a very original way to interpret the rules without violating them. Don't be afraid to find loopholes and take advantage of them. Visualize different solutions on your own and determine the pros and cons of how they comply with the rules.
3. **Building the robot** This phase will very likely present some surprises to you. Implementing your desired strategy, you'll discover new constraints and opportunities you hadn't thought of while imagining your robot. As for programming, we strongly suggest that you stay with simple but solid strategies. Only when you're sure the basic behaviors work as expected should you add the more sophisticated components, making sure not to introduce bugs in the previous code. You won't believe how many matches people win by keeping it simple!
4. **Attending the contest** This is the most exciting moment—on the field, testing your ability against your competitors! It's also the moment to learn: Study the other robots and their strategies; observe the course of the matches. Don't be frightened to ask for explanations and details; most of the builders are usually more than happy to describe their creatures. All that you learn will be useful for other contests, whether run on the same set of rules or not. One last suggestion: Never throw in the towel before the end, because anything can happen during the event. The strongest competitors aren't always crowned the winners. Learn from each match. Would your robot do better if you moved its starting point or direction slightly while still complying with the rules? In a head-to-head competition, pay attention to other matches and learn how to best position your robot to beat other robots.

Use the Internet to search for other MINDSTORMS fans. One popular resource is the LEGO Users Group Network (LUGNET), which lists dozens of local groups. Many of them also have their own Web sites, which shouldn't be difficult to find using any search engine. Once you've found a group, or some individual users, there's no certainty that anyone's going to leap up and organize a robotics contest from time to time. But you, yes you, can be the one to get the ball rolling (or robots, rather).

LUGNET is the best place to find information about contests of all sorts, as most local groups advertise the contests they organize there. Usually they refer you to a Web site where you can find all the details about the time, place, and rules of the contest. Some contests require a small admission fee for each robot, which funds the prize for the winner. Events are characterized by a very friendly atmosphere, and you'll be welcomed even if you just go to watch and learn.

Optimizing Speed

The first challenge described here concerns pure speed. Don't make the mistake of thinking speed is purely trivial and poses few challenges in terms of robotics. We've been proven wrong on this score ourselves. Even a straight-out speed race promises surprises.

Drag Racing

"A starting line; a finish line; the fastest robot to cover the distance wins." Described in these terms, the race sounds boring. But stay tuned, and take a closer look at the implications of this definition.

The speed of a vehicle is affected by a number of factors: motor power, gear ratio, mass, and friction. Using electric motors, the maximum power you can apply to your race car depends on the kind and number of motors, and the current you supply them. With the addition of the new NXT motors, three types of motors are widely available. Also available are the traditional MINDSTORMS "gray" motors, as well as the black "RC Buggy" motors that are now available as a separate motor pack (LEGO set #8287) and as part of kits such as the yellow crane (LEGO set #8421). Of course, the rules of the competition will probably specify the allowed motor type and a restriction on power source.

For the purposes of this chapter, it is assumed that the competition limits the vehicle to two NXT motors and an NXT as the power source. Even with these limits, there are still many variables to consider in your design process.

The gear ratio and mass will have a strong influence on the acceleration rate of your vehicle; here is a short list of tips:

- The shorter the gear, the shorter the time it takes to reach the maximum speed. The problem is that a short gear also has low top speed. You have to balance the two effects, and the optimal choice depends also on the length of the race: Favor acceleration on short tracks, and maximum top speed on longer ones.
- Build your robot in a way that allows easy replacement of the gears, so you can experiment with different ratios in a time-efficient manner.
- Keep the gearing pared down to the essentials. Remember that each stage adds some friction. There's no need for a differential gear, because the dragster travels on a straight run.
- The diameter of the wheels has its role in the conversion of power to speed. If you substitute the wheels of your car with ones half the diameter in size, you get the same effect as though you had reduced the gear ratio by a factor of two.
- Acceleration is also influenced by the mass you have to move: Under the same power, higher mass equates to lower acceleration. This is due to inertia (see Chapter 6), which explains why it's harder to get a car rolling than it is to push a child in a stroller. So, a very important thing to do is to keep the mass at a minimum. Build a lightweight structure.
- Another factor related to mass is the center of gravity of the vehicle. As with a top fuel racer that goes 300+ miles per hour, your vehicle should center the weight almost over the drive axle at the rear of the vehicle. The center of gravity should be just far enough in front of the rear axle to keep the vehicle from lifting its front wheels off the ground for a significant amount of time.

At this point, you haven't yet considered the modes of operation allowed by the NXT.

Up to this point, the challenge is essentially electro-mechanical. There's no need for an NXT; a vehicle supplied by a battery box would perform the same, or even better (recall that the RCX has an inner current-limiting device, and the battery box doesn't). To create the necessity of at least a few lines of code, we suggest that the dragsters be run down a narrow corridor with a three-quarter-inch black line down the center. Just as top fuel dragsters need to stay in their lanes, our robotic counterparts will incur their own time penalty by bumping the walls. Of course, there will need to be a rule against the vehicle intentionally riding against the wall.

Combining Speed with Precision

When you move from races based on pure speed to those that require additional skills, your projects become more complex, and more than likely, the resulting vehicle will move slower. All the considerations listed in the preceding section still apply—batteries, motors, gear ratio, mass—but you must also take new variables into account. Speed will actually become what makes your task more difficult: When you design a robot for yourself, you usually feel satisfied when it works; but when you have to build and program it to be as fast as possible, some techniques that worked at a slower pace prove unsuccessful at higher speeds.

Sometimes you reach a point where you cannot increase speed without compromising the reliability of your robot. This is the time when a further improvement can come only from a *paradigm shift*, a change from one way of thinking to another. This principle can be summarized in a few words: Don't set your heart on a particular solution. Try to look at the problem from different angles and keep your mind open to any idea, even those which initially seem strange or impractical may lead to a winning configuration.

Line Following

Don't worry, this chapter doesn't start discussing line following again! Jump back to Chapter 14 if you feel compelled to revise some of those concepts. Line following just couldn't be ignored in a chapter that talks about races against time, because it presents many interesting discussion points.

If you are the one who decides the rules, don't underestimate the importance of the details. State the number and kind of the allowed parts—motors and sensors in particular. More important, be very precise regarding the nature of the path, informing competitors about the width of the line and the minimum radius of the turns, the latter having a strong influence on the structure of the robots.

Line-following contests are usually judged by speed alone. Evaluating accuracy, though theoretically possible, is not a very practical option. However, if you want to try this option, you can use a paper pad and attach a pen to each robot so that they draw a line as they move. At the end of each run, measure the maximum distance between the course of the robot and the main line, and apply greater penalties to greater distances.

Line following allows for many interesting variations, including these:

- **Round trip** When the line ends, the robots must return to the starting point.
- **Short interruptions in the line, specified by number and length** For the robots, it's like hanging in midair for a while. The restart point of the line might even be offset from where the line broke.
- **Small obstacles to overcome** The robots should detect these with bumpers, suspend line following, pass the obstacle, and resume line following again.
- **Obstacle removal** Similar to the preceding variation, except that objects of a specific size and shape must be removed instead of climbed over.
- **Specific robotic architecture** Specifying that a particular type of architecture be incorporated into the robot design. For example, all the robots must use legs instead of wheels.

Wall Following

Conceptually similar to line following, in this challenge, the competing robots must follow a wall instead of a line. The software is actually very similar to what works for line following, with only a few adjustments to reflect the difference in sensors.

If you decide to organize a wall-following competition, remember that the walls used need not be real walls. You can create temporary walls with wood, cardboard, or any other material of your choice. Wall following can be as simple to set up as having the robot find its way around the perimeter of a large cardboard box. As with all competitions, it's important that you put a lot of care in specifying the details, including the following:

- The height of the walls, their color, and the material they are made of
- The color of the floor and the material it is made of
- Whether the robots are required to remain in constant contact with the wall, or if they can move apart from it for a while
- The shape of the course, or at least what kind of angles the robots should expect
- Whether the robots are allowed to "hook" the upper edge of the walls

Moving to the point of view of the participant, the hardware configuration required to follow walls can be very similar to that of a wall-maze-solving robot (maze solving actually being a sophisticated variant of wall following). However, this is one of those cases where an increase in speed brings new difficulties. Similar to what happens in high-speed line following, the critical factor here is the reaction time of the robot. In fact, anytime it loses contact with the wall and needs to undertake a corrective action, that longer reaction time entails a stronger correction.

As mentioned in Chapter 14 when discussing how to optimize line following, this is easier said than done. To recapitulate, the elements you have to consider include:

- **The mechanical configuration of your robot** These include the type of drive, number of motors, position of the sensors, gear ratio, and backlash within gears.
- **The firmware you installed on your NXT** Alternative firmware installations are available and the variants that are available are constantly changing. Some alternatives support the standard NXT-G language as well as other programming environments. You need to choose a firmware and environment that you can use effectively to provide the fastest response time for your robot.
- **The algorithms used in the software** This comprises strategies adopted to keep the robot on course as much as possible.

The mechanical configuration of your robot is something you have to experiment with. The optimal solution depends on the set of rules with which you'll use to race. As for the firmware options, this is an opportunity to study a new language and install a new system, though not everyone will want to do that just to attend a contest.

As for the strategies, some of you may recall that Chapter 12 introduced hysteresis as a technique aimed at improving the efficiency of a system, because it reduces the number of corrections it has to make. It was definitely an interesting option for line following, but is it applicable to wall following too? The answer depends on the configuration of your robot. If it relies on a touch sensor to "feel" the wall, hysteresis will be of no help, because all you can determine from the robot is whether it's touching the wall. To take advantage of hysteresis, you need finer information—you need to know the *distance* from the wall, so you can make your robot decide when and how much to correct the route. This implies that you have to replace the touch sensor with a more sophisticated device. For example, you could arrange a bumper, or antenna, connected to a rotation sensor in such a way that the count of the sensor is proportional to the distance. Or you may be able to use the Ultrasonic Sensor to detect a distance from the wall—like the Maze Runner of Chapter 17.

Other Races

Many other type of contests require your robot to perform some action as quickly as possible. As we explained in the introduction, most of them require some additional capability rather than just speed. In Chapter 20, we will describe contests in which speed is important, but this is usually in the background when compared to other factors, such as efficiency in finding and gathering objects. In the following list, we suggest a few ideas for competitions in which speed is the most important component:

- **Car racing** Car racing is similar to drag racing, but the robotic cars run on a circuit that is more complex than just a straight track. The circuit may be delimited with colored tape on the floor, or with side walls. Avoid reducing the contest to line or wall following; instead, design the circuit so that a robot that follows one of the sides takes a longer route than those that run inside the track. If the circuit is delimited with real walls, encourage the competitors to use sophisticated detection techniques, such as proximity sensing, by applying a penalty for every collision with a wall. Another approach to the car racing track was developed by the Lafayette LEGO Robotics Club. They use an oval track that is about 2 feet wide and has a black-to-white gradient across the track.
- **Fast painting** Each robot is equipped with a felt-tip pen and is asked to paint a given area on a sheet of paper. The robot that covers the surface fastest wins. Consider basing the results of each competitor on a combination of the elapsed time with the comprehensiveness of the coverage. The panel could be provided with a robot designed to scan the sheet and evaluate the result!

- **Wall climbing** Prepare a climbing wall equipped with special holds that a robot can seize (this could be as simple as a grid of horizontal bars); the fastest robot to reach the top wins. You can keep the competition open to ideas, allowing any kind of technique to reach the top, including lifting mechanisms and the launching of ropes. Be sure to provide a soft surface under the wall as you don't want anyone to break his NXT if it were to fall off the wall.

Summary

This chapter introduced you to the world of contests which represent a great opportunity to expand your knowledge, stimulate your creativity, and compare your ideas with others.

Even races that seem the least "robotic" of all the possible types of competitions can spur you to find new solutions or improve old ones. During contests, the details are very important. Your robot should not only work, but work better than its competitors. For this reason, an apparently simple task such as going straight and fast requires thoughtful planning of your project: batteries, motors, geartrains, wheels, the weight of the vehicle, and the center of gravity… all of these elements are crucial to success.

When you move to contests that involve highly specialized capabilities, such as navigation, the problems become much more complex. Tasks as simple as line following and wall following require a tremendous effort when your purpose is to design, build, and program a robot tuned for optimal performance. This is a process which proceeds by trial and error, and which will test your skills, your experience, your creativity, and, most of all, your patience!

We encourage you to participate in contests. They can really be a great experience. Be humble enough to learn from your mistakes, or from more effective techniques rather than completely different approaches adopted by other robots. Take everything very seriously during preparation: Try different solutions; perfect the details; test your program thoroughly until you feel satisfied. But don't take the final rankings too seriously—remember, it's all in fun!

Chapter 20

Hand-to-Hand Combat

Solutions in this chapter:

- Building a Robotic Sumo
- Attack Strategies
- Getting Defensive
- Testing Your Sumo

Introduction

The contests described in other chapters are more specific to those where each competitor has its turn, and the results compare the individual performances. In this chapter, we'll talk about competitions where the rival robots fight face to face in a more spectacular way.

In our experience, sumo is one of the most suitable kinds of competition for small robots, offering the opportunity to test an incredible range of techniques that may prove useful in all your projects, not just during contests. We will take a look at variations on some familiar solutions—such as bumpers and the use of the ultrasonic sensor—and we will introduce some new ones. For example, we will illustrate a transmission, which behaves like a sort of automatic gear switch.

Although the technical aspects of building a successful sumo robot are important, the design requires much more than simply putting together a few mechanical solutions: It requires a strategy. Will your robot be very aggressive, or do you prefer a defensive approach? It could be robust and slow, or lightweight and fast. It could be designed to actively search out its opponent, or to react when it's under attack. You cannot work at the mechanical configuration and decide how the robot should behave after it's finished. On the contrary, you have to pick up a strategy and design both the mechanics *and* the program according to it. This principle applies to any robot, but it is particularly important for sumo robots, and it is the key to understanding this chapter: We want you to devote the proper attention to the connections between the planned behavior of your robot and the solutions you can adopt to effectively implement it.

Building a Robotic Sumo

We explained in this chapter's introduction that when you start building a robot for a sumo contest, you must have a strategy in mind. The process starts before you build your robot. It begins by examining the rules carefully, understanding what you can and cannot do, and deciding your line of action. You must try to imagine what the opponents' strategies can be, and plan your robot to be able to resist their attacks and take advantage of their weak points. Obviously, you cannot really know how the other competitors will strategize and behave, but this exercise helps you to focus on a well-defined strategy. Remember that any strategy is better than no strategy at all! Figure 20.1 shows a simple sumo robot ready for action.

Figure 20.1 A Simple Sumo Robot

This section starts by describing a typical set of rules, which will help you in framing what a sumo contest is, and provide a starting point in case you want to organize your own. Then we'll describe how you can tune your robot to produce maximum force, which is undoubtedly a very important component in a sumo competition. We will also explain how to configure your robot to take advantage of some important offensive and defensive behavioral strategies.

Setting the Rules

In a typical sumo competition, you will receive two sets of similar rules. The first set of rules states that the robots can be made out of any original LEGO pieces, in any desired quantity, but that they must be within a maximum size of 32 x 32 studs and a maximum weight of 1.5kg (3 lbs). In the alternative set of rules, which we call Mini Sumo, each robot may be built using only parts from a single MINDSTORMS NXT set; there is therefore no need for size and weight constraints.

For most other aspects the two sets of rules are almost the same:

- The field is a circular or square pad with a contrasting external strip of 20 cm (8 inches). Usually the pad is white and the strip black, or vice versa. There is also an optional 2-inch warning line that warns the robot as it approaches the edge of the ring.

- Only two robots can fight on the field at a time. Should one robot for any reason find itself outside the field boundaries, that is, any portion of it touches a point beyond the external strip, the robot loses the round. If neither robot is eliminated within a chosen time limit (e.g., three minutes), the match ends in a draw.
- A robot may also be eliminated if it is overturned by its opponent or it finds itself in a situation where it can no longer maneuver.
- No "violent" behaviors are allowed. A robot can only push or lift its opponent. It is in no way allowed to damage its opponent's structure or parts. This will be left to the judge's discretion.
- A robot cannot drop any part or subsystem in the field either deliberately or involuntarily. Any part found loose on the field will be removed by a member of the panel.
- The robots must be fully autonomous; any kind of remote control is forbidden.
- Every robot must comply with the limits in size and weight at the beginning of a match, but once the match starts, it can modify its own structure, perhaps extending parts itself so its dimensions become larger than the initial specified size limits.

There are many other, less important, rules covering items such as batteries, the composition of the panel, the prematch test time, and more. Some sumo competitions require that your robot pass an admissions test: It should be able to push a block of wood out of the fighting ring. If it can't beat a block of wood, it has little chance against another robot, and this rule is meant to screen out robots that are too weak to enter the contest. This rule is not that commonly enforced, but it is a good exercise, and it may help in filtering out weaker robots to make large tournaments quicker.

Maximizing Strength and Traction

The making of a strong sumo robot requires much more than just brute force, but we cannot deny that maximizing the generated push will increase your chances of winning some matches and, maybe, the tournament.

When optimizing the pushing power of your robot, the first thing you need is an objective way to measure it. Without measuring the force, the improvements you make are subjective and, as a result, are very inaccurate. During preparation for one of the first robotic sumo contests, a friend and robot builder suggested a simple trick based on a very common object: scales, such as those used in many kitchens to weigh flour, sugar, and other ingredients.

You have to place the scale on its side, on the table or the floor, possibly removing the upper tray, and hold it firmly while your robot pushes against it. You're not interested in the absolute value that the scale indicates, but rather in comparing the push produced by different setups.

Many factors affect this force; you can imagine a sort of path of power that goes from the batteries to the wheels, passing through the motors and the gearing, decreasing in accordance with the variables that affect each part along the path (see Figure 20.2).

Figure 20.2 Limitations on Force

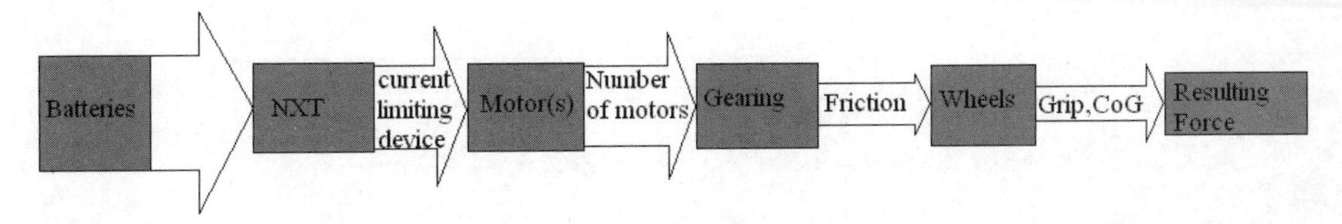

The rules will hopefully specify that all competitors use the same kind of commercial batteries. Between the batteries and the motors, there's the NXT. It's worth reminding you once again that the NXT incorporates a current-limiting device to protect the motors connected to its output ports. If the rules allow the use of custom parts and you have them or are willing to make them, you can consider the use of a Motor Multiplexer from Mindsensors or HiTechnic, or the use of a homemade motor hub made by Philo (Appendix A).

The number of motors influences the generated power. Simply use the maximum allowed by the rules and by your own inventory. As for the mobility configuration, the differential drive allows for the highest combination of maneuverability and simplicity. The fact that it doesn't go perfectly straight is not relevant to sumo fighting, and the dual differential drive has no advantages in this case. On the contrary, the capability to use one motor to turn and the other to move reduces the maximum generated force.

The optimal gearing is, as always, easier to determine by experiments than by calculations. Generally speaking, the higher the reduction ratio, the higher the push, but this doesn't mean you should gear down too much. Speed has its importance (we'll explain why later in the chapter), and very high reduction ratios introduce too much friction, which uses up precious power.

Now we come to the part where you have to convert the produced torque into the actual push. The wheels are a critical component: If they don't grip the pad well, the rest of your efforts will prove fruitless. This is when the scales we mentioned earlier prove to be an enormous benefit. By testing different kinds of LEGO wheels, you'll discover that there are significant variations in grip. The ones from the 8462 Tow Truck work particularly well, as well as the large spoke wheels from the W979648 Education Resource Set and many others.

On no account should you use tracks. They offer extremely low grip, and almost no grip at all in the direction perpendicular to its motion. You'd have little hope at all if your opponent broadsided you—an eventuality more probable than a head-on collision.

If possible, try to test your robot on a surface similar to the contest's official pad. Different materials require different wheels. For example, the wheel having the best grip on a smooth tabletop is not necessarily the one with the best grip on a rough plywood surface.

The position of the center of gravity is also very important when it comes to friction and your wheels. Keep the center of gravity (COG) as close as possible to the main drive axles.

Designing & Planning…

When Reinforcement Really Counts

Remember that reinforcement is very important to any successful robot. This applies especially to sumo robots because they have to withstand the pressure of other robots. So where do you start to reinforce your robots?

Generally, you want to reinforce the motors the most to ensure that the drive motors don't have much friction so that they can reach their optimum efficiency. You'll also notice that the orange rotating piece on the motor is rather loose. That's the best place to start reinforcing. Take advantage of every hole on the motor. Anywhere that can be cross-braced should be. We won't tell you how many matches were lost due to a robot falling apart!

Attack Strategies

We anticipated that force wouldn't always make the difference in a robotic sumo contest. Many different strategies can affect the result and cause a robot to win out against a more powerful competitor. These include finding the enemy first, using speed as a force, using a gear switch for maximum speed and push, and other offensive tricks.

Finding the Enemy

A very important rule is to find your enemy before he finds you. This basic military principle applies to sumo robots as well, for the simple fact that the first one to engage the other has a good chance of attacking it on a weak side. Sumo robots are generally designed to push forward, and they offer much less resistance when attacked from the side or rear. In fact, they often don't even realize they're under attack, because often they're not designed to detect the enemy from behind or from the side. In such cases, you can say that three sides out of four are generally weak.

When you start planning, the most logical sensor to use to find the enemy is the ultrasonic sensor. Spin, go toward the closest object, and repeat. It seems perfect, but there are some loopholes. Among them is the fact that they interfere with each other. If both you and your opponent used the ultrasonic sensor at the same time, your robot would start reading your opponent's signals and get confused. You could solve this problem by using Guy Ziv's Ultrasonic ping method (see Appendix A).

Another issue is that you're trying to find a pile of LEGO pieces, and the ultrasonic sensor is meant to find large solid objects. Possibly, you could mount the ultrasonic sensor on its side so that you have a more narrowed range but a better resolution. This can be left as an exercise to the reader.

A technique that is simpler but just as effective employs contact sensors, in the form of either bumpers or antennas. Bumpers don't require any particular trick. You simply program your robot to turn toward the obstacle instead of avoiding it. Design compact and smooth bumpers devoid of any unnecessary protrusion, to reduce the chances of getting caught on an enemy robot and dragged off the playing surface. With antennas you can use either touch or rotation sensors, the latter being able to tell you more about the direction of the opponent. But with the latter, you will have to use the legacy rotation sensor because the NXT motors have too much internal gearing to move from a light push. And you would have wasted an entire motor!

Using Speed

Speed is an extremely important factor in the search for the enemy. Imagine two robots running freely on the sumo field, simply going straight until they find the border and change direction randomly. Supposing that they have different speeds, the faster of the two has a much greater chance of intercepting the other. For this reason, it's important not to have too slow a robot. Find a compromise between pushing capability and speed.

The robots built around speed all have a common strategy: Crash into the opponent repeatedly and use its momentum instead of its strength. The resulting energy makes the opposing robot lose contact with the ground, which gives your robot time to rear up and assault again. One charge later, the enemy is often found helpless. Though these robots may seem cheap, you have to appreciate how much experimentation it takes so that they do not illegally destroy their opponents.

NOTE

Momentum is a physical quantity defined as the product of mass times velocity. You can understand what it means through an example: You face a person of your same weight and build that's trying to knock you down. If you're both stationary, you have a good chance to resist. If, on the other hand, you are stationary and the other person is running toward you, you will very likely go down.

Using a Transmission

Other robots use a transmission to get the best of both worlds: fast speed during the search phase, and maximum push after the engagement. Sure, some robots will use a special transmission ring included in some vehicle sets, but it was soon proven during a contest that it's possible to make a sort of automatic gear shift even inside the strict rules of Mini Sumo. Look at the assembly in Figure 20.3. It's not very solid, but it explains the principle: The wheel on the right in the picture is geared with a shorter ratio than the main one, and during normal motion it slips a bit because the robot is moving faster than the speed of the idler wheel. When the robot slows down or stops for any reason, the faster wheel slips, and at that point, the slowest one grips. Because it's mounted on a short independent beam with a free end, part of the torque pushes the wheel down and consequently lifts the robot. This mechanism is very fascinating to watch, but it's very difficult to understand just by looking at the picture, so we encourage you to build it and try it out. Just remember that you need two of these assemblies, one for each side, and a supporting wheel. The wheels in the NXT kit can replace the ones in the picture.

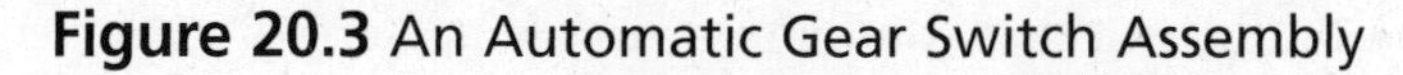

Figure 20.3 An Automatic Gear Switch Assembly

Other Sumo Tricks

Many other tricks prove useful during a sumo contest. The ones most often used are meant to lift the opponent, thus getting two positive effects: reducing or canceling the grip of its wheel and transferring part of its weight on your robot. This class of method includes at least two large families, one based on inclined planes and the other on counter-rotating wheels.

An inclined plane works like a wedge that slips under the enemy robot. It can have the shape of some small slopes placed at the front side of the robot, or of a large inclined surface that covers the whole robot. In the latter case, a LEGO baseplate is the better choice: Mount it studs-down and you'll have a very smooth top surface to wedge under your opponent.

Counter-rotating wheels are very effective too, but they require an additional motor to operate them. Be sure they don't touch the ground, though; otherwise, they'll counteract the forward motion of your own robot! The combined effects of the front wheels with the push of the robot may even overturn the opponent, a spectacular but rare event.

Getting Defensive

So far we have discussed attack strategies, but protecting the weak sides of your robots is important as well.

Every active defense system relies on the fact that you know what's happening around you, and require some sensor to detect a possible attack. Depending on the rules of the tournament, you might find yourself dealing with a limited number of input ports, requiring that you carefully plan how to allocate them in regard to your navigation, attack, or defense subsystems. The simplest detecting system is a sort of large bumper that covers a whole side of the robot. With this "ring" bumper, one touch sensor will be able to detect an attack from any direction (see Chapter 9).

Another method of detecting the enemy is to check whether your motors have been stalled. This is easy to do with the built-in rotation sensor of the servo motors. If you run your motors, you can monitor their speed and rotation values. If the motors are under stress, through software you can detect a stall condition and assume you are jammed against something—in this case, the opponent. Then you can test each direction to know from which side the opponent got you.

When you detect that you've been tackled, you have the option of either escaping or facing your enemy. The first choice is best when fighting a slow, strong opponent, whereas the second works well when it's your robot that has a strong push (though it's not always easy to turn in place when being pushed). Some rules allow competitors to use more than one program. Take advantage of this opportunity by preparing different versions to implement different strategies, and then select the one most suitable when you know which robot you'll be facing in a given match.

Also consider passive defense systems, the kind that don't require any sensor or port. The more obvious defense mechanisms pertain to the shape and size of the robot itself. A smaller robot offers less surface area to an opponent than a larger one, and though a triangular shape is more difficult to build, it's also more difficult to catch. Make the perimeter somehow convex if you can, so as not to offer any holds that will help your opponent. Clearance from the ground is important for the same reason: It reduces your enemy's chances of wedging itself under your robot.

More sophisticated passive defenses include protruding beams or axles meant to keep the enemy away from your robot's vital organs, freewheeling vertical wheels on the sides to neutralize lifting wheels, and free horizontal wheels to allow your robot to slip away when engaged on one side.

Testing Your Sumo

This phase is crucial to a good result. Start testing your robot on a pad similar to the tournament's to make sure it doesn't do senseless things in the most common situations. It should detect the edge of the field when reaching it from any angle: You can't imagine how many robots won a match because their opponents killed themselves!

When everything works well, you can start more advanced testing. You really need a sparring partner, but it need not be a second robot. Many reasons suggest you use a fake robot as a sparring partner, something you can move by hand to create any situation you want. (Using a real robot, you'd end up testing both instead; plus, you risk not being able to control specific scenarios.) A simple box does the trick, or a heavy book. Start by leaving the fake robot still and in the middle of the field, and see what happens. Your robot should find it, sooner or later, and push it off the pad. When this works, move the fake robot yourself to test the defensive strategy of your robot, and its behavior at the edge of the pad, the most dangerous area.

Remember that the perfect robot doesn't exist. For any winner of a contest, it's possible to design an "antidote" robot capable of beating it. You just have to accept some compromises in your project and make some assumptions about your opponents, hoping they won't prove too far from reality.

Summary

If you have no previous experience in robotic sumo, you may think of it as a competition based solely on brute force. We must confess that we also had many preconceptions our first time out at a competition of this kind, but we had to change our minds. Force is indeed important, but it typically proves useless when you're up against a good deal of intelligence.

These competitions have nothing in common with the kinds of events that feature radio-controlled machines, called "robots," that try to destroy each other. These are not robots, simply because they totally lack a distinctive robot property: autonomy.

The first important lesson that this chapter taught is that you must design your robot with a strategy in mind, choosing the configuration that best suits your goal. Start examining the rules, and then make a hypothesis about your opponents and devise a strategy to beat them. Your opponents may be very different from how you imagined them, but this is not important—what's important is that you build and program your robot to be consistent with the strategy you chose. A perfect robot doesn't exist; in fact, situations in which Robot A beats Robot B, which then beats Robot C, which in turn actually beats Robot A, are very common in contests. And they're what make contests so interesting and instructive.

We hope you also understand the second important message of this chapter: When building and programming your robot, make reliability your first priority. If you can beat a block of wood in a sumo match, you're halfway to success!

Chapter 21

Searching for Precision

Solutions in this chapter:

- **Precise Positioning**
- **Shooting with Precision**
- **Fine Motor Skills of Your Robot**
- **Fire Fighting in a Maze**
- **Playing Soccer**

Introduction

This chapter is dedicated to contests based on some specific capability. Occasionally, speed is important, too, but not so much as in the competitions described in Chapter 19, and although two or more robots may perform at the same time on the same field, physical contact is not the main goal, as was the case in the competitions discussed in Chapter 20.

These capabilities include what we described in Part I as the most challenging tasks for NXT robots: finding and grabbing objects (Chapter 11), and knowing precise positioning (Chapter 13). The need to use them in a contest makes your mission even more demanding: You must consider the interference that comes from sharing the playing field with other robots that may voluntarily or involuntarily disturb the action of your robot. The recipe for success is the same as that proposed in the previous two chapters. This applies to any kind of contest: Study the rules, define a strategy, make a few assumptions about the opponents, build a prototype, experiment with it, test the software carefully … and rebuild everything from scratch until you are satisfied. In other words, you need some ideas, some skills, and lots of patience!

The last challenge described in this chapter—RoboCup Junior—shows an interesting variation on the theme of object finding: It is the object itself—the ball—that guides the robot to its position, through the emission of IR light. You will discover that this change in the nature of the problem is enough to simplify the robot's requirements considerably, to the point where its software isn't so different from that used to implement the simple light-following algorithm.

Precise Positioning

The challenge of precise positioning requires that your robot go, or return, to a specific point. The robot whose degree of error is smallest, wins. You can define many implementations of that simple statement, each one with its own peculiarities. As always, even a small change in the rules can have radical effects on the difficulty of the challenge. A very simple version is: Starting from a predefined point, the robots must move forward until they hit an obstacle, then turn in place 180 degrees and return to the spot where they began. The obstacle will be the same, at the same distance from the start for all the robots, and the contest may require many runs with different distances. It's important that the rules specify that the robots must turn 180 degrees before returning to the starting point; otherwise, most of them will simply go in reverse!

If you're the one who decides the rules, calibrate the difficulty of the contest by setting the limit on the number of parts admitted. For example, a dual differential drive can be very precise, but requires two differential gears. Limiting the equipment to just the NXT set will make the contest fair to a larger number of participants, but more difficult.

Have you any initial ideas about how you would make a precisely turning robot with only NXT parts? At this point in the book, you should have many ideas. However, let's do this exercise together. Starting from the mobility configuration, you can proceed by a process of elimination: A simple steering is easy to make with parts from the standard NXT set, but the small radius of turn and the lack of Ackerman's steering correction introduce unpredictable slippage, which is very bad for precise positioning. A differential drive won't work because of the lack of a differential gear in the standard set. So, you end up using a tribot with one caster wheel.

A tribot with a caster wheel requires that you use a synchronized feature of the NXT servo motor to run the robot along a straight line, relying on the caster to turn smoothly while one of these motors rotates at a different speed to turn the robot.

With any solution, you may manage to go straight, but you still have to turn precisely 180 degrees. This is the most critical point, because even a small error in the angle will leave your robot very far from the starting point. Do you remember what we said about tuning the turning capability of robots in prior chapters? Use the distance between the wheels to adjust the turning angle so that you have predictable encoder values of the NXT servo motor to make a U-turn. Take some trial runs, and get a feel for the encoder values you will need to make such a turn. Monitor the encoder and control this rotation of the motor to ensure precise turning. Thorough testing is, as always, your ticket to success.

A challenge based on positioning may be made significantly more complex by simply adding more segments and checkpoints to the route. For example, instead of a round trip, you can prepare a triangular path—ask the robots to stop in any vertex and measure the deviation between their actual positions and the expected ones. Each robot should have an easily identifiable part to use as a reference point for measuring the starting and ending points of the journey—for example, a vertical axle with one end very close to the ground.

Shooting with Precision

In summer 2006, NXTasy.org created and hosted an innovative contest called "Throw Me!". The contest required participants to throw a blue plastic ball from the NXT set as far as possible. The contest was worldwide and was open to all users with an NXT set. There wasn't any specific place for the event and the participants had to send documentation of their robot, including video and pictures showing the robot's performance. The participants prepared their own field which, among other things, was required to have a calibrated line to measure how far the ball landed.

The challenge was simple, but required the ball to land within about 5 centimeters of the line! Think about it—when the robot threw the ball, any small deviation in the throw was going to be magnified multifold when it landed. To land close to the line, the throw had to be reproducible with near-zero deviation! The challenge was a feat of robot precision and strength at the same time.

A few months later, Mr. Barak took the prize with a whooping score of 11 meters. The robot used a rotating arm that hit a free-falling ball. The rotating arm was allowed to build momentum before the ball fell. The positioning of the ball, its fall under gravity, and the point at which it touched the rotating arm were adjusted so that it would be thrown as far as possible. It landed 11 meters from the robot—right on the 11-meter marker! You can find more details, images, and video at the NXTasy.org Web site (see Appendix A).

Fine Motor Skills of Your Robot

First Lego League, popularly known as FLL, has used innovative ideas for creating contests. These contests motivate participants to refine their robots' *fine motor skills*. The Nano Quest theme of the 2006 FLL season is another noteworthy contest. Participants have several missions to accomplish, and limited time to accomplish them (for details, see Appendix A). It's a daunting task to make a robot that can accomplish all the missions in this contest. Chances are the winning team will have accomplished all of them, but rookie teams will probably plan which missions to accomplish, and if time permits, they may stretch beyond to accomplish additional missions. Several times, the robots would have interchangeable modules to accomplish specific missions. Time is critical, and thinking in terms of a single module to perform more than one mission helps significantly.

Removing the Bricks

For the Individual Atom Manipulation mission from this contest, the robot has to remove at least one white brick from a blue platform without removing any red bricks. The platform has eight red bricks and eight white bricks. To get any points, the robot has to leave all eight red bricks behind; leaving behind fewer than eight red bricks results in no points!

This is a complex challenge that requires that your robot be able to:

- Navigate the field and find the platform.
- Ascertain the correct bricks.
- Remove at least one brick from the platform.

Moreover, the platform is suspended on four rubber bands, so a hastily approaching robot could knock everything down and lose points. Fortunately, plenty of aid is available for the robot to reach the platform without bumping into it. All the navigation techniques described in prior chapters would be an asset here. The field setup table has an edge, which is distinctly black. The field mat has several lines and colored markings which represent landmarks for absolute positioning, as described in Chapter 13. The positioning of sensors on the robot needs to be well thought out, as there is not much space to maneuver around the platform.

The next difficulty comes from the bricks. The white bricks are in a predictable position and they are tall, attributes on which a robot could strategize. But in order to rely on that, the robot needs to be precisely positioned on the field. Any misalignment quickly knocks the bricks over as the robot reaches them. If a brick falls off the platform, the team gets the point. But if the robot just knocks it over and it remains on the platform, the brick is almost impossible to remove (see Figure 21.1).

Figure 21.1 The Atoms Platform

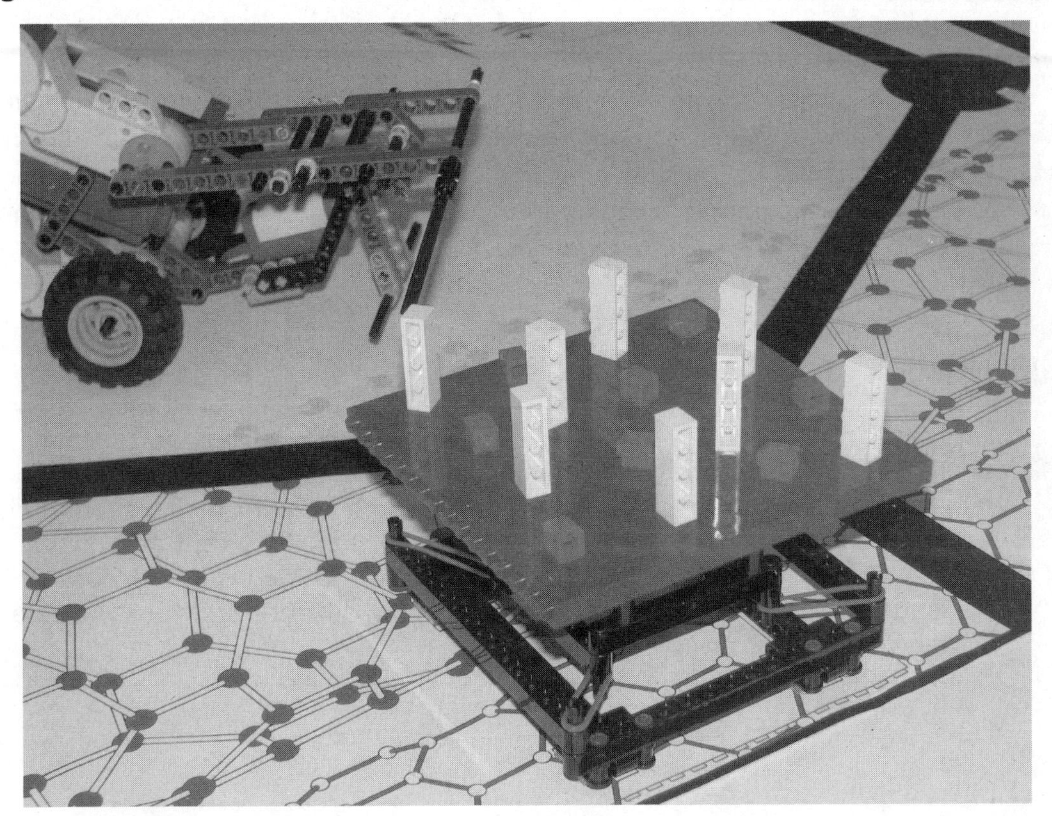

Freeing the Magnets

The Atomic Force Microscopy is another challenging mission. The robot needs to free a stuck nanotip. The nanotip has a magnet and so does the base, and they are stuck together, forcing the nanotip to be bent into position.

Again, the robot needs to navigate the field, find the nanotip, locate its stuck magnets, and free them. The navigational challenges are the same as those discussed in the preceding section. Moreover, the robot has to be fitted with a module to separate the two magnets. You may think that you could use the same module that removed the bricks to separate the magnets, but alas! The base magnet is attached to the bottom mat, and with that technique, instead of separating the magnets, the mat lifts up. You really need a two-hands approach to hold the base magnet down while lifting the top magnet.

The team NANO RATS, which took the winner's trophy at the Virginia State Championship this year, had some simple strategies for dealing with these challenges. Sally and Bill Sylvester, coaches of the team, said their team made a robot with attachments on both sides, and to achieve some missions, the robot ran in reverse. That saved them some time in terms of exchanging the attachments. They reduced their initial four minutes of run time to accomplish six of nine missions to two and a half minutes. For the state championship, they had their robot optimized so well that they had a few seconds left to do one more mission and get additional points.

The team relied quite heavily on synchronized motors to run their robot in a straight line. Remember how to compute distance traveled based on number of rotations? The one with 'PI' in the formula? That was the method they used to ensure that their robot stopped at the right place. They launched the robot from the base, pointing to the mission, and programmed the robot to run straight to the precise location and then operate on the mission (see Figure 21.2).

NOTE

Another factor that plays silently in robot precision is battery voltage. If you're running your robot using timing, it is important to maintain consistent battery voltage. You can do this by fully charging up your rechargeable battery before each run or maintaining a supply of batteries kept at a consistent voltage. Fortunately, the new NXT servo motors include built-in encoders that you can use to ensure greater precision for distance and turns, even at varying battery strength.

Figure 21.2 The NANO RATS' Robot Approaching the Stuck Nanotip

Fire Fighting in a Maze

Trinity College in Connecticut hosts a Fire Fighting Contest which is open to all kinds of robots. The participating robot needs to find its way through a maze that represents a house, find a lit candle that represents a fire, and extinguish the candle. Points are awarded based on several criteria, including shortest time (for details, see Appendix A). Unlike the FLL field, for the most part, the maze flooring is a featureless terrain and has no markings to aid in navigation. Instead, there are walls, so you could use several of the techniques we discussed in Chapter 17 to navigate the maze. Simply sliding the robot along the walls would work, but there are penalty points for that. Several of the robots use ultrasonic or infrared sensors to determine distance from walls.

Finding and extinguishing the candle is also challenging. The robot could use a light sensor to detect the flame, but it needs to be fairly close to the candle before it can reliably detect it. In doing so, there is always a fear of bumping into the candle and tipping it over. The candle is encircled with a line at a 30 cm radius, which could aid in determining when to stop. In fact, per the contest rules, the robot must step inside that arc before extinguishing the candle. Blowing air to extinguish the candle is another option. That would work for a candle, but it's not a reliable proposition for a real-life fire.

Because the competition is open to all, we should mention the difference between NXT robots and other robots. In general, non-NXT robots have options for integrating high-powered batteries and motors with sophisticated processors and sensors, which helps to improve precision and speed. On the other hand, the ease of building robots using the NXT is undeniably an advantage for first-time entrants.

Playing Soccer

RoboCup Junior is a simplified form of soccer, a rather suitable game for small robots. The robots don't actually kick the ball, but instead push it toward the goals. The required capabilities are similar to finding an object (a ball), and knowing where you need to take it to reach the opponent's goal (see Appendix A for Web site details). The field is covered with a simple linear gradient, black at one end and white at the other, and there are walls all around, with goal posts. All of this can be used in navigation.

The ball is a special, active ball: Made of clear plastic, it is filled with infrared LEDs and batteries to be detected by light sensors. With a light sensor, the robot can turn in place until the sensor reads the expected value, which would be the direction of the ball.

The field mat gradient can be used with the techniques we described in Chapter 13. This geometric pattern has the property that, if you follow the darkest path from any point, you arrive at the black edge, whereas if you choose the lighter path, it drives you to the white edge. Using this gradient, it's very easy for robots to reach their desired goals by employing a very simple navigation algorithm.

The program is not too difficult to write. Make your robot turn in place, searching for the ball until it finds it (the algorithm is actually very similar to a light-following algorithm). If it doesn't find it, make it move a bit in any random direction and look around again. When you find the ball, move the robot forward to catch it and then start moving toward the opponent's net.

Summary

The competitions we talked about in this chapter require some capabilities that we described in Chapter 1-13 of the book as being the most challenging to implement: finding objects, directing with precision, knowing where you are, and navigating precisely.

If these activities prove difficult to implement when you build a robot for yourself, situating them in the context of a competition makes your mission even more difficult. This happens because you must push the performance of your robot to its maximum. You have to consider all the details, optimize the software, and reach the highest possible level of reliability and precision.

The soccer competition we described in this chapter is a good example of how a few changes can radically affect the solution to a problem. It also shows the practical application of two techniques described in Chapter 13 regarding absolute positioning: the use of an IR beacon, and a pad with a special pattern that eases navigation.

These challenges require complex behavior composed of many different actions that need to be coordinated together well. If you decide to take up the challenge, we suggest that you think of both your hardware and your software in terms of subsystems. This way, they will be easier to test, debug, and maintain. You also could design your hardware modules to be interchangeable during the competition to perform different missions. In addition, you could make your program modular with a top-level program that manages small subroutines corresponding to the basic actions the robot has to perform: navigation, object detection, and object collection. Mastering this kind of challenge won't be easy, but as with most difficult things in life, your satisfaction will be directly proportional to the effort you expend!

Appendix A

Resources

Introduction

There are quite a few reference materials to be found regarding MINDSTORMS inventions, including some very good books, and hundreds of Internet sites, and numerous blogs that cover specific topics and show interesting models. In this appendix, you'll find a section about books, another one about links of general utility, and a section specific to each chapter of this book (many of the quoted sites pertain to more than a single chapter topic, so browse through them all). We apologize in advance for the significant number of interesting sites that we surely (and unintentionally) omitted from the list.

Every link of this appendix has been checked, but as you know, the Internet is a dynamic animal, so we cannot guarantee that all the links will be still valid at the time you read the book. If you find any broken links, use the descriptive information we provided about each site address to hunt for it using your favorite search engine.

A few of the links point to commercial sites or to sites that not only provide information about the making of some custom part but also sell a kit or the finished product. We have no direct or indirect interest, nor any connection with them; we included the links simply as a help to the reader.

Bibliography

The Unofficial Guide to LEGO MINDSTORMS Robots, by Jonathan B. Knudsen; O'Reilly & Associates, 1999. The first to appear on the market, Knudsen's book is still a very good resource for introducing readers to the MINDSTORMS world. It covers many topics, ranging from construction techniques to programming with different languages.

Dave Baum's Definitive Guide to LEGO MINDSTORMS, by Dave Baum and Rodd Zurcher (Illustrator); Apress, 1999. Baum is the creator of NQC, the most successful alternative programming environment for the RCX. In this book, he not only explains how to use NQC but also explores many building and programming techniques.

Extreme MINDSTORMS: An Advanced Guide to LEGO MINDSTORMS, by Dave Baum, Michael Gasperi, Ralph Hempel, Luis Villa; Apress, 2000. Four gurus of the independent MINDSTORMS community introduce you to the secrets of NQC, legOS, pbForth, and to the making of custom sensors.

Creative Projects with LEGO MINDSTORMS, by Benjamin Erwin; Addison-Wesley, 2001. Erwin invites the reader to be creative, to explore different approaches, and even use different materials. He also covers topics like ROBOLAB, not covered in any other book.

Joe Nagata's LEGO MINDSTORMS Idea Book, by Joe Nagata; No Starch Press, 2001. Nagata is without a doubt a great designer. In his book, he steers you step by step through the building of some instructive and efficient models.

LEGO MINDSTORMS: The Master's Technique, by Jin Sato; No Starch Press, 2001. This is a great book, containing both general building suggestions and programming tips. It also includes step-by-step instructions on how to replicate MIBO, his famous robotic dog.

General Interest Sites

LEGO MINDSTORMS (http://mindstorms.lego.com)

The first site to mention is, of course, the LEGO MINDSTORMS official site. It contains technical tips, a gallery of inventions, events, contests, answers to frequently asked questions (FAQs), and more. The official LEGO MINDSTORMS FAQ site is: http://mindstorms.lego.com/products/whatis/faq.asp.

LUGNET (www.lugnet.com)

The LEGO Users Group Network (LUGNET) is the most comprehensive Internet resource for LEGO, and it's difficult to describe in a few words. It features a database containing all the LEGO sets ever released, as well as a reference list citing all the single LEGO parts. But, more important, its newsgroups are the meeting point of LEGO fans of any age and from any part of the world, and it's one of the friendliest places on the Internet. Don't miss the LUGNET newsgroup (http://news.lugnet.com/robotics), the place where you can ask any number of questions and have them answered with completeness, competence, and patience.

LEGO Set/Parts Reference (www.peeron.com)

A valuable resource for sourcing sets and part numbers/images for LEGO sets.

NXTasy.org

An unofficial LEGO MINDSTORMS blog focused on LEGO Robotics, including the NXT.

The NXT STEP (http://thenxtstep.blogspot.com/)

An unofficial LEGO MINDSTORMS blog focused on LEGO Robotics, including the NXT.

NXTbot.com

An unofficial blog focused on various robotics interests, including the LEGO Robotics and the NXT.

BNXT (www.bnxt.com)

An unofficial LEGO MINDSTORMS blog focused on LEGO Robotics, including the NXT.

Brickshelf (www.brickshelf.com)

Brickshelf is a site that offers everybody the extraordinary opportunity of having free space to show off his or her own LEGO models.

NOTE

The official Web site for this book is at www.syngress.com/solutions. Check it out for lots of additional features and resources.

Chapter 1 Understanding LEGO Geometry

LEGO Geometry (www.brickshelf.com/gallery/GJansson/Geometry/legogeometry.doc)

Gustav Jansson created a document that introduces the basics of LEGO geometry.

LEGO on My Mind (http://homepages.svc.fcj.hvu.nl/brok/legomind)

Don't miss Eric Brok's site; it is filled with explanations and suggestions. LEGO geometry is just one of the many topics covered here.

Studless building (http://thenxtstep.blogspot.com/2006/06/studless-part-ii-hybrid-templates.html)

Brian Davis explains some concepts of studless building geometry and how they relate to studded beams.

Chapter 2 Playing with Gears

Rice University (www.owlnet.rice.edu/~elec201/Book/legos.html)

A LEGO Tutorial proving a variety of information on the fundamentals of LEGO design and gearing.

Technica (http://isodomos.com/technica/registry/gear/gear_2.php)

This site contains a repository of the various LEGO gears that have been produced over the years. It includes dates and model numbers.

Sergei Egorov's LEGO Geartrains (www.malgil.com/esl/lego/geartrains.html)

Egorov's site provides a table listing useful ways to position LEGO gears so that they mesh properly.

Chapter 3 Controlling Motors

Philippe (Philo) Hurbain (www.philohome.com/)

Hurbain has hacked the NXT and provides a look at things like motor internals, hardware interfacing, custom sensors, etc.

Brickshelf Gallery (www.brickshelf.com/cgi-bin/gallery.cgi?f=226241)

This site contains recent introductions of Power Functions by LEGO, such as contain motors, IR controller, and battery box.

NXT Motor Characteristics (http://web.mac.com/ryo_watanabe/iWeb/Ryo%27s%20Holiday/NXT%20 Motor.html)

Visit this site for a mathematical analysis of NXT motor's performance by Ryo Watanabe.

Chapter 4 Reading Sensors

PlastiBots (www.plastibots.com)

Dave Astolfo's LEGO MINDSTORMS robotics site contains pictures and documentation on all of his robots (including the BrickSorter on the front cover of this book). There are also a number of reviews on aftermarket sensors with additional information and pictures.

Techno-Stuff (www.techno-stuff.com)

Pete Sevcik creates and sells LEGO MINDSTORMS compatible sensors and controls for both the NXT and RCX systems. Check it out if you are looking to expand your robot's abilities.

Mindsensors (www.mindsensors.com)

You can find a large variety of aftermarket sensors for the NXT, RCX and VEX systems.

HiTechnic (www.hitechnic.com)

A variety of LEGO NXT and RCX compatible sensors can be purchased from HiTechnic.

Vernier (www.vernier.com/nxt/)

Vernier provides an adapter for the LEGO NXT to be able to use more than 40 of its own analog sensors.

MindStorms RCX Sensor Input Page (www.extremenxt.com/lego.htm)

Michael Gasperi's site about LEGO MINDSTORMS NXT and RCX sensors—the starting point for any investigation about this component. It also contains information on creating your own homebrew sensors.

Sivan Toledo (www.tau.ac.il/~stoledo/lego/)

Toledo provides some interesting creations on his site, where he pushes the limits of the NXT. He also provides some excellent tutorials on I2C interfacing and creating your own custom sensors with it.

Krystian Majewski (http://kisd.de/~krystian/nxt/)

This site features some interesting NXT projects. Specifically, the JennToo radar where Majewski uses the NXT ultrasonic sensor to map the area as it scans. While it's mapping and scanning, it displays the results on the screen as a visual representation of the surroundings.

Philippe (Philo) Hurbain (www.philohome.com/sensors.htm)

Hurbain has a variety of sensors that can work with both the NXT and RCX. Be sure to browse his site as there are many interesting things there.

Chapter 5 What's New with the NXT

Ross Crawford's Tic-Tac-Toe (www.br-eng.info/words/index.php/category/moc/nxt/)

There are few other games on this site, too.

NXT Mobile Applications (http://mindstorms.lego.com/overview/Mobile%20Application.aspx)

The software runs on a Java-capable mobile phone to use as a remote control for NXT.

NeXTScreen by John Hansen (http://bricxcc.sourceforge.net/)

The "Programmable Brick Utilities" section on this site provides a link to information about several NXT utilities, including NeXTScreen.

Wiimote Controlled Robot by Jose Bolaños (www.techeblog.com/index.php/tech-gadget/wiimote-controlled-robot)

The Wiimote and NXT are connected via a computer running a .NET program.

Coco5 by Martyn Boogaarts (http://mindstorms.lego.com/MeetMDP/Martyn.aspx)

This robot has a camera mounted on it, and it can take pictures when it detects someone nearby.

NXTiiMote Controller by Philo (http://philohome.com/nxtiimote/ni.htm)

This controller transmits spatial information to a remote NXT using Bluetooth.

Mini Block Library (http://mindstorms.lego.com/support/updates/)

From this site, you can download several of software updates to MINDSTORMS NXT.

Paul Tingey's Wii-like Controller (www.mindsensors.com/Wii_like_Controller.htm)

This controller is wired to the NXT and functions like Wiimote, driving the RoboArm T-56.

Chapter 6 Building Strategies

LEGO MINDSTORMS NXT Building Instructions (http://mindstorms.lego.com/buildinginstructions/)

LEGO provided building instructions and tips for robots such as a classic clock and a sound robot.

Ldraw.org (www.ldraw.org)

LDraw is a freeware program that can create LEGO models in 3D on your computer screen. Did you ever dream of working with an unlimited supply of any LEGO part in any color?

MLCad (http://mlcad.ldraw.org)

Michael Lachmann's MLCad is a great (and free) CAD program for creating LEGO-like building instructions of your own models. The MLCad site has been recently incorporated into the Ldraw.org domain.

Chapter 7 Programming the NXT

LEGO NXT-G (http://mindstorms.lego.com)

LEGO's MINDSTORMS NXT product, including NXT-G programming graphical interface.

LEGO Robolab (www.legoeducation.com)

LEGO Education RobotC (www.robotc.net)

Developed by the Robotics Academy at Carnegie Mellon University, RobotC is an easy-to-use text-based programming language with many powerful features.

Ralph Hempel pbLUA (www.hempeldesigngroup.com/lego/pbLua/)

Ralph Hempel is the sole developer of the pbLUA application. Having developed a version of the FORTH language (pbFORTH) for the RCX, Hempel is well experienced with the MINDSTORMS products.

John Hansen—NBC & NXC (http://bricxcc.sourceforge.net/nbc/)

Next Byte Codes (NBC) is a simple language with an assembly language syntax that can be used to program LEGO's NXT programmable brick. Not eXactly C

(NXC) is a high-level language, similar to C, built on top of the NBC compiler. It can also be used to program the NXT brick.

LeJOS NXJ (http://lejos.sourceforge.net/)

LeJOS NXJ is currently in alpha release and is a subset of the Java language for the NXT. A core team of about four developers is working on the leJOS for the NXT. Most of the members of this team were already developers of the leJOS version for the RCX.

NXTasy.org (www.nxtasy.org)

For a more thorough list of current development environments, visit NXTasy.org.

Chapter 8 Playing Sounds and Music

Bricx Command Centre (bricxcc.sourceforge.net/)

This site provides the brick piano tool for creating NXT Melody files, as well as the WAV2RSO and MIDIBatch and RMDPlayer utilities.

Sivan Toledo's Clap Counter (www.tau.ac.il/~stoledo/lego/ClapCounter/)

This site includes a car that makes musical sounds through a speaker connected to the output port of an RCX.

Katherine Anderson's SnackBot (mindstorms.lego.com/MeetMDP/KatherineAnderson.aspx)

On this site you'll find a robot that scoops dog food into a bowl then emits a high-pitched tone to let the family dog know it's time to eat.

Note Names, MIDI Numbers and Frequencies (www.phys.unsw.edu.au/~jw/notes.html)

This site contains a table that gives the frequency of any standard keyboard note and its MIDI number.

Chapter 9 Becoming Mobile

Bryan Bonahoom www.funtimetechnologies.com/teamb2

This site is home of some neat NXT creations, including the W.O.P.R Tic-Tac-Toe playing robot.

Laurens Valk (www.freewebs.com/laurens200/)

This site has a collection of robots built with LEGO MINDSTORMS. Some of them include building instructions.

Steve Hassenplug (www.teamhassenplug.org/)

Hassenplug is the designer of many excellent robots. Check out http://mindstorms.lego.com/MeetMDP/SteveH.aspx for details on his holonomic platform—OMNI.

The Straight and Narrow (www.oreillynet.com/pub/a/network/2000/05/22/LegoMindstorms.html)

Jonathan Knudsen's article is about using a differential drive to go straight. Even though this is an older article, the concepts are still the same with the NXT.

Doug's LEGO Robotics Page (www.visi.com/~dc/index.htm)

Although this site contains RCX-based content, there are many nice robots here that will inspire you to try something with the NXT.

Ackerman Steering (http://www.nationaltbucketalliance.com/tech_info/chassis/ackerman/Ackerman.asp)

A nice write-up of Ackerman Steering by George Barnes.

Chapter 10 Getting Pumped: Pneumatics

Ralph Hempel: www.hempeldesigngroup.com/

Christopher R. Smith: www.brickshelf.com/cgi-bin/gallery.cgi?m=Littlehorn

C. S. Soh: www.fifth-r.com/cssoh1/

C. S. Soh's site is subtitled "…where air is power." This is the most important reference for LEGO pneumatics on the Web.

Kevin Clague: www.kclague.net/

Clague is known in the LEGO AFOL community as one of the leading experts on pneumatics. Check out his site for some of his amazing creations.

TECHNIC Double-Acting Compressor (www.hempeldesigngroup.com/lego/compressor/index.html)

This site is the home page of Ralph Hempel's famous double-acting compressor. The same site also contains his Pressure Switch (www.hempeldesigngroup.com/lego/pressureswitch/index.html).

Sergei Egorov's LEGO Pneumatics Page (www.malgil.com/esl/lego/pneumatics.html)

This is a nice page with detailed plans for a double-acting compressor and pneumatic switch.

LEGO Construction Site—Ideas (www.telepresence.strath.ac.uk/jen/lego/ideas.htm)

It's difficult to find a place in this appendix for Jennifer Clark's wonderful site because it covers so many aspects of robotics. Her page of ideas contains many useful suggestions about pneumatics, but don't miss the other tips and her models as well!

Alex Zorko (www.nicjasno.com/)

Zorko has some inspiring model cars built—many of which use pneumatic engines. Look at his site to find many great samples of pneumatic building, instructions, and videos.

Chapter 11 Finding and Grabbing Objects

LEGO MINDSTORMS NXTLog

This site is a central repository for fan-built LEGO MINDSTORMS NXT Robots.

Chapter 12 Doing the Math

Numerical Methods (http://tonic.physics.sunysb.edu/docs/num_meth.html)

This Web site covers all aspects of numerical analysis, although finding what you're looking for may require some time.

Introduction to Time Series Analysis (www.itl.nist.gov/div898/handbook/pmc/section4/pmc4.htm)

An index page from the NIST/SEMATECH Engineering Statistics Internet Handbook about the methods used to analyze time series. It includes moving averages and exponential smoothing.

What's Hysteresis? (www.lassp.cornell.edu/sethna/hysteresis/WhatIsHysteresis.html)

Jim Sethna explains hysteresis in laymen's terms and provides some examples.

Chapter 13 Knowing Where You Are

Probabalistic Localization with the RCX by Dr. Lloyd Greenwald (www.cs.hmc.edu/roboteducation/itcsl_RCXparticlefilteringWkshp.pdf)

Workshop materials prepared by Dr. Lloyd Greenwald (with help from Babak Shirmohammadi), for Thinking Outside the (Desktop) Box, National Science Foundation Workshop, University of Mississippi.

Where Am I (www-personal.engin.umich.edu/~johannb/position.htm)

The site where you can download the not-to-be-missed "Where am I?—Systems and Methods for Mobile Robot Positioning" by J. Borenstein, H. R. Everett, and L. Feng.

Using PID-Based Technique for Competitive Odometry and Dead Reckoning (www.seattlerobotics.org/encoder/200108/using_a_pid.html)

An excellent article written by G. W. Lucas about using the proportional, integral, and derivative (PID) approach in odometry.

JP Brown's Serious LEGO (http://jpbrown.i8.com/)

Here, Jonathan Brown describes the Laser Target we mentioned in Chapter 13. Although much of his work is done with the RCX, it is well worth a visit as the ideas and principles behind LEGO MINDSTORMS robot building are still similar. Don't miss his wonderful creations, especially his world-famous Rubik's Cube solver.

Robotics Introduction (www.restena.lu/convict/Jeunes/RoboticsIntro.htm)

Boulette's Robotics Page is one of those sites difficult to classify because it contains useful tips and interesting projects in many different areas. We chose to place it here for its discussion on positioning and for its description of highly specialized sensors used for the task: laser emitters and decoders, compasses, and infrared-ultrasonic beacons.

Chapter 14 Classic Projects

Line Following Samples (www.bnxt.com/paper/line_follower)

This site provides a few different approaches to line following algorithms in NXT-G.

Dead Reckoning Wiki (http://en.wikipedia.org/wiki/Dead_reckoning)

This site contains a Wiki explanation of dead reckoning with lots of information and links.

Chapter 15 Building Robots That Walk

Kevin Clague (www.kclague.net/)

One of the most well known pneumatic/biped robot builders in the community, Clague has been involved with numerous books and is respected for his experience in bipeds. You can also find information on his site about tools he has authored for LPUB, a tool for the rendering of LEGO for building instructions.

Joe Nagata Walker ND1 (http://web.mac.com/joenagata/iWeb/MindstormsNXT/Welcome.html)

Joe Nagata has published a number of MINDSTORMS RCX and NXT books and has published some NXT creations on his site. You will have to bear through some translation issues, but using the Google Translate tool should provide enough assistance for you to follow along.

Miguel Agullo (http://technicpuppy.miguelagullo.net/)

A nicely done site with some excellent creations, videos, instructions, etc.

Chapter 16 Robotic Animals

Mac Ruiz (http://mobildefencelab.blog.homepagenow.com/?txnid=77e675814ba675d5741a3632aa05a756)

Ruiz is the creator of a number of interesting robots, including the monkey and mouse in Chapter 16 of this book.

LEGO.com's NXTLOG listing of animals (http://mindstorms.lego.com/nxtlog/projectlist.aspx?SearchText=animal)

Yoshihito Isogawa (www.isogawastudio.co.jp/legostudio/index.html)

You will find some excellent ideas at his site (use the translator option at the bottom).

Chapter 17 Solving a Maze

Guy Ziv's MazeRunner solving a maze built of piles of books:

http://nxtasy.org/2007/03/21/nxt-mazerunner/

Maze Solving Algorithm (www.lboro.ac.uk/departments/el/robotics/Maze_Solver.html)

A description of the Bellman flooding algorithm.

Micromouse: Maze Solving (www.cannock.ac.uk/~peteh/micromouse/maze_solving.htm)

This site is dedicated to Micromouse maze-solving competitions. The page we mention is specifically about maze-solving algorithms.

Think Labyrinth: Maze Solving (www.astrolog.org/labyrnth/algrithm.htm)

This site provides a nice variety of maze types and algorithms/ideas for how to go about solving them.

Chapter 18 Drawing and Writing

Logo Foundation (http://el.www.media.mit.edu/groups/logo-foundation/index.html)

The Logo Foundation Web site: a place to find information and resources useful in learning and teaching Logo.

Haiku Program (http://severed.tentacle.net/rpeake/archives/programming/haiku.html)

C source for an automatic Haiku writer.

Chapter 19 Racing Against Time

The Northeast Indiana Robot Games (NEIRG) (www.sciencecentral.org/NEIRG.htm)

ChiBots, Chicago Area Robotics Groups (www.chibots.org/index.php)

Central Illinois Robotics Club – (http://circ.mtco.com/)

Lafayette LEGO Robotics Club (http://cobweb.ecn.purdue.edu/~andy/LAFLRC/BLOCKSandBOTS.html)

RTLToronto (http://peach.mie.utoronto.ca/events/lego/)

Chapter 20 Hand-to-Hand Combat

No Screwdrivers Needed (http://stage6.divx.com/No-Screwdriver-Needed/)

A video site with excellent videos of NXT and RCX Sumo matches.

NXTasy (www.nxtasy.org)

NXTasy is a blog that contains a NXT-G code repository for numerous custom blocks, including an Ultrasonic Ping Block created by Guy Ziv. It allows you to run 2 Ultrasonic sensors that don't interfere with each other.

Lugnet (www.lugnet.com)

This discussion site for all things LEGO includes a way to find a Sumo contest and other events in your area.

Philo Hurbain (www.philohome.com)

This site contains many example robots and hacks, including a way to expand port A for more motors.

Chapter 21 Searching for Precision

Throw Me Contest at NXTasy.org

(www.nxtasy.org/challenges/challenge-no-1-throw-me/)

The rules, winners, images, and videos of the contest held in the summer of 2006.

Fire Fighting Robot Contest at Trinity College

(www.trincoll.edu/events/robot/default.asp)

The robots must navigate through a maze, find a lit candle, and extinguish the flame.

Playing Soccer at RoboCup Junior (http://rcj.sci.brooklyn.cuny.edu/)

A team of autonomous mobile robots plays simplified soccer games in an enclosed field.

First LEGO League (www.firstlegoleague.org/)

On this site you can apply math, science, and technology to solve real-world problems and have fun.

Maxwell's Demons—Official Rules
(http://news.lugnet.com/org/us/smart/?n=22)

David Schilling's original post about the rules concerning his Maxwell's Demons competition.

LEGO Robots: Challenge
(www.cs.uu.nl/~markov/lego/challenge/index.html)

The account of a soda can retrieval challenge at the Department of Computer Science at Utrecht University (in the Netherlands).

Appendix B

Matching Distances

Legend:

- Each cell of the table contains three data: the distance in LEGO units (studs), the quality of the matching, and the resulting angle in degrees.
- Distances are measured *excluding* the starting point. (For example, if one peg is in the first hole of a beam and another is in the tenth, the distance is nine units.)
- The quality of the matching is expressed with a symbol that reflects the difference between the actual distance and the closest perfect match, expressed in LEGO units, according to the following scheme:

Symbol	Meaning	Maximum Tolerance
P	Perfect match	0.00 studs
V	Very good	0.02 studs
G	Good	0.04 studs
N	Not so good	0.06 studs
B	Bad	0.08 studs

Height in Bricks and Plates	Base in Studs																				
	0	1	2	3	4	5	6	7	8	9	10	11	12	13	14	15	16	17	18	19	20
1										9 B 8°	10 B 7°	11 B 6°	12 N 6°	13 N 5°	14 N 5°	15 N 5°	16 N 4°	17 N 4°	18 G 4°	19 G 4°	20 G 3°
1 1/3																	16 B 6°	17 B 5°	18 B 5°	19 B 5°	20 B 5°
1 2/3	2 P 90°																				
2																					
2 1/3		3 G 70°																			
2 2/3						6 B 33°															
3							7 V 31°														
3 1/3	4 P 90°			5 P 53°				8 B 30°	9 N 27°												
3 2/3					6 N 48°					10 V 26°	11 B 24°										
4						7 B 44°						12 V 24°	13 B 22°								
4 1/3				6 V 60°			8 B 41°						13 B 23°	14 V 22°	15 B 20°						
4 2/3			6 N 70°					9 G 39°							15 B 22°	16 V 20°	17 N 19°				
5	6 P 90°								10 P 37°									18 G 19°	19 G 18°	20 B 18°	
5 1/3				7 B 65°						11 N 35°										20 N 19°	21 V 18°
5 2/3							9 B 49°					13 B 32°									
6								10 N 46°					14 V 31°								
6 1/3									11 G 44°					15 N 30°	16 B 28°						
6 2/3	8 P 90°	8 B 83°			9 N 63°		10 P 53°			12 N 42°						17 P 28°					
7								11 B 50°			13 N 40°						18 B 28°	19 G 26°			
7 1/3			9 G 77°																20 G 26°	21 B 25°	

Continued

Height in Bricks and Plates	Base in Studs																				
	0	1	2	3	4	5	6	7	8	9	10	11	12	13	14	15	16	17	18	19	20
7 2/3					10 G 67°		11 V 57°							16 B 35°							22 V 25°
8				10 N 73°											17 G 34°						
8 1/3	10 P 90°	10 N 84°														18 G 34°					
8 2/3							12 V 60°											20 B 31°			
9			11 V 80°							14 N 50°									21 V 31°		
9 1/3											15 V 48°				18 B 39°					22 N 31°	23 B 29°
9 2/3				12 V 75°			13 N 63°					16 V 47°				19 G 38°					
10	12 P 90°	12 N 85°				13 P 67°				15 P 53°			17 G 45°				20 P 37°				
10 1/3					13 G 72°						16 B 51°			18 G 44°				21 N 36°			
10 2/3			13 N 81°												19 G 42°						
11								15 N 62°		16 G 56°						20 V 41°					24 G 33°
11 1/3				14 B 78°													21 V 40°				
11 2/3	14 P 90°	14 G 86°																22 G 39°			
12					15 N 74°			16 V 64°		17 V 58°									23 N 39°		
12 1/3			15 B 82°				16 G 68°						19 N 51°			21 B 45°					
12 2/3						16 V 72°								20 V 49°			22 B 44°				
13										18 V 60°					21 G 48°			23 B 43°			
13 1/3	16 P 90°	16 G 86°											20 P 53°			22 B 47°					
13 2/3														21 B 52°							
14				17 B 80°						19 N 62°											

Continued

Height in Bricks and Plates	Base in Studs																				
	0	1	2	3	4	5	6	7	8	9	10	11	12	13	14	15	16	17	18	19	20
14 1/3									19 G 65°				21 G 55°								
14 2/3					18 N 77°			19 N 68°													
15	18 P 90°	18 G 87°					19 G 72°														
15 1/3						19 B 75°			20 B 67°		21 N 61°		22 G 57°					25 N 47°			
15 2/3				19 G 81°				20 B 70°								24 N 51°			26 G 46°		
16																	25 V 50°			27 V 45°	
16 1/3					20 V 78°						22 V 63°		23 V 59°					26 N 49°			28 V 44°
16 2/3	20 P 90°	20 G 87°								22 B 66°						25 P 53°					
17						21 V 76°											26 B 52°				
17 1/3				21 V 82°				22 N 71°			23 B 64°		24 V 60°		25 B 56°						
17 2/3							22 G 74°			23 G 67°						26 G 55°					
18					22 G 80°				23 G 70°												
18 1/3	22 P 90°	22 G 87°											25 N 61°		26 B 58°					29 B 49°	
18 2/3						23 N 77°						25 N 64°				27 N 56°					30 G 48°
19				23 V 83°															29 N 52°		
19 1/3							24 G 75°													30 V 51°	
19 2/3					24 B 80°							26 G 65°		27 N 61°		28 G 58°					31 B 50°
20	24 P 90°	24 G 88°						25 P 74°			26 P 67°								30 P 53°		

Appendix C

Note Frequencies

The following table contains note frequencies rounded to the nearest whole number.

Octave	C	C#	D	D#	E	F	F#	G	G#	A	A#	B
1	33	35	37	39	41	44	46	49	52	55	58	62
2	65	69	73	78	82	87	92	98	104	110	117	123
3	131	139	147	156	165	175	185	196	208	220	233	247
4	262	277	294	311	330	349	370	392	415	440	466	494
5	523	554	587	622	659	698	740	784	831	880	932	988
6	1047	1109	1175	1245	1319	1397	1480	1568	1661	1760	1865	1976
7	2093	2218	2349	2489	2637	2794	2960	3136	3322	3520	3729	3951
8	4186	4435	4699	4978	5274	5588	5920	6272	6645	7040	7459	7902

Appendix D

Math Cheat Sheet

Sensors

Raw values to percentage (light sensor):

percentage = 146 – raw value / 7

Raw values to temperatures, in C° (temperature sensor):

C° = (785 – raw value) / 8

Conversion of Celsius to Fahrenheit degrees:

F° = C° x 9 / 5 + 32

Averages

Simple average:

$$A = (V_1 + V_2 + \ldots + V_n) / n$$

Weighted average:

$$A = (V_1 \times W_1 + V_2 \times W_2 + \ldots + V_n \times W_n) \;/\; (W_1 + W_2 + \ldots + W_n)$$

Exponential smoothing:

$$A_n = (V_n \times W_1 + A_{n-1} \times W_2) \;/\; (W_1 + W_2)$$

Interpolation

Linear interpolation: Find the value of the dependent variable Y for a given value of the independent variable X, knowing that for X equal to X_a, Y is Y_a, and for X equal to X_b, Y is Y_b.

$$(Y - Y_a) / (Y_b - Y_a) = (X - X_a) / (X_b - X_a)$$
$$Y = (X - X_a) \times (Y_b - Y_a) / (X_b - X_a) + Y_a$$

Equation of the straight line which connects the points (X_a, Y_a) and (X_b, Y_b):

$$m = (Y_b - Y_a) / (X_b - X_a)$$
$$b = Y_a - m \times X_a$$
$$Y = m \times X + b$$

Gears, Wheels, and Navigation

Output angular velocity of the body of a differential gear Oav, given the input angular velocity of the two axles Iav_1 and Iav_2:

$$Oav = (Iav_1 + Iav_2) / 2$$

Distance, Time, Speed:

distance = speed x time

speed = distance / time

Circumference C of a wheel, given the diameter D:

$$C = D \times \square$$

$$\square = 3.1415926...$$

Increment in rotation sensor count I that corresponds to a turn of the wheel, given R the resolution of the sensor and G the gear ratio between the wheel and the sensor:

$$I = G \times R$$

R = 16 (for Lego rotation sensors)

Conversion factor F which measures the traveled distance of a wheel for any single increment in the count of a rotation sensor:

$$F = C / I = (D \times \square) / (G \times R)$$

Actual traveled distance, given F and the count of the sensor:

$$T = Count \times F$$

Traveled distance T_C of a differential drive robot's centerpoint, given the traveled distances T_L and T_R of its left and right drive wheels:

$$T_C = (T_R + T_L) / 2$$

Change of orientation $\square O_R$, in radians, of a differential drive robot, given the traveled distances T_L and T_R of its left and right drive wheels, and the distance B between the wheels:

$$\square O_R = (T_R - T_L) / B$$

New orientation O_i of a robot after a change in orientation □O from the previous orientation O_{i-1}:

$O_i = O_{i-1} +$ □O

New position of a robot (x_i, y_i) of a robot after having covered a distance T_C in direction O_i from position (x_{i-1}, y_{i-1}):

$x_i = x_{i-1} + T_C \times \cos O_i$

$y_i = y_{i-1} + T_C \times \sin O_i$

Conversion of radians to degrees:

Degrees = Radians x 180 / □

Required increment in rotation sensor count for a given change of orientation □O_R in radians or □O_D in degrees:

Count = T / F = (□O_R x B / 2) / F = □O_R x B / 2F

Count = □O_D x π x B / (360 x F)

Index

A

B

C

D

G

H

N

Q

R

S

T

U

V

W

Z

Syngress: *The Definition of a Serious Security Library*

Syn•gress (sin-gres): *noun, sing.* Freedom from risk or danger; safety. See *security*.

AVAILABLE NOW
order @
www.syngress.com

Cyber Spying: Tracking Your Family's (Sometimes) Secret Online Lives

Dr. Eric Cole, Michael Nordfelt, Sandra Ring, and Ted Fair

Have you ever wondered about that friend your spouse e-mails, or who they spend hours chatting online with? Are you curious about what your children are doing online, whom they meet, and what they talk about? Do you worry about them finding drugs and other illegal items online, and wonder what they look at? This book shows you how to monitor and analyze your family's online behavior.

ISBN: 1-93183-641-8

Price: $39.95 US $57.95 CAN

Stealing the Network: How to Own an Identity

AVAILABLE NOW
order @
www.syngress.com

Timothy Mullen, Ryan Russell, Riley (Caezar) Eller, Jeff Moss, Jay Beale, Johnny Long, Chris Hurley, Tom Parker, Brian Hatch

The first two books in this series "Stealing the Network: How to Own the Box" and "Stealing the Network: How to Own a Continent" have become classics in the Hacker and Infosec communities because of their chillingly realistic depictions of criminal hacking techniques. In this third installment, the all-star cast of authors tackle one of the fastest-growing crimes in the world: Identity Theft. Now, the criminal hackers readers have grown to both love and hate try to cover their tracks and vanish into thin air...

ISBN: 1-59749-006-7

Price: $39.95 US $55.95 CAN

AVAILABLE NOW
order @
www.syngress.com

Software Piracy Exposed

Paul Craig, Ron Honick

For every $2 worth of software purchased legally, $1 worth of software is pirated illegally. For the first time ever, the dark underground of how software is stolen and traded over the Internet is revealed. The technical detail provided will open the eyes of software users and manufacturers worldwide! This book is a tell-it-like-it-is exposé of how tens of billions of dollars worth of software is stolen every year.

ISBN: 1-93226-698-4

Price: $39.95 U.S. $55.95 CAN

SYNGRESS®

Syngress: *The Definition of a Serious Security Library*

Syn•gress (sin-gres): *noun, sing.* Freedom from risk or danger; safety. See *security*.

AVAILABLE NOW
order @
www.syngress.com

Phishing Exposed
Lance James, Secure Science Corporation,
Joe Stewart (Foreword)

If you have ever received a phish, become a victim of a phish, or manage the security of a major e-commerce or financial site, then you need to read this book. The author of this book delivers the unconcealed techniques of phishers including their evolving patterns, and how to gain the upper hand against the ever-accelerating attacks they deploy. Filled with elaborate and unprecedented forensics, Phishing Exposed details techniques that system administrators, law enforcement, and fraud investigators can exercise and learn more about their attacker and their specific attack methods, enabling risk mitigation in many cases before the attack occurs.

ISBN: 1-59749-030-X

Price: $49.95 US $69.95 CAN

Penetration Tester's Open Source Toolkit

AVAILABLE NOW
order @
www.syngress.com

Johnny Long, Chris Hurley, SensePost,
Mark Wolfgang, Mike Petruzzi

This is the first fully integrated Penetration Testing book and bootable Linux CD containing the "Auditor Security Collection," which includes over 300 of the most effective and commonly used open source attack and penetration testing tools. This powerful tool kit and authoritative reference is written by the security industry's foremost penetration testers including HD Moore, Jay Beale, and SensePost. This unique package provides you with a completely portable and bootable Linux attack distribution and authoritative reference to the toolset included and the required methodology.

ISBN: 1-59749-021-0

Price: $59.95 US $83.95 CAN

AVAILABLE NOW
order @
www.syngress.com

Google Hacking for Penetration Testers
Johnny Long, Foreword by Ed Skoudis

Google has been a strong force in Internet culture since its 1998 upstart. Since then, the engine has evolved from a simple search instrument to an innovative authority of information. As the sophistication of Google grows, so do the hacking hazards that the engine entertains. Approaches to hacking are forever changing, and this book covers the risks and precautions that administrators need to be aware of during this explosive phase of Google Hacking.

ISBN: 1-93183-636-1

Price: $44.95 U.S. $65.95 CAN

Syngress: *The Definition of a Serious Security Library*

Syn•gress (sin-gres): *noun, sing.* Freedom from risk or danger; safety. See *security*.

AVAILABLE NOW
order @
www.syngress.com

Cisco PIX Firewalls: Configure, Manage, & Troubleshoot

Charles Riley, Umer Khan, Michael Sweeney

Cisco PIX Firewall is the world's most used network firewall, protecting internal networks from unwanted intrusions and attacks. Virtual Private Networks (VPNs) are the means by which authorized users are allowed through PIX Firewalls. Network engineers and security specialists must constantly balance the need for air-tight security (Firewalls) with the need for on-demand access (VPNs). In this book, Umer Khan, author of the #1 best selling PIX Firewall book, provides a concise, to-the-point blueprint for fully integrating these two essential pieces of any enterprise network.

ISBN: 1-59749-004-0

Price: $49.95 US $69.95 CAN

Configuring Netscreen Firewalls

Rob Cameron

Configuring NetScreen Firewalls is the first book to deliver an in-depth look at the NetScreen firewall product line. It covers all of the aspects of the NetScreen product line from the SOHO devices to the Enterprise NetScreen firewalls. Advanced troubleshooting techniques and the NetScreen Security Manager are also covered..

ISBN: 1--93226-639-9

Price: $49.95 US $72.95 CAN

AVAILABLE NOW
order @
www.syngress.com

Configuring Check Point NGX VPN-1/FireWall-1

Barry J. Stiefel, Simon Desmeules

Configuring Check Point NGX VPN-1/Firewall-1 is the perfect reference for anyone migrating from earlier versions of Check Point's flagship firewall/VPN product as well as those deploying VPN-1/Firewall-1 for the first time. NGX includes dramatic changes and new, enhanced features to secure the integrity of your network's data, communications, and applications from the plethora of blended threats that can breach your security through your network perimeter, Web access, and increasingly common internal threats.

ISBN: 1--59749-031-8

Price: $49.95 U.S. $69.95 CAN

SYNGRESS®

Syngress: *The Definition of a Serious Security Library*

Syn•gress (sin-gres): *noun, sing.* Freedom from risk or danger; safety. See *security*.

AVAILABLE NOW
order @
www.syngress.com

Syngress IT Security Project Management Handbook
Susan Snedaker

The definitive work for IT professionals responsible for the management of the design, configuration, deployment and maintenance of enterprise-wide security projects. Provides specialized coverage of key project areas including Penetration Testing, Intrusion Detection and Prevention Systems, and Access Control Systems.

ISBN: 1-59749-076-8

Price: $59.95 US $77.95 CAN

Combating Spyware in the Enterprise
Paul Piccard

AVAILABLE NOW
order @
www.syngress.com

Combating Spyware in the Enterprise is the first book published on defending enterprise networks from increasingly sophisticated and malicious spyware. System administrators and security professionals responsible for administering and securing networks ranging in size from SOHO networks up to the largest enterprise networks will learn to use a combination of free and commercial anti-spyware software, firewalls, intrusion detection systems, intrusion prevention systems, and host integrity monitoring applications to prevent the installation of spyware, and to limit the damage caused by spyware that does in fact infiltrate their networks.

ISBN: 1-59749-064-4

Price: $49.95 US $64.95 CAN

Practical VoIP Security
Thomas Porter

After struggling for years, you finally think you've got your network secured from malicious hackers and obnoxious spammers. Just when you think it's safe to go back into the water, VoIP finally catches on. Now your newly converged network is vulnerable to DoS attacks, hacked gateways leading to unauthorized free calls, call eavesdropping, malicious call redirection, and spam over Internet Telephony (SPIT). This book details both VoIP attacks and defense techniques and tools.

ISBN: 1-59749-060-1

Price: $49.95 U.S. $69.95 CAN

SYNGRESS®

Syngress: *The Definition of a Serious Security Library*

Syn•gress (sin-gres): *noun, sing.* Freedom from risk or danger; safety. See *security*.

AVAILABLE NOW
order @
www.syngress.com

Snort 2.1 Intrusion Detection, Second Edition

Jay Beale, Brian Caswell, et. al.

"The authors of this *Snort 2.1 Intrusion Detection, Second Edition* have produced a book with a simple focus, to teach you how to use Snort, from the basics of getting started to advanced rule configuration, they cover all aspects of using Snort, including basic installation, preprocessor configuration, and optimization of your Snort system."

—Stephen Northcutt
Director of Training & Certification, The SANS Institute

ISBN: 1-931836-04-3
Price: $49.95 U.S. $69.95 CAN

Ethereal Packet Sniffing

AVAILABLE NOW
order @
www.syngress.com

Ethereal offers more protocol decoding and reassembly than any free sniffer out there and ranks well among the commercial tools. You've all used tools like tcpdump or windump to examine individual packets, but Ethereal makes it easier to make sense of a stream of ongoing network communications. Ethereal not only makes network troubleshooting work far easier, but also aids greatly in network forensics, the art of finding and examining an attack, by giving a better "big picture" view. *Ethereal Packet Sniffing* will show you how to make the most out of your use of Ethereal.

ISBN: 1-932266-82-8
Price: $49.95 U.S. $77.95 CAN

AVAILABLE NOW
order @
www.syngress.com

Nessus Network Auditing

Jay Beale, Haroon Meer, Roelof Temmingh, Charl Van Der Walt, Renaud Deraison

Crackers constantly probe machines looking for both old and new vulnerabilities. In order to avoid becoming a casualty of a casual cracker, savvy sys admins audit their own machines before they're probed by hostile outsiders (or even hostile insiders). Nessus is the premier Open Source vulnerability assessment tool, and was recently voted the "most popular" open source security tool of any kind. *Nessus Network Auditing* is the first book available on Nessus and it is written by the world's premier Nessus developers led by the creator of Nessus, Renaud Deraison.

ISBN: 1-931836-08-6
Price: $49.95 U.S. $69.95 CAN

Syngress: *The Definition of a Serious Security Library*

Syn•gress (sin-gres): *noun, sing.* Freedom from risk or danger; safety. See *security*.

AVAILABLE NOW
order @
www.syngress.com

Buffer OverFlow Attacks: Detect, Exploit, Prevent

James C. Foster, Foreword by Dave Aitel

The SANS Institute maintains a list of the "Top 10 Software Vulnerabilities." At the current time, over half of these vulnerabilities are exploitable by Buffer Overflow attacks, making this class of attack one of the most common and most dangerous weapons used by malicious attackers. This is the first book specifically aimed at detecting, exploiting, and preventing the most common and dangerous attacks.

ISBN: 1-932266-67-4

Price: $34.95 US $50.95 CAN

Programmer's Ultimate Security DeskRef

James C. Foster

AVAILABLE NOW
order @
www.syngress.com

The Programmer's Ultimate Security DeskRef is the only complete desk reference covering multiple languages and their inherent security issues. It will serve as the programming encyclopedia for almost every major language in use.

While there are many books starting to address the broad subject of security best practices within the software development lifecycle, none has yet to address the overarching technical problems of incorrect function usage. Most books fail to draw the line from covering best practices security principles to actual code implementation. This book bridges that gap and covers the most popular programming languages such as Java, Perl, C++, C#, and Visual Basic.

ISBN: 1-932266-72-0

Price: $49.95 US $72.95 CAN

AVAILABLE NOW
order @
www.syngress.com

Hacking the Code: ASP.NET Web Application Security

Mark Burnett

This unique book walks you through the many threats to your Web application code, from managing and authorizing users and encrypting private data to filtering user input and securing XML. For every defined threat, it provides a menu of solutions and coding considerations. And, it offers coding examples and a set of security policies for each of the corresponding threats.

ISBN: 1-932266-65-8

Price: $49.95 U.S. $79.95 CAN

SYNGRESS®